# Galbraith's Building and Land Management Law for Students

# Galbraith's Building and Land Management Law for Students

Sixth edition

Michael Stockdale, PhD, LLB
Stephen Wilson, BA
Rebecca Mitchell, LLB, Solicitor
Russell Hewitson, LLB, Solicitor
Mick Woodley, BA, MSc
Simon Spurgeon, BA, Solicitor

ELSEVIER

AMSTERDAM • BOSTON • HEIDELBERG • LONDON • NEW YORK • OXFORD
PARIS • SAN DIEGO • SAN FRANCISCO • SINGAPORE • SYDNEY • TOKYO
Butterworth-Heinemann is an imprint of Elsevier

Butterworth-Heinemann is an imprint of Elsevier
The Boulevard, Langford Lane, Oxford OX5 1GB, UK
30 Corporate Drive, Suite 400, Burlington, MA 01803, USA

First published as *Building Law for Students* 1989
Second edition 1991
Reprinted 1992
Third edition 1993
Reprinted 1994
Fourth edition 1998
Fifth edition 2005
Sixth edition 2011

**British Library Cataloguing-in-Publication Data**
A catalogue record for this book is available from the British Library

**Library of Congress Cataloging-in-Publication Data Control Number**
A catalog record for this book is available from the Library of Congress

ISBN: 978-0-08-096692-2

For information on all Butterworth–Heinemann publications visit our website at elsevierdirect.com

Typeset by MPS Limited, a Macmillan Company, Chennai, India
www.macmillansolutions.com

Printed and bound in the United States of America
11 12 13  10 9 8 7 6 5 4 3 2 1

Working together to grow
libraries in developing countries

www.elsevier.com | www.bookaid.org | www.sabre.org

ELSEVIER    BOOK AID International    Sabre Foundation

# Contents

# Preface to the sixth edition

The sixth edition of *Galbraith's Building and Land Management Law for Students* has been written by a large team of subject specialists to reflect the broad scope of the topics covered.

The new edition has provided the authors with an opportunity to update the existing chapters of the book. Examples of significant new developments that are dealt with include material on the creation of the Supreme Court of the United Kingdom, changes to the institutions of the European Union, the new tribunal structure created by the Tribunals Courts and Enforcement Act 2007, changes to the constitution of the House of Lords and the structure of local government, the Equality Act 2010, the Planning Act 2008 and the redrafting of JCT standard building contracts.

The law is stated as of 31 April 2010.

*Michael Stockdale*
*Stephen Wilson*
*Rebecca Mitchell*
*Russell Hewitson*
*Mick Woodley*
*Simon Spurgeon*

# Table of statutes

**Regulations**

# Table of cases

# 1
# The nature of law

## Introduction

Any society quickly finds that it needs rules or laws to enable it to function smoothly. Consider the results if you had decided to drive to work today on the right instead of the left. People need to be able to live and go about their business in certainty, knowing that they can expect others to abide by the same rules. It is probably true that most people obey most of the rules most of the time. They may not have any precise knowledge of the actual rules which they are obeying, but that precise knowledge only becomes important when disputes arise or problems have to be solved.

As society evolves and becomes more complex, the number of laws increases and the types of laws become more sophisticated. Initially people need laws to protect their person or their property. Some rules accord with basic ideas of morality, e.g. not killing or not stealing, but it is not essential for all the rules to have a moral base. For instance, when we create a rule for motorists that they must drive on the left the rule commands obedience simply because it is necessary and certain.

Although certainty is an important principle in any legal system, the rules of law should also aim to achieve justice. This is sometimes accomplished by giving a discretion. Some areas of English law are subject to discretionary principles known as equity. It should be remembered that laws are equally binding even if they are thought to be unjust. The remedy then is to seek to reform the law, not to violate it.

### Sanctions

If a law is disobeyed, there must be some effective means by which it can be enforced. This may be described as a sanction and could take the form of a fine, imprisonment, an award of damages or an injunction, which is an order of the court forbidding certain actions or behaviour. Although such sanctions may deter people from breaking the law, in fact the reasons for obeying the law may be far more complex. A person may obey the law because he or she fears the disapproval or hostility of others, or because he or she believes in the 'rightness' of the rule, or because he or she sees obedience of the law as a duty owed as a citizen.

Galbraith's Building and Land Management Law for Students. DOI: 10.1016/B978-0-08-096692-2.00001-4
Copyright © 2010 by Elsevier Ltd

## Divisions of the law

It should be noted that all the contents of this book relate to English law. Although it is appropriate to talk about the British Constitution it should be remembered that Scotland has a different system of law to England. Whereas some of the rules are common to both countries, there are still significant differences. These are most apparent in areas like the system of land holding and in the structures and procedures of the courts.

English law may be subdivided in a number of ways. A simple division would be into criminal and civil law. It is quite common for the layman to believe that the bulk of our law consists of criminal law. This inevitably results from the publicity given in the media to major criminal cases. In this sense the law is like an iceberg, the criminal law is the part seen above the water and the civil law is the mighty bulk that lies below. The state has an interest in preserving our society and upholding law and order, and it makes criminal laws to secure those objectives. A person who infringes those laws commits a criminal offence for which he or she can be prosecuted by the state. The principal objectives of the criminal law may, therefore, be seen to be to deter and to punish. The civil law protects rights and creates obligations between individuals, although sometimes those individuals will be large public corporations, government departments, local authorities, etc.

There is no simple way to classify civil law, but it may be subdivided into the law of contract, the law of tort (i.e. civil wrongs), the law relating to property and the law relating to persons. The rules of civil law must deal with matters as diverse as contracts for the sale of goods, actions for damages for negligence, planning applications and compulsory purchase, divorce, tenancy, creation of companies, contracts of insurance, making a will, money lending, recovery of debts, defective workmanship and unfair dismissal.

## Language of the law

Every profession has its language or jargon, which is often a convenient shorthand method of communication between people engaged in that profession. With its long history, the language of the law is especially rich as it frequently uses Latin expressions such as *res ipsa loquitur* (the facts speak for themselves) or *caveat emptor* (let the buyer beware). Different terminology applies to criminal and civil law. In criminal proceedings a person is arrested, charged with an offence and prosecuted in summary or indictable proceedings by a prosecutor. If the accused person is found guilty he or she will be sentenced and punished. If found not guilty, he or she will be acquitted. In a civil action a claimant (formerly known as a plaintiff) sues a defendant. If the defendant is found to be liable he or she may be ordered to pay damages or an order of the court such as an injunction may be made against him or her. Where the terminology is correctly used it is possible to tell whether a case concerns criminal or civil law.

Although it is convenient to break down the mass of English law into subdivisions for the purposes of study, it should be realized that these various divisions of the law are not mutually exclusive. It is possible for one set of facts to give rise to both civil and criminal liability. The simplest example is the motorist who drives carelessly and injures a pedestrian. A motorist may both be prosecuted in criminal proceedings for an offence under the Road Traffic Acts and be sued in civil proceedings by the injured pedestrian hoping to recover compensation. Similarly, an employee injured at work may wish to bring civil proceedings against his or her employer to recover damages where the employer has been negligent in caring for the employee's safety. Arising out of the same set of facts, the Health and Safety Executive may prosecute the employer for breaches of the Health and Safety at Work Act 1974.

## Making the law

Unless some catastrophic event like a war or a revolution occurs, which may cause a country to adopt a completely new system of law, the law-making processes will have developed over a period of centuries. Many of the rules of English law which are still in force are of considerable antiquity. Laws do not become ineffective merely because they are very old. Indeed, some old rules have been used very imaginatively by the judges in the courts to create principles relevant to modern life.

In the early stages of law making, custom will usually play a large part. As the needs of a society become more sophisticated, custom as a source of law tends to be superseded by more formal sources. Local custom may still be upheld as a valid part of the law where it can be shown that the custom is reasonable and certain and has been continued without interruption since time immemorial. For practical purposes this normally means that the custom must be shown to have existed during living memory. Occasional examples still come before the courts.

The two main sources of law in our system today are legislation (laws made or approved by Parliament) and judicial precedents (binding decisions of the judges). Both of these sources of law must now take account of the fact that the United Kingdom is a member of the European Union (see Chapter 2) and note the impact of the Human Rights Act 1998 (see later in this chapter).

# New legislation

Law-making authority is vested in the Crown and Parliament, although the role of the Crown is now almost entirely formal. Our complex modern society requires rule-making techniques which are capable of coping with economic, social and welfare problems. Parliament, consisting of the House of Lords and the House of Commons, is said to be supreme, but in practice its ability to make rules must be viewed subject to a number of limitations. By virtue of the European Communities Act 1972 a new element of European Community law was introduced into our system. European Community law takes precedence over the national law of any member state. This, therefore, imposes a limit on the ability of Parliament to make whatever laws it wishes. If English law is found to conflict with European Community law the Community law will prevail. The United Kingdom is a signatory to the European Convention on Human Rights and is obliged by international law to ensure that English law does not contravene the Convention. Under the Human Rights Act 1998 the main provisions of the European Convention on Human Rights became part of UK law and thus enforceable through the English courts. Other factors which would limit the power of Parliament are: the existence of an opposition party, or parties, whose duty is to seek to curb or limit government proposals; Parliamentary question time which is held daily, when ministers, including the Prime Minister, must justify their activities; the two-chamber system whereby the House of Lords can at least delay the passing of legislation; and public opinion, pressure groups, freedom of speech and publicity through the media. Once laws are enacted by Parliament the interpretation of those laws is carried out by the judges in the courts. If Parliament has passed a law which is seen to be too harsh or too extensive in its application it may be possible for the judges, by restrictive interpretation of the words of the statute, to limit the scope of the new rule.

Acts of Parliament, which can also be called statutes, contain the main laws made by Parliament, acting in its legislative role. Until the statute or Act has passed through all its stages in both Houses and received the Royal Assent, it is referred to as a bill.

The inspiration for new legislation may come from a number of sources. A new government will have made manifesto commitments and will have outlined its own policies. During its first year or so in office it will be keen to push through those changes which were outlined in its election manifesto. Inevitably this source of new legislation becomes comparatively less significant as the government's term of office, a maximum of five years, progresses. Each of the major government departments will have a programme of legislation which it would like to introduce. Parliamentary time is at a premium and this may well produce competition between departments. Where a small measure is needed a department may seek to persuade a private member who has won a high place in the private members' ballot to introduce a bill on its behalf.

New laws may also be needed to implement recommendations of the Law Commission, a body established in 1965 to review the law with a view to its systematic development and reform. Occasionally the government of the day sets up Royal Commissions or other special committees of enquiry. These are usually established to investigate one specific topic area and they will be disbanded once their reports are published. Many of these reports gather dust and do not result in their recommendations being implemented in legislation. In other cases the recommendations may be implemented.

Examples include the Robens Report on Health and Safety, which very quickly became the Health and Safety at Work etc. Act 1974, and the Royal Commission on Criminal Justice, appointed in the wake of miscarriages of justice such as the 'Birmingham Six', which resulted in changes to the law (see the Criminal Appeal Act 1995). There is a limited scope for individual politicians to introduce private members' bills, by which means small changes may be introduced into our law. Occasionally new laws will be required to meet a sudden emergency and, where necessary, Parliament can act with considerable speed. It is possible for an Act of Parliament to pass through all its stages in both Houses and receive the Royal Assent in a day if necessary.

Once an idea for a piece of legislation has been accepted by the Legislation Committee of the Cabinet there will usually follow an intensive period of deliberation and consultation. Parliamentary Counsel (the draftsmen of the bills) will then be required to draw up the bill clause by clause. Many bills will be significantly redrafted before they are enacted in their final version. The problems which beset the draftsmen include: the lack of precision of our language; trying to reconcile many conflicting demands; attempting to cover situations which can be envisaged but which have not yet arisen in practice; and pressure of time. Not surprisingly many Acts of Parliament are subsequently found to create interpretative difficulties which must be resolved in the courts.

Acts of Parliament begin life as bills, which may be either public, private or private member's bills. Public bills comprise the vast majority passed in each parliamentary session and are general in their operation. An example of a private bill may be one promoted by a local authority to authorize some activity specifically in its own local area. Private members' bills provide the limited opportunity available in each parliamentary session for an MP to introduce some proposals for change in the law on a topic of his or her choice; private members' bills are public bills. Public bills are usually introduced by government departments, and may be introduced in either House. The likelihood is that most of these public bills will be passed. A government, particularly one with a large majority, has an effective stranglehold on procedures in Parliament. Non-controversial bills are often introduced in the House of Lords but all money bills must be introduced in the Commons.

The normal procedure in both Houses is that when the bill is introduced it will have its first reading. Normally the sponsors of the bill present it in dummy form at the table of the House and one of the clerks reads out the title. The bill is then deemed to have been read for the first time. It is next ordered to be printed and published, and a date fixed for the second reading. Its second reading is the occasion for a parliamentary debate on the principles of the bill. The bill is not considered clause by clause at this stage. There will be a vote at the end of the debate and the bill could be rejected at this stage. Assuming that the bill is not lost it is then referred to a committee, to undergo its committee stage. It will usually go to a standing committee of between 16 and 50 members, chosen to reflect the relative strength of the political parties in the House, and having regard to the special qualifications, concerns and interests of the MPs in question. These standing committees will consider the bill in detail, examining it clause by clause, trying to produce a result which is unambiguous. The committee will also deal with proposed amendments to the bill. Where a bill is of major constitutional importance, the committee stage may be taken in a Committee of the Whole House. However, the drawback of this system is that, while it sits as a committee, no other business can be conducted by the House.

Many of the amendments proposed to the bill at the committee stage may be put down by the minister in overall charge of the bill himself. These proposed amendments may reflect afterthoughts by him/her or officials, or they may be the result of concessions to outside pressure groups. Once the committee stage is completed, the bill is then reported back to the House (the report stage) when the changes introduced in committee will be outlined. There is then a third reading of the bill, often done immediately after the report stage is concluded, and the debate on the third reading may culminate in a vote.

Where the measures proposed are controversial, both the government and the opposition will be anxious to have as many of their supporters voting as possible, and each may have issued a three-line whip, which is an instruction to their supporters to attend and vote in accordance with instructions. The whip system is the means by which political parties control and organize their members in Parliament. There is a system of whipping in the Lords, but its principal importance is in relation to the activities of the House of Commons. The name comes originally from the world of hunting. The whipper-in is a hunt official charged with the control of the hounds. The word has now transferred itself to the parliamentary context. A whip in Parliament is a person whose job it is to give information to his or her party members and to maintain discipline among them.

Once the bill successfully passes all its stages in both Houses, it receives the Royal Assent and becomes an Act of Parliament. It takes effect from the date on which it receives Royal Assent, or from the date of commencement set out in the Act itself, or from a date to be fixed in the future. Power to fix that date will usually by given to an appropriate minister. The Act then remains in force until it is repealed. Repeals are effected by exactly the same process. It is usual to refer to the Act by its short title, for example, the Sale of Goods Act 1979, or the Arbitration Act 1996. Unless the Act states to the contrary, it will apply throughout the UK.

Nowadays Acts of Parliament follow a fairly standard pattern. Before setting out any of the main text there is a long title which establishes the purpose of the Act. For example, the Arbitration Act 1996 states that it is 'an Act to restate and improve the law relating to arbitration pursuant to an arbitration agreement; to make other provision relating to arbitration and arbitration awards; and for connected purposes.' The main text is then divided into sections and subsections and, if appropriate, the whole Act will be set out in parts. Where the Act contains detailed lists, these will often be contained in a schedule to the Act. Most modern Acts have sections which deal with definitions, repeals, date of commencement and area of operation. The interpretation section in an Act of Parliament is important because within it the draftsman of the Act can set out the precise meaning of a word for the purpose of that Act only. This can be of considerable help to judges when they subsequently need to interpret the exact meaning of the legislation. The repeals section will indicate which earlier Acts or regulations or parts of Acts and regulations have been repealed by the present Act.

## Pressure groups

It is undoubtedly true today that in drawing up and seeking to implement its legislative programme, no government can afford to ignore the views of pressure groups. Joining a pressure group is one of the ways in which individuals have an opportunity to influence government decisions and future legislation. There are two main types of pressure groups – interest groups, where the group exists primarily to promote the interests of its own members (for example, the Automobile Association or the National Farmers Union or the various trade unions), and cause groups, which are usually formed for the purpose of lobbying for a specific cause (for example, Help the Aged, The National Society for the Prevention of Cruelty to Children and the Campaign for Nuclear Disarmament).

The word 'lobbying' means influencing members of the legislature. The word itself derives from the practice of people meeting their MPs in the lobby of the House of Commons. The process of lobbying can be extended over a very long period and requires considerable energy, organization and resources to carry it out effectively. A large organization may employ professional lobbyists to pursue its cause.

The government of the day is likely to be influenced by pressure groups for a number of reasons – the pressure group may well be expert within its particular field or the government may need active cooperation from the members of the pressure group in order to implement its policies. Some pressure groups are immensely wealthy, and can promote their cause by advertising and other forms of publicity, while other pressure groups are powerful in terms of the significant proportion of the community which they represent.

Pressure groups can seek to exercise influence in a number of ways and at a number of points in the legislative process. Before any legislation is introduced, members of pressure groups may be represented on advisory bodies connected with appropriate government departments. If legislation is proposed, consultation with appropriate groups will frequently take place before a bill is introduced. A pressure group will always be working to advertise and promote its cause, possibly by liaising with sympathetic MPs, even when no immediate legislation is proposed. Once a bill has been introduced, MPs sympathetic to particular causes may participate at the detailed committee stage, when numbers of amendments to a bill may be proposed. At that point, the pressure group will wish to stay in touch with its 'tame' MPs. Even after an Act has been passed, a watching and monitoring role can be played by pressure groups who may seek changes by regulations if an Act is seen to have loopholes. Much of the success of a pressure group may depend on making and maintaining good personal contact with MPs, peers who sit regularly in the Lords, civil servants and ministers.

## Delegated legislation

It can easily be seen that pressure on parliamentary time is so great that all the rules necessary in a complex modern society cannot be made in the laborious and time-consuming way previously outlined. It is therefore common for Parliament to lay down a framework in an Act (an 'enabling Act') and then grant power to some other person or organization to make the detailed regulations. For example, by power given under the Building Act 1984 the Secretary of State has made the Building Regulations 2000. (Note that the Building Regulations have been amended by later delegated legislation.)

Delegated legislation can take one of three forms:

1    *Orders in Council* – where the rule-making power is vested in the Privy Council. This is normally confined to emergency situations and was much used during the Second World War.
2    *Statutory instruments* – this is the form commonly used when ministers make regulations by virtue of delegated powers. For example, the Building Regulations are in this form.
3    *Bye-laws* – made by local authorities or other public bodies under powers granted to them by Parliament.

Although delegated legislation can be seen to be necessary and to have some advantages as a rule-making technique, there must be adequate control of the process. The advantages include:

- *speed* – where there is an emergency or a temporary or rapidly changing situation;
- *flexibility* – outdated rules can be more easily changed;
- *expertise* – consultation with outside bodies may harness the experience of technical and other experts.

Despite these advantages, it will be realized that rule making by these methods gives great power to persons who have not been elected. Critics of delegated legislation would argue that it vests too much power in the hands of the executive, and that scrutiny and control of the process are vital. That control may come through the procedures themselves, for example by delegating powers on strict terms to prevent abuse, or by subjecting any regulations proposed to an affirmative resolution procedure of one or both Houses of Parliament, or by requiring prior consultation with an advisory body.

If the power which had been granted is exceeded, control can be exercised by the application of the *ultra vires* principle. This would require a person aggrieved by the abuse of power to claim in court that the rule maker had acted outside the scope of his or her authority, i.e. that the action has been *ultra vires*. If so found by the court, the *ultra vires* action will be declared to be void, i.e. of no legal effect. Parliament itself is aware of the danger of abuse of power and has established committees to scrutinize statutory instruments. Parliament might pass only 50 Acts of Parliament in any one year, yet in the same period more than 3000 Statutory Instruments might be made, so the potential size of the problem can be seen.

## Statutory interpretation

When an Act of Parliament comes into force it may cause disputes as to its true meaning. The written word is not an exact form of communication and it will ultimately be the duty of the judges, when cases arise on disputed points in an Act, to determine the 'intention of Parliament'.

Faced with such a problem, the judge starts by looking at the Act itself. Most Acts contain a definition section, in which words will be given a specific meaning for the purposes of that Act only. For example, in s.14 of the Unfair Contract Terms Act 1977, definitions are given of the words 'business', 'goods' and 'notice', thereby giving those words a specific and restricted meaning confined to that Act. 'Business' is defined there as including a profession and the activities of any government department or local or public authority.

Where no definition is given within the Act itself, judges may take the meaning of a word from a dictionary, but if they do so, they must bear in mind the principle expressed in Latin, *noscitur a sociis*, which means 'a word is known by the company it keeps'. The meaning of a word may well be governed by the context in which it is used, e.g. in the phrase used in another Act, 'interest, annuities or other annual payments', the word 'interest' means annual interest. The judge may also turn to the Interpretation Act 1978, which gives meanings for standard phrases and expressions like day, week and month. And again, the word or phrase in question may have been interpreted and applied in an earlier case heard by the court, so that the present judge would be obliged by the rules of judicial precedent to accept that meaning in the case before him or her.

Assuming that these aids do not solve the problem, the judge may then adopt one of the following approaches to the question:

1   *The literal approach* – this approach dictates that words should be given their ordinary, plain and natural meaning, even if this produces an absurd result. Judges favouring this approach take the view that it is then Parliament's responsibility to put the situation right by amending the law. In using this approach, two general rules of construction should be remembered:

   (a)   The *euisdem generis* rule – when specific words are followed by general words, the general words must be read in the light of the specific words, for example 'on a pond, lake, river, canal or other water'. By applying this rule, the words 'other water' would be read to include a reservoir but not the open sea. It is clearly intended by the specific words to confine the stretches of water to inland water.

(b) The rule *'expressio unius est exclusio alterius'* – if one thing is specifically mentioned in the Act, the Act is presumed not to apply to another, e.g. some sections of an Act may refer to the seller and the buyer. A subsequent section mentioning only the seller is presumed not to apply to the buyer.

The literal approach to interpretation is not favoured by some judges. One judge has said: 'When a defect appears a judge cannot simply fold his hands and blame the draftsman.'

2   *The golden rule* – this allows the judge to interpret so as to avoid the absurdity, or to avoid a result which is inconsistent with the Act itself or common law principles. This approach was used in *Re Sigsworth*, in a case where a man claimed his mother's property, despite having murdered her. Under the Act which governed the distribution of her property, this ought to have been shared amongst her children. The son was the only child. The judge held that the common law rule that a murderer cannot benefit from the person he has murdered prevailed over the clear words of the Act.

3   *The mischief rule* – another approach adopted where the judge looks to see which defect of the law is intended to be corrected by the Act and then interprets to correct the defect.

An important development in the interpretation of statutes was seen in *Pepper v. Hart*. In this case the House of Lords changed the existing judicial practice of not allowing reference to be made to parliamentary materials in the interpretation of statutes. The use of Hansard (a record of debates in Parliament) is now allowed within the following limits:

(a)   the legislation must be ambiguous or unclear or a literal interpretation would lead to absurdity;

(b)   the material relied upon consists of statements made by the minister or other promoter of a bill together with such other parliamentary materials as is necessary to understand such statements; and

(c)   the statements relied upon are clear.

This allows the courts to collect the mischief from another source outside the words of the legislation. The judges also sometimes refer to a purposive approach to interpretation, meaning that the courts' task is to give effect to Parliament's purpose, i.e. the object of the legislation.

A simple example from a case may illustrate some of the problems of interpretation faced by judges. In a case concerning single payments under the old Supplementary Benefit scheme, a claimant requested a single payment to help towards the purchase of wallpaper to redecorate her lounge and kitchen. The wording of the appropriate regulation required that payments could only be made in respect of expenses of *essential* internal redecoration. The court had to consider what was the meaning of the word 'essential'. It was not defined anywhere in the Supplementary Benefits Act 1976 or the relevant Regulations and it fell, therefore, to be given its ordinary meaning in everyday use. When a dictionary was consulted, the *Shorter Oxford English Dictionary* gave two rather differing meanings of the word 'essential'. It can mean 'material' or 'important' in the sense of 'your work is essential to the success of this project', or it can mean 'indispensably requisite', as for example 'water is an essential ingredient of a cup of tea'. Relying on the stricter meaning of the word, the benefit officer had refused a single payment. On appeal to the tribunal, however, the judge favoured the less strict interpretation of the word. He was guided in doing so by the fact that the same word 'essential' was used elsewhere in the Regulations, where its meaning tended to be 'material' and 'important' rather than 'indispensably requisite'.

Following the introduction of the Human Rights Act 1998 the courts are, so far as it is possible to do so, to interpret Acts of Parliament and delegated legislation to ensure compatibility with Conventions Rights (see page 25). It is not possible for the courts to declare legislation invalid on the ground of incompatibility.

No one instructs a judge as to which of these approaches should be used. It is clear that the process of interpreting and applying Acts of Parliament gives considerable power to the judges. But their role in the law-making process is not limited to the interpretation of statutes because the other principal source of law is judicial precedent (binding decisions of the judges).

## Judicial precedent

When a judge gives his decision in a case before him, this has two elements:

1    The actual decision affecting the parties, e.g. X is found to be liable and must pay damages to Y of £1000.
2    The principles of law which have caused the judge to arrive at that decision.

Where the case concerns facts or situations which are comparable with earlier cases heard by the courts, the judge will normally apply the principles from the earlier cases. In one well-known case, *Donoghue v. Stevenson*, a soft-drinks manufacturer was held to be liable to the ultimate consumer who had suffered physical harm by drinking a bottle of the manufacturer's ginger beer which was contaminated by the remains of a decomposed snail. If, on a future occasion, a court hears a case brought by a person made ill by eating a pie containing a dead mouse, the same principles can be applied. By analogy these principles may be extended further. The principle in the 'ginger beer' case was actually applied to a later case where a person contracted dermatitis because of wearing underpants where chemicals had been left in the material during their manufacture.

Although many systems of law in practice apply this principle of deciding cases in the same way as previous cases of a similar nature, the English system goes further. The judge is bound by the earlier decisions, hence the system is often referred to as the system of binding precedent. In the English legal system there is a hierarchy of courts, so the Supreme Court of the United Kingdom (formerly the House of Lords) binds all the courts beneath it by its decisions. The Court of Appeal, both Civil and Criminal Divisions, is bound by the Supreme Court and binds all the courts below. The High Court is bound by the Supreme Court and the Court of Appeal and binds all the courts beneath it. Finally, the Crown Court, magistrates' courts and the county courts are bound by all the courts previously mentioned but do not bind any courts. Also note that some courts are bound by their own previous decisions and some are not. For example, the Supreme Court is not strictly bound by its own previous decisions, but usually follows them, whereas the Court of Appeal is bound by its own previous decisions, subject to certain exceptions.

One difficulty presented by the doctrine of precedent lies in finding the actual part of the case which is binding. When a judge gives a decision, his legal reasoning, when set out in the law reports, can be several pages long. Much of what is said may relate to why he or she rejected other arguments put by counsel, or what would have influenced him or her if the facts had been slightly different. These parts of the judgment are not binding in later cases. They are 'said by the way' or in legal terms, *obiter dicta*. The judge in a later case must search to find the rule of law upon which the decision is based, properly called the *ratio decidendi*.

The process of identifying the ratio is a vital part of the training of a lawyer. In each case the first step is to establish the material facts, meaning facts relevant to the decision. For example, Mrs Smith, a mother of four aged 45, knocked down and injured Mr Brown last Friday. The accident was caused because Mrs Smith had gone to sleep at the wheel. The only material facts here are that a motorist injured a pedestrian by falling asleep at the wheel of a car. Once the material facts have been established, if they correspond with the material facts of an earlier decision, the judge will usually be bound by the principle of law in the earlier case. However, judges are normally only bound by decisions given in a court superior to their own. This allows the system to be flexible (as it is open to a higher court to overrule decisions of lower courts) but certain (as it obliges lower courts to follow the decisions of higher courts). The structure of the courts is considered in Chapter 3.

## Advantages and disadvantages of the system of precedent

One of the greatest advantages of the system of binding precedent is that the rules have evolved from real-life cases and are, therefore, essentially practical. Again, the 'binding' feature of the system makes it reasonably certain. However, two major criticisms are often levelled against the system. The first is that it creates a bulky system of very detailed rules which requires the production, at considerable expense, of large numbers of law reports to enable the previous cases to be checked and quoted. A more significant problem is that the system can be very uncertain, due to the powerful role played by the judges. Consider the case where a judge is faced with a precedent which in theory is binding on him, but which he prefers not to apply. There is a technique called 'distinguishing' one case from another which may allow him to avoid applying the earlier decision. The judge is allowed to distinguish (and therefore not apply) any case where the material facts are not the same as in the case before him. When this technique is properly used it allows the law to be flexible. When improperly used it results in hair-splitting distinctions, which can leave two apparently conflicting decisions reported in the casebooks.

So that the system of binding precedent can function effectively there must be a well-organized procedure for reporting cases and there must be an established hierarchy of the courts. Generally, lower courts are bound by their own previous decisions and by those of higher courts.

If a person is dissatisfied with the outcome of a case he or she may have grounds for appeal. The final court of appeal in the English system is the Supreme Court of the United Kingdom (formerly the House of Lords). In relation to European Community law, the decisions of the Court of Justice of the European Communities (the European Court) bind the Supreme Court and all other English courts. Until 1966, even the House of Lords was bound by its own previous decisions, but it was then announced that: 'Their Lordships propose to modify their present practice, and while treating former decisions of this House as normally binding, to depart from a previous decision when it appears right to do so.' An example of such a departure can be seen in *Murphy v. Brentwood D.C.* in 1990 (see page 260).

## Applying a precedent

Examples taken from actual cases demonstrate how the case law system works. To see how the normal processes of applying a precedent operate, look first at a case from 1932, *Donoghue v. Stevenson*. A manufacturer of ginger beer which was marketed in an opaque bottle was held to be liable to the ultimate consumer of the drink in the tort of negligence. (A tort is a civil wrong.) The consumer had become ill on discovering the remains of a decomposed snail in the bottle. If this precedent had subsequently been confined by judges to sets of facts involving drinks sold in opaque bottles its value as a precedent would have been very limited. However, in 1936, another court seized the opportunity to extend the scope of the precedent. In the case of *Grant v. Australian Knitting Mills*, the claimant, Dr Grant, had purchased a pair of long woollen underpants, manufactured by the defendants. He contracted a severe form of dermatitis after wearing them, shown to have been caused by excessive amounts of sulphur which had not been successfully washed out during the manufacturing process. Were the defendants liable? The judges examined the Donoghue case, and extracted from it the following statement of principle:

A manufacturer of products which he sells in such form that he intends them to reach the ultimate consumer with no reasonable possibility of intermediate examination, and where the absence of reasonable care in the preparation of the products will result in injury to the consumer, owes a duty to the consumer to take reasonable care.

Undoubtedly a manufacturer of underpants must realize that potential customers cannot carry out scientific tests to measure sulphur levels before purchasing, and the courts had no difficulty in applying the Donoghue principle. If the court in the Grant case had thought the principle was unsound, it could have sought to distinguish the two cases on the ground that the material facts were not similar. For example, the snail case concerned goods to be consumed internally, whereas the underpants case concerned goods to be worn externally. Such fine distinctions could have been justified if the earlier precedent had been unpopular for any reason, but the law develops best when the scope of a sound principle is gradually extended.

## Common law and equity

The expression 'common law' is frequently used and can have a number of meanings. It may mean the law which is common throughout England. Indeed, the phrase was originally used in that sense to distinguish it from local rules and customs. The expression may also mean rules developed through precedents, rather than created by Acts of Parliament. It may further be used to mean rules which are not derived from equity. An unusual system emerged in England between the twelfth century and 1875, whereby a completely separate system of courts developed, administering quite separate rules – the principles of equity. These rules were usually developed to meet situations where the common law had no remedy to offer or where the common law was in some way deficient or unjust. The equitable principles are discretionary in nature, but are part of a body of case law which operates according to the doctrine of precedent. Since 1875, the principles of equity have been administered side by side with the rules of common law. If the two sets of rules should ever conflict, the principles of equity will prevail. Examples of equitable principles are promissory estoppel and undue influence; an equitable remedy is specific performance.

## The European Convention on Human Rights

The United Kingdom (UK) has been a signatory to the European Convention on Human Rights since 1950 and UK citizens have, since 1966, been able to take cases to the European Court of Human Rights in order to obtain a remedy therefrom. Moreover, human rights arguments based upon the Articles of the Convention have long been deployed before UK courts. It was only in October 2000, however, when the majority of the provisions of the Human Rights Act 1998 came into force, that the Convention became part of domestic UK law.

Before considering the significance of the Human Rights Act 1998, it is first necessary to briefly consider both the nature of the rights that the Convention's Articles protect and that of the European Court of Human Rights itself.

## The Convention rights

### Article 1

Article 1 of the Convention requires the contracting parties (including the UK) to secure the rights that the Convention protects to those within their jurisdiction. Article 1 has not been incorporated into UK law but, given the obligations imposed upon public authorities by the Human Rights Act 1998, this does not appear to be problematic.

### Article 2

Article 2 requires that the right to life must be protected by law. Thus, as a general rule, the state (e.g. the police) should not intentionally deprive the citizen of his life. Further, as the decision of the European Court of Human Rights in *Osman v. UK* made clear, the state must take appropriate steps to protect life (e.g. where the life of a person is at risk from the unlawful acts of a third party). Moreover, the decision of the European Court of Human Rights in *McCann v. UK* demonstrates that where death results from the use of force by agents of the State, Article 2 implicitly requires an effective official investigation into the killing.

The intentional deprivation of life by the state does not contravene Article 2 where the use of force which resulted in the deprivation of life was absolutely necessary in order to defend a person from unlawful violence, in order to effect a lawful arrest or prevent a person who had been lawfully detained from escaping or for the purpose of lawfully quelling a riot or insurrection. Whilst Article 2 does not prevent the execution of a lawfully imposed death penalty following a criminal trial, the UK is a party to the Convention's Sixth Protocol, which requires the abolition of the death penalty, and this has been incorporated into the Human Rights Act 1998.

### Article 3

Article 3 prohibits torture or inhuman or degrading treatment or punishment. Thus, for example, the state (e.g. the police) must not torture a person suspected of a crime when they are interviewing him. Equally, the state must protect persons from torture, e.g. by ensuring that the law prevents parents from violating the Article 3 rights of their children (*A v. UK*). The Article 3 right is an absolute right, from which derogation is not permitted (see Article 15 below). Thus, for example, even in the context of the fight against terrorism, the Article 3 rights are absolute (*A and Others v. UK*).

## Article 4

Article 4 prohibits the holding of persons in slavery or servitude and also provides that persons shall not be required to perform forced or compulsory labour. Article 4 does not, however, prohibit what would otherwise be termed forced or compulsory labour where it is required in the ordinary course of detention or conditional release from detention, where it amounts to service of a military character or is required of a conscientious objector as an alternative to compulsory military service, where it is required in the context of an emergency or a calamity which threatens the life or well-being of the community, or where it forms part of normal civic obligations. Like the Article 3 right, the Article 4 right is an absolute right, from which derogation is not permitted (see Article 15 below).

## Article 5

Article 5 protects the person's right to liberty and security of person, providing that a person shall only be deprived of his or her liberty in any of a number of cases specified by Article 5 and only in accordance with a procedure prescribed by law. The requirement that the procedure must be prescribed by law means that the domestic law must permit the arrest or detention, must be sufficiently precise to enable persons to understand when they might be affected by it and must be accessible (e.g. must not be secret) (*Hashman and Harrup v. UK*). Because the Article 5 right is subject to specified exceptions, however, the right is not an absolute right but, rather, is a limited right. The situations in which Article 5 does not prevent a person's liberty from being taken away comprise: lawful detention after conviction by a criminal court; lawful arrest or detention for non-compliance with a court order or to secure the fulfilment of legally prescribed obligations; lawful arrest or detention on reasonable suspicion of committing an offence to bring a person before the competent legal authority or where reasonably necessary either to prevent the commission of an offence or to prevent flight after its commission; lawful detention of a minor for educational supervision or to bring him before the competent legal authority; lawful detention to prevent the spreading of disease or of persons of unsound mind, alcoholics, drug addicts or vagrants; or lawful arrest or detention to prevent unauthorized entry into the country or in the context of action to deport or extradite a person. These six exceptions to the Article 5 right only apply where detention is 'lawful', however, and, consequently, even where detention is in accordance with national law and one of the six exceptions to the Article 5 right would otherwise seem to apply, detention will still give rise to a violation of Article 5 if it is arbitrary (*Saadi v. UK*). Article 5 also, essentially, gives a person who is arrested or detained the right to be informed of the reasons for his arrest and of the charges against him; the right to be brought promptly before a judge or a person authorized to exercise judicial authority and to be tried within a reasonable time and released on bail (*Wemhof v. Germany*), the right to take proceedings in court to speedily determine the lawfulness of the detention and the right to compensation where arrested or detention is in contravention of Article 5.

## Article 6

Article 6 guarantees the right to a fair trial. Thus, in the context of determination of his civil or criminal rights, a person is entitled to a fair and public hearing which takes place within a reasonable time and is held by an independent and impartial tribunal that has been established by law. Article 6 requires that the court must pronounce its verdict in public but permits the exclusion of the press or public from the trial for a variety of specified reasons, e.g. in the interests of national security, in the interests of juveniles or to protect the private lives of the parties. Article 6 expressly provides that a person charged with a criminal offence shall be presumed innocent until proven guilty in accordance with law. Article 6 also expressly gives a person charged with a criminal offence a number of minimum rights, namely: to be informed promptly of the nature and cause of the charge; to have adequate time and facilities to prepare a defence; to defend him- or herself in person or through chosen legal advisers or, if he or she cannot afford legal assistance, to be provided with free legal assistance where this is required by the interests of justice; to examine the witnesses against him or her or to have them examined and to secure their attendance and examination under the same conditions as these witnesses are required to attend and are examined; and to have an interpreter free of charge if he or she cannot understand the court or cannot speak the language used. In determining whether there has been a violation of Article 6, the issue is whether the proceedings as a whole, including the possibility that an appeal may have cured an earlier defect, were fair (*Edwards v. UK*). A trial may be unfair even though there has been no violation of the minimum rights which the Article guarantees (*Edwards v. UK*). Thus, for example, although these are not expressly mentioned by Article 6, aspects of a fair criminal trial include the right of silence and the right not to incriminate oneself (*Saunders v. UK*), and the right of the defence in criminal proceedings to have disclosure of material in the hands of the prosecution (*Edwards and Lewis v. UK*). In determining whether a trial was fair within the meaning of Article 6, it is necessary to consider the issue of 'equality of arms'. The significance of 'equality of arms' is, essentially, that the conditions under which a party is permitted to present his case in court should not place him or her at a substantial disadvantage when compared to the conditions under which the other party is permitted to present their case (*Neumeister v. Austria*).

## Article 7

Article 7 provides both that a person shall not be found guilty of a criminal offence in consequence of an act or omission which did not amount to a criminal offence at the time when the person acted or omitted to act and that a person who is convicted of an offence shall not receive a heavier sentence than could have been imposed at the time the offence was committed. Article 7 does not prevent the trial and punishment of a person where, at the time he or she acted or omitted to act, the general principles of law recognized by civilized nations would have classified this act or omission as criminal. Moreover, Article 7 does not prevent the normal development of the law via case law provided that such development (e.g. a change in the common law to the effect that a man can rape his wife) is reasonably foreseeable (*SW v. UK*).

## Article 8

Article 8 concerns the right of persons to respect for their private and family life, for their home and for their correspondence. Article 8 does, however, permit interference with the exercise of the Article 8 right by a public authority if such interference is in accordance with the law and is necessary in a democratic society for any of a number of specified reasons. These reasons comprise the interests of national security, the interests of public safety, the country's economic well-being, the prevention of disorder or crime, the protection of health or morals, and the protection of the rights and freedoms of others. The requirement that the interference must be in accordance with law means that domestic law must permit the interference, must be sufficiently precise to enable persons to understand when they might be affected by it and must be accessible (e.g. must not be secret) (*Sunday Times v. UK*). The requirement that the interference must be necessary in a democratic society means that it must correspond to a pressing social need and must be proportionate to the legitimate aim that it pursues (an example of the application of the concept of 'proportionality') (*S and Marper v. UK*). In assessing whether this is so, the domestic authorities are allowed a 'margin of appreciation' (whereby the European Court will respect the domestic authorities' decision making and effectively grant them discretion as to how to give effect to the rights), though the margin of appreciation will tend to be narrow if the effective enjoyment by an individual of intimate or key rights is at stake (*S and Marper v. UK*). Thus, the Article 8 right was violated where, for example, personal data, in the form of fingerprint and DNA information relating to persons who had not been convicted of offences, was retained by the police under powers which the European Court of Human Rights regarded as having a 'blanket and indiscriminate nature' (*S and Marper v. UK*).

## Article 9

Article 9 protects a person's right to freedom of thought, freedom of conscience and freedom of religion, and encompasses both a person's right to change religion or belief and the freedom, whether alone or with others and whether in private or in public, to manifest this religion or belief via worship, teaching, practice or observance. Article 9 provides, however, that a person's freedom to manifest religion or beliefs may be subject to limitations which are prescribed by law and are necessary in a democratic society for any of a number of specified reasons. These reasons comprise the interests of public safety, the protection of public order, health or morals, and the protection of the rights and freedoms of others. The fact that a Christian employee was required by her contract of employment to work on Sundays and was dismissed when she refused to do so did not violate her Article 9 right because the matter was contractual, she had been free to resign and she had not been dismissed from her employment because of her Christian beliefs (*Stedman v. UK*). The meaning of 'prescribed by law' is the same as that of 'in accordance with the law', which was considered in the context of Article 8 above.

## Article 10

Article 10 concerns the right to freedom of expression, including freedom to hold opinions and receive or impart information and ideas in the absence of interference by public authorities and regardless of frontiers. Article 10 does not prevent states from requiring broadcasting, TV or cinema enterprises to be licensed. The exercise of the Article 10 freedoms may be subject to formalities, conditions, restrictions or penalties which are prescribed by law, provided that they are necessary in a democratic society for one or more of a number of specified reasons. These reasons comprise: the interests of national security, territorial integrity or public safety; the prevention of disorder or crime; the protection of health or morals; the protection of the reputation or rights of others; prevention of the disclosure of information received in confidence; and maintaining the authority and impartiality of the judiciary. An aspect of the right of freedom of expression is the right of journalists to protect their sources (*Goodwin v. UK*), though a court order requiring a journalist to disclose a source will not violate Article 10 if it is carried out in accordance with law and is necessary in a democratic society for one or more of the reasons specified above. Even where interference with the Article 10 right serves a legitimate aim, however, the interference will not be necessary in a democratic society, and thus will not be justified if, in the circumstances, the method used to achieve the legitimate aim was not a proportionate response to the legitimate aim (*Goodwin v. UK*). The domestic authorities do possess a 'margin of appreciation' in determining whether there is a 'pressing social need' for imposing a restriction upon the right to freedom of expression but, in the context of disclosure of journalistic sources, the interests of maintaining a free press in a democratic society possesses considerable weight when the European Court of Human Rights is determining whether an interference with the right of freedom of expression which had a legitimate aim was proportionate to that legitimate aim (*Financial Times and Others v. UK*). The meaning of 'prescribed by law' is the same as that of 'in accordance with the law', which was considered in the context of Article 8 above.

## Article 11

Article 11 protects a person's right to freedom of assembly and his right to freedom of association, including the right to form or join a trade union. Restrictions may be placed on the exercise of the Article 10 rights provided they are prescribed by law and are necessary in a democratic society for one or more of a number of specified reasons. These reasons comprise the interests of national security or public safety, the prevention of disorder or crime, the protection of health or morals and protection of the rights or freedoms of others. Article 11 does not prevent members of the police, the armed forces or the administration of the state from imposing lawful restrictions on the exercise of the Article 11 rights. The Article 11 rights appear to encompass the right not to join an association, such as a trade union (*Young v. UK*) and the right of peaceful protestors to state protection (*Plattform 'Arzte Fur Das Leben' v. Austria*). The meaning of 'prescribed by law' is the same as that of 'in accordance with the law', which was considered in the context of Article 8 above.

## Article 12

Article 12 concerns the right of men and women of marriageable age to marry and found a family according to domestic laws which govern the exercise of the Article 12 right. It appears that the Article 12 right does not apply to marriages between persons of the same sex or to transsexuals (*Cossey v. UK*), though the European Court of Human Rights has recognized the right of a transsexual to marry in her new gender (*Goodwin v. UK*).

## Article 13

Article 13, which concerns the right to an effective remedy when a person's Convention rights and freedoms have been violated, has not been incorporated into UK law, though sections 7 and 8 of the Human Rights Act 1998 (which are considered below) were intended to give effect to the obligations imposed by Article 13.

## Article 14

Article 14 provides that persons are entitled to enjoy the Convention rights regardless of discrimination on grounds such as, but not restricted to, sex, race, colour, language, political or other opinion, national or social origin, association with national minorities, birth or status of some other sort. Article 14 does not have an independent existence but, rather, only applies in the context of other Convention rights (*Burden v. UK*). Article 14 may be violated where people in similar situations are treated differently, but differences in treatment do not violate Article 14 if they pursue a legitimate aim and the method used to achieve the legitimate aim is a proportionate response to the legitimate aim (*Burden v. UK*). In assessing whether differences are justified, the national authorities are allowed a 'margin of appreciation' which is usually wide in the context of matters of economic or social strategy (*Burden v. UK*).

## Article 15

Article 15 gives a 'High Contracting Party' (e.g. the UK) limited power to derogate from Convention obligations during a war or public emergency (e.g. in the context of terrorism) which threatens the life of the nation. Derogation from Articles 3, 4(1) and 7 is not permitted, however, and derogation from Article 2 is only permitted in relation to deaths resulting from lawful acts of war. The European Court of Rights has recognized with regard to the operation of Article 15 that the Contracting States have a wide 'margin of appreciation' when determining whether derogation is necessary but has also made clear that they do not have an unlimited discretion in this regard (*A and Others v. UK*).

## Article 16

Article 16 permits a High Contracting Party (e.g. the UK) to impose restrictions on the freedom of expression, freedom of assembly and freedom of association of foreign nationals.

## Article 17

Article 17 essentially provides that the Convention rights and freedoms do not give any state, group or person the right to perform acts which are either aimed at the destruction of the Convention rights and freedoms or which are aimed at limiting those rights and freedoms to a greater extent than the Convention itself makes provision for.

## Article 18

Article 18 provides that the restrictions that the Convention permits to the Convention rights and freedoms must not be used for any purpose other than those which the Convention prescribes.

## First Protocol

The effect of Article 1 of the Convention's First Protocol is, essentially, that persons (including non-natural persons such as companies) are entitled to the peaceful enjoyment of their possessions, shall not be deprived of their possessions other than where this is in the public interest and shall only be so deprived subject to the conditions for which domestic law and the general principles of international law provide. Article 1 of the First Protocol does not, however, restrict the right of a state such as the UK to enforce the use of those domestic laws that it believes necessary either to control the use of property in the general interest or to secure the payment of taxes, etc.

Deprivation of possessions, interference with peaceful enjoyment of possessions or control of possessions by the state can only be justified where there is a fair balance between the interest of the community and the interests of individuals (*Sporrong v. Sweden*), and the domestic law which provides for such deprivation, interference or control must be accessible, must be sufficiently certain and must provide protection against arbitrary decisions. In determining whether such a fair balance has been achieved in the context of a deprivation of property, one matter which will be of relevance will be the adequacy of the compensation, if any, payable to the property owner, though it appears that the public interest is capable of justifying a level of compensation below market value (*Lithgow v. UK*). Examples of matters which are potentially capable of giving rise to issues under Article 1 of Protocol 1 in appropriate circumstances include planning decisions, compulsory purchase or nationalization of property and statutory controls restricting the ability of landlords to recover possession of rented property.

Article 2 of the First Protocol concerns the right to education, providing that no one shall be deprived of that right and that the state must respect the parental right to ensure that education and teaching conform with their religious and philosophical convictions. The right to education does not require the state to provide education but, rather, imposes conditions where the state does provide education. A UK reservation concerning Article 2 of Protocol 1 (which has never been tested by the European Court of Human Rights) means that the right to education has only been accepted by the UK to the extent to which it is compatible with efficient instruction and training and avoiding unreasonable public expenditure.

Article 3 of the First Protocol requires the High Contracting Parties (e.g. the UK) to hold free elections by secret ballot at reasonable intervals under conditions which ensure the free expression of the opinion of the people. This has led, for example, to prisoners being able to successfully challenge the blanket ban on their right to vote in elections (*Hirst v. UK*).

The Convention's other Protocols are not further considered in this book.

## The European Court of Human Rights

The European Court of Human Rights has one judge from each of the High Contracting Parties (one of which is the UK) who are elected from a list of three candidates nominated by the High Contracting Party and sit independently, i.e. not as a representative of their country. They must be persons of high moral character who are either qualified for high judicial office or are juriconsults of recognized competence. They sit for six years and may be re-elected, but their terms of office expire at the age of 70. When Protocol No. 14 has been ratified by all the High Contracting Parties (see below), judges will be elected for a nine-year period but when the nine-year period has ended, re-election will not be permitted.

For the purpose of considering cases, the European Court sits in Committees of three judges, Chambers of seven judges and a Grand Chamber of 17 judges. The full Court elects its President, Vice President, Presidents of Chambers, etc. Committees may, upon a unanimous vote, finally declare applications from individuals, groups of individuals or non-governmental organizations inadmissible, or strike them out, where they can make such a decision without further examination. Unless a Committee declares such an application inadmissible or strikes it out, a Chamber will decide both on the admissibility and on the merits of the application. A Chamber also determines the admissibility and merits of applications by one High Contracting Party against another. Unless a party objects, a Chamber may relinquish jurisdiction to the Grand Chamber where a serious question concerning the interpretation of the Convention or protocols arises or where the result of the case might be inconsistent with an earlier decision of the Court. A party to the judgment of a Chamber may, within three months, request the referral of the case to the Grand Chamber. The request will be considered by a panel of five judges of the Grand Chamber and will be accepted if it raises a serious question concerning the Convention's interpretation or application or concerning the interpretation of application of a protocol or if it raises a serious issue of general importance. If the request is accepted the judgment of the Grand Chamber will determine the case. The Grand Chamber may also give advisory opinions on legal questions concerning the interpretation of the Convention and protocols where requested to do so by the Committee of Ministers.

The European Court of Human Rights can only receive applications from a person, group of persons or non-governmental organization claiming to be the victim of violation of Convention rights by a High Contracting Party (e.g. the UK). The applicant may be a person who has been affected by the violation or is under a real risk of being so affected. Thus, the family of a deceased person may be a victim (*McCann v. UK*) but it appears that a pressure group is not a victim if not itself a victim.

The jurisdiction of the Court encompasses all matters concerning the interpretation and application of the Convention and its protocols. The Court can only deal with a matter if all domestic remedies have been exhausted and within six months of the making of the final decision. It will not deal with anonymous applications, with applications that are substantially the same as matters that the Court has already examined or with applications that have been submitted to other forms of international investigation or settlement and do not contain new relevant information. It will declare an application inadmissible if it is incompatible with the Convention, is manifestly ill-founded or is an abuse of the right of application. It may reject an application which it considers inadmissible at any stage of the proceedings. It may strike an application out if the applicant does not intend to pursue the application or the matter has been resolved or if for any other reason the examination is no longer justified.

Where an application is admissible the Court will pursue the examination of the case together with the parties' representatives and undertake an investigation if necessary. It will also place itself at the disposal of the parties in an attempt to secure a friendly settlement. If a friendly settlement is effected the Court will strike the case out. A hearing will normally be in public but proceedings to secure a friendly settlement will be confidential.

Where no friendly settlement is reached, if the Court finds that the Convention or the Protocols have been violated and the relevant domestic law (e.g. UK law) only allows for partial reparation, the Court will, if necessary, afford 'just satisfaction' (i.e. compensation). A judgment of a Chamber becomes final either when the parties declare that they will not request referral to a Grand Chamber or if no such request has been made within three months of the judgment or when the request is rejected by the panel of the Grand Chamber.

In order to assist the European Court of Human Rights to cope with its increasing workload, a number of changes will be made by Protocol No. 14, once this has been ratified by all of the High Contracting Parties. Two of the main changes that Protocol No. 14 will make concern applications from individuals, groups of individuals or non-governmental organizations. First, a single judge will be able to declare such applications inadmissible, or strike them out, if the judge can make such a decision without further examination (though a single judge will not be permitted to examine an application against the High Contracting Party that elected the single judge). Secondly, Committees of three judges will be able to declare a case admissible and render a judgment on the merits if the underlying question in the case is the subject of well-established case law of the European Court of Human Rights. Whilst, at the time of writing, one of the High Contracting Parties had still not ratified Protocol No. 14, the provisions of Protocol No. 14 concerning single judges and the competence of Committees of three judges are already applicable on a provisional basis in relation to applications brought against High Contracting Parties who have consented to be bound by Protocol No. 14 bis (the UK having so consented). Probably the most important change which Protocol No. 14 will make when all of the High Contracting Parties have ratified it is a new admissibility criterion, the effect of which will be that even where there has been a violation of a Convention right, cases that have already been considered by a domestic tribunal will be rejected if 'the applicant has not suffered a significant disadvantage unless respect for human rights ... requires an examination of the application on the merits'.

A principle which the European Court of Human Rights applies in appropriate circumstances when determining whether an Article of the Convention has been violated is that of the 'margin of appreciation'. This essentially means that when determining matters such as whether interference with a Convention right is 'necessary in a democratic society', the European Court of Human Rights recognizes that the authorities of the individual state concerned are better placed than the European Court to determine whether this is so in the circumstances of their own nation (*Handyside v. UK*). Whilst the European Court recognizes that this is so it also recognizes, however, that the final decision is one for the Court, not for the state, and that the principle of the margin of appreciation does not give the state unlimited power to determine what is best for its people. The principle of the margin of appreciation is not a principle to be applied by the UK courts when deciding whether there has been a violation of Convention rights. The UK courts will, however, be required to apply the concept of proportionality (examples of the application of which were seen above) and, in so doing, in a democracy, may, at times, find it appropriate to defer to the opinion of an elected body or individual (*R v. DPP ex parte Kebelene*). Examples of the application of this principle were identified in the context of specific Articles of the Convention, which were examined above.

It should also be noted that the Convention is a 'living instrument' which must be interpreted in the light of current values and of social change (*Tyrer v. UK; Cossey v. UK*). Consequently, there is no strict doctrine of precedent in Convention law and a previous decision of the European Court of Human Rights will not be followed by the European Court on a subsequent occasion if it does not reflect current values. This is a matter which the UK courts are required to take into account when interpreting the Convention and considering earlier decisions of the European Court of Human Rights.

## The Human Rights Act 1998

The majority of the provisions of the Human Rights Act 1998 (HRA) came into force in October 2000. Essentially, its effect was to make the Convention part of UK law, though, as was seen above, not all of the Convention rights embodied in the Convention's Articles and Protocols became part of UK law. Thus, for example, s.1 of the HRA provides that, for the purposes of the HRA, the 'Convention rights' do not include the rights set out in Article 13.

When the UK courts are required to interpret Convention rights, the effect of s.2 of the HRA is essentially that they must take into account relevant decisions of the European Court of Human Rights, decisions of the European Commission of Human Rights (which no longer exists) and decisions of the Committee of Ministers. The Act does not require the UK courts to follow decisions of the European Court of Human Rights, though where they do not do so an aggrieved party will still have the option of taking the matter to the European Court of Human Rights itself.

Section 3(1) of the HRA requires the UK courts, so far as this is possible, to read both primary and subordinate legislation in a way which is compatible with the Convention rights. This means that unless it is impossible to interpret legislation in a way that is compatible with the Convention rights, the UK courts must interpret it so as to give effect to those rights. Consequently, the operation of s.3(1) may require the courts to adopt an interpretation of unambiguous legislation which is contrary to its literal meaning (*R v. A*).

At times the courts may find that legislation is irredeemably incompatible with a Convention right. In the case of primary legislation (i.e. Acts of Parliament plus certain other types of legislation), the court, under HRA s.3(2), must give effect to the legislation even though it is incompatible with the Convention right. In the case of subordinate legislation (i.e. many but not all forms of delegated legislation), the courts, again under HRA s.3(2), must give effect to the legislation if primary legislation prevents the removal of the incompatibility. If this is not the case, however, then the validity, continuing operation or enforcement of subordinate legislation may be affected by its incompatibility with a Convention right.

Where a court determines that primary legislation is incompatible with a Convention right, the court may, under HRA s.4, make a declaration of incompatibility. The Court may also make a declaration of incompatibility where subordinate legislation is incompatible with a Convention right, if primary legislation prevents the removal of the incompatibility. Not every court can make a declaration of incompatibility; in English law the power to do so is restricted to the Supreme Court, the Judicial Committee of the Privy Council, the Courts-Martial Appeal Court, the Court of Appeal and the High Court. When the Court is considering making a declaration of incompatibility the Crown must be notified and a Minister of the Crown, a person nominated by a Minister or some other person permitted by s.5 of the HRA is entitled to be joined as a party to the proceedings. Where such a person is joined as a party to criminal proceedings and a declaration of incompatibility is made, the person may, with leave, appeal to the Supreme Court against the declaration of incompatibility. Between October 2000 and January 2009, 26 declarations of incompatability were made, of which 16 had become final and were not subject to further appeal (*Responding to Human Rights Judgments Cm 7524*).

A declaration of incompatibility does not affect the validity, continuing operation or enforcement of the relevant legislation and does not bind the parties to the relevant proceedings. Rather, the significance of a declaration of incompatibility is that, under HRA s.10, if no appeal is being made from the relevant decision or after any appeal has been determined, the Government may, by making a 'remedial order', amend the relevant primary legislation so as to remove the incompatibility (or so as to allow the incompatibility between subordinate legislation and a Convention right to be removed). A remedial order may also be made where, following a decision of the European Court of Human Rights in proceedings brought against the UK, it appears that UK legislation is incompatible with UK obligations under the Convention. A remedial order may be made with retrospective effect, i.e. it may take effect from a date earlier than the date on which it was made. The use of a remedial order may be appropriate where legislation requires minor amendment but may not be appropriate if substantial amendment is required to primary legislation or the issue is controversial. This is so because whilst remedial orders are subject to Parliamentary scrutiny, they cannot be amended in Parliament. Thus, by January 2009, the Government was disappointed that it had not been able to make greater use of remedial orders (*Responding to Human Rights Judgments Cm 7524*).

HRA s.6 makes it unlawful for a 'public authority' to act in a way which is incompatible with a Convention right. Section 6 also makes a failure to act in a particular way unlawful where this failure is incompatible with a Convention right. A failure to introduce or lay proposed legislation before Parliament or to make primary legislation or a remedial order will, however, never be unlawful under s.6. Moreover, it is not unlawful under s.6 for a public authority to act in a way which is incompatible with a Convention right if in consequence of primary legislation the public authority could not have acted differently or if the public authority was acting so as to give effect to or enforce legislative provisions which could not be read or given effect to in a way which was compatible with the Convention rights. A court or a tribunal is a public authority for the purposes of s.6 and even where a person or body performs some functions of a private nature and some of a public nature the person or body will still be a public authority for the purposes of s.6 so far as acts which he/it performs which are of a public nature are concerned. The Houses of Parliament are not public authorities for the purposes of s.6, however, and neither are persons who exercise functions in connection with Parliamentary proceedings.

Where a public authority fails to comply with its s.6 duty (i.e. when it acts unlawfully in the s.6 sense) it does not commit a criminal offence (unless its conduct is criminal under the general criminal law). Rather, the potential effects of an act which is (or would be if it took place) unlawful under s.6 are that, under s.7, a person who is or would be a victim of the relevant act (in the Article 34 sense, seen above) may either bring proceedings against the public authority under the HRA (e.g. via an application for judicial review) or may rely on the relevant Convention rights in the course of other legal proceedings, whether civil or criminal. When a court (including a tribunal) finds that an act or a proposed act of a public authority is unlawful, the court, under s.8, may grant any remedy etc. which is within its powers but damages may only be awarded by a court which possesses the power to award damages or order the payment of compensation in civil proceedings (which does not include the Crown Court). Moreover, the court may only award damages if, in all the circumstances and taking into account the principles applied by the European Court of Human Rights in this context, the award of damages is necessary to award 'just satisfaction'. In practice, where human rights have been violated and a civil claim for damages results, it seems that, where this is possible, it would be sensible for the victim to rely on 'traditional' common law or statuory rights (such as a civil claim in tort) as well as relying on section 7 of the 1998 Act, because the level of damages awarded under section 8 may not equate with that which might be awarded in the context of a claim based on such rights.

Where a person's Convention rights are violated not by a public authority but, rather, are violated by a private person or body, the conduct of the person or body will not be unlawful under HRA s.6. Moreover, the victim will not, under s.7, be entitled to bring proceedings under the HRA. This does not mean, however, that the violation of Convention rights will be of no significance. This is so because if the dispute between the parties comes before a court of law or a tribunal, the court or the tribunal will be under the s.3(1) duty to read legislation where possible in a Convention compatible way and the s.6 duty not to act unlawfully. Thus, it appears that the violation of Convention rights may still be of some relevance even though the rights were not violated by a public authority (*Venables v. News Group Newspapers*).

HRA s.12 concerns those circumstances in which the court is considering granting any relief which might affect the Convention right to freedom of expression (guaranteed by Article 10). Essentially, s.12 requires the court to pay particular regard to the importance of the Convention right as well as to other specified matters. The section also provides that relief should not be granted in the absence of the respondent unless either the applicant has taken all reasonable steps to notify him or there are compelling reasons for not notifying him. Moreover, s.12 also provides that relief must not be granted so as to restrain pre-trial publication unless the applicant is likely to establish that publication should not be allowed. One particular problem which the courts have had to deal with in the context of s.12 is that of balancing the right to freedom of expression against the right to respect for private and family life etc. (guaranteed by Article 8), the House of Lords having recognized that neither right takes precedence over the other, the court being required to balance the proportionality of interfering with one right against the proportionality of interfering with the other (*Campbell v. MGN Ltd*).

Finally, HRA s.13 provides that when a judicial determination of a question under the HRA might affect the exercise of the Convention right to freedom of thought, conscience and religion by a religious organization or its members (guaranteed by Article 9), the court must have particular regard to the importance of the Convention right.

# 2

# The United Kingdom and the European Union

The European Economic Community (EEC) was brought into being by the Treaty of Rome 1957. Initially there were six members of the EEC: France, Luxembourg, Belgium, West Germany, Italy and the Netherlands. The purpose behind the formation of the EEC was the promotion of the economic integration of the member states for their mutual benefit. At the time of writing, the European Union (EU), as it is now known, comprises 27 member states, the United Kingdom having been a member since 1973.

The European Union now has a wide variety of objectives which are listed in the Treaty on European Union and the Treaty on the Functioning of the European Union (TFEU; the EC Treaty was renamed as the TFEU following the Lisbon Treaty 2009). The list of current objectives extends well beyond the original purpose for which the EEC was formed. These objectives include promoting peace, the values of the EU and the well-being of its peoples, offering its citizens an area of freedom, security and justice without internal frontiers, establishing an internal market, establishing economic and monetary union based on the euro, and upholding and promoting its values and interests in its relations with the wider world. More specifically, the second of these objectives involves ensuring the free movement of persons whilst taking appropriate measures as regards border controls, asylum, immigrations and preventing and combating crime. Under the third of these objectives the EU will work for sustainable development (upon the basis of balanced growth and price stability), a highly competitive social market economy (aiming towards full employment and social progress), and high-level protection and improvement of environmental quality, will promote scientific and technological advance, will combat social exclusion and discrimination, promoting social justice and protection, equality between male and female, solidarity between generations and protecting children's rights, will promote economic, social and territorial cohesion and solidarity amongst member states, will respect cultural and linguistic diversity, and will ensure that the cultural heritage of Europe is safeguarded and enhanced. The fifth of these objectives requires the EU, in the context of its relations with the wider world, to contribute to peace and security,

Galbraith's Building and Land Management Law for Students. DOI: 10.1016/B978-0-08-096692-2.00002-6
Copyright © 2010 by Elsevier Ltd

sustainable development, solidarity and mutual respect, free and fair trade, eradication of poverty, the protection of human rights (particularly those of children), and strict observance and development of international law.

The expansion of the EU's role and influence has, at times, brought it into conflict with some of the member states who are concerned that their national identity and/or sovereignty will be subsumed into a federal European state somewhat akin to the USA. For example, the UK opted out of monetary union (i.e. the UK did not adopt the euro) and also opted out of the relaxation of cross-border immigration controls.

# The institutions of the EU

## The European Council

The European Council has existed for more than 20 years but has only recently become an institution of the European Union. Its membership comprises the Heads of Government of the member states (their Foreign Ministers will also attend) plus its own President and the President of the Commission. The European Council does not legislate, but provides an impetus for the development of the EU and defines the EU's general political directions and priorities. The European Council meets twice every six months but special meetings may also be arranged, when this is required. It takes its decisions by consensus, unless the Treaties provide to the contrary.

## The Council of the European Union

The Council (often referred to as the Council of Ministers) consists of one government minister from each member state. The identity of the government minister varies depending on the nature of the meeting taking place.

The Council has legislative and budgetary functions, which it performs jointly with the European Parliament. The Council also has policy-making and coordinating functions (relating to, for example, economic policy, foreign and security policy, policing and criminal justice). The Council meets in various configurations, the Presidency of the Council's configurations (other than the Foreign Affairs Council) rotating between the member states. The High Representative for Foreign Affairs and Security Policy, who chairs the Foreign Affairs Council, is appointed by the European Council, with the agreement of the President of the Commission.

Unless the Treaties provide otherwise, the Council acts via a 'qualified majority'. From 2014 this will require at least 55% of the members of the Council and must comprise at least 15 members whose represent at least 65% of the EU's population. In the absence of a 'blocking minority' of at least four members of the council, a qualified majority will be deemed to have been attained.

## The Commission

Essentially, the Commission, which acts as the EU's 'civil service', devises policy, proposes new EU legislation, ensures that EU legistlation is applied throughout the EU (if necessary, bringing a member state which fails to implement its EU obligations before the European Court of Justice) and can make 'regulations' (see below), which can become law in a member state. The Commission is required to promote the general interests of the EU, to ensure the application both of the treaties and of measures adopted by the instututions in pursuance thereof, to oversee the application of EU law under the control of the European Court of Justice, to execute the EU's budget and manage programmes, to exercise coordinating, executive and management functions as required by the treaties (apart from foreign and security policy or as otherwise specified in the Treaties) to ensure the external representation of the EU and to initiate the annual and multiannual programming of the EU in order to achieve inter-institutional agreements.

The Commission consists of Commissioners, drawn from the member states, there currently being a Commissioner from each member state. The Commissioners are appointed for a five-year period and their appointment can be renewed. Once confirmed in post by the Council, the Commissioners become officers of the EU and must act in its interests. Any previous national and political affiliations must be put aside. The Commission is led by a President who is proposed by the European Council and elected by the European Parliament. The Commissioners are each given a designated area of responsibility by the President and must promote the EU's aims within that area. They are assisted by a large number of administrative staff, both dedicated to them and general support staff who work for the Commission (known as the Commission Services and organized into Directorate Generals).

## The European Parliament

The European Parliament exercises legislative and budgetary functions jointly with the Council. It merely has the right to be consulted in relation to a range of matters but has the right to veto decisions of the Council in relation to others and can amend or even reject the EU's draft budget or parts thereof. It also elects the President of the Commission.

The Parliament currently consists of 736 directly elected representatives of the member states, members being elected for five years. These can be identified in the United Kingdom by the designation MEP after their name. The Parliament is split along political lines but the MEPs vote on individual bases and much of the Parliament's work is conducted via Parliamentary Committees. The Parliament elects its own President and officers.

## The Court of Justice of the European Union

The Court of Justice of the European Union comprises the European Court of Justice (ECJ), the General Court and specialized courts and is required to ensure that the law is observed in the interpretation and application of the Treaties. In performing its role, the Court may have to adjudicate on disputes between a member state and the EU, a dispute between two or more of the EU's institutions or between an individual and a member state.

The Court of Justice is comprised of one judge from each member state who has a tenure of six years, which can be renewed. The judges are selected by the member states and must be independent and impartial arbiters of EU law disputes. The President is elected by the other judges. There are, currently, also eight 'Advocates General' who must give an impartial summary at the end of a case and a reasoned opinion which the Court considers before reaching its decision. Normally the Court sits in chambers of three or five judges, though in certain circumstances a Grand Chamber of 13 judges or even the full Court may sit. Where the ECJ makes a decision, only one opinion is given and no dissenting judgments are published.

Some actions or questions are determined by the General Court (previously called the Court of First Instance), which has its own judges, rather than by the Court of Justice. The General Court may also hear actions against decisions made by the specialized courts (the Council and the Parliament being empowered to legislate so as to establish specialized courts). There is a right to appeal to the Court of Justice against certain decisions made by the General Court and certain other decisions made by the General Court may be subject to review by the Court of Justice.

### The European Central Bank

The role of the European Central Bank relates to the 'euro area' (i.e. the 16 countries in the European Union which have adopted the euro). The United Kingdom is not one of these countries.

### The Court of Auditors

The Court of Auditors has a member from each member state. The members are appointed for six years and elect their President. Essentially the role of the Court of Auditors is to audit the accounts of the EU, assessing the collection and expenditure of EU funds. The members of the court of auditors must be independent of governments or other bodies and must act in the interests of the EU.

## Sources of EU law

The primary sources of law are the treaties (for example, the Treaty on European Union and the Treaty on the Functioning of the European Union), which prevail over any other legislation. The treaties can provide individuals with enforceable rights. The Council (at times in conjunction with the Parliament) and the Commission have the power to make various types of delegated legislation, which are discussed below.

## Regulations

A regulation will set out general rules. 'It shall be binding in its entirety and directly applicable in all member states.' This means that once a regulation is made, it applies to all member states without the need for any action by the domestic legislature of those states.

## Directives

Directives apply to those specific member states to which they are addressed. They set out an end to be achieved but leave the member state in question an element of discretion as to how they amend their law to achieve the end. Member states are under an obligation to enact national legislation which gives effect to the directive. They must do so by the deadline laid down in the directive.

## Decisions

Decisions apply to member states or citizens. They are binding in their entirety and thus need no further implementation. Where a decision so specifies, it will only be binding upon those to whom it is addressed.

## Recommendations and opinions

These are the final two types of EU legislation and are not binding. They do not provide individuals with enforceable rights.

# EU law in United Kingdom courts

Where an issue of EU law is raised in a United Kingdom court, the court can deal with it itself or may prefer to refer the case on a point of law to the European Court of Justice. If the case is being heard by the House of Lords then the issue must be referred. The referral will be in the form of a specific question which the ECJ will answer. The case is then adjourned in the United Kingdom court pending the European Court of Justice's decision. Once the European Court of Justice's decision has been communicated to the national court, it must then be applied by the national court to the instant case. This is known as the preliminary rulings procedure.

# The supremacy of EU law

The United Kingdom became members of what is now the EU by virtue of the European Communities Act 1972. Section 2 of the Act is intended to bring about the harmonization of United Kingdom and EU law. This created some problems initially. Traditionally British law has been founded on the principle of parliamentary sovereignty, i.e. Parliament is the supreme legislative authority. This caused a problem after the 1972 Act because the EU institutions were making law rather than Parliament.

This meant there was a situation where the European Court of Justice was promoting and enforcing the supremacy of EU law, whilst the British courts were promoting the supremacy of UK Acts of Parliament. At some point there was going to be a conflict in this area. The pivotal case was *R. v. Secretary of State for Transport ex parte Factortame*. In this case the House of Lords disapplied an Act of Parliament because of a conflict with EU law, thus confirming the supremacy of EU law. Thenceforth it appears that EU law is supreme.

# 3
# Settlement of disputes

In any situation where things go wrong, causing a dispute or conflict between people or organizations, the parties involved may think of turning to the 'law' for a remedy or a solution. So if you have bought faulty goods, or want to claim against a motorist who has damaged your car, or are owed money and want to recover the debt, you may wish to take proceedings in the courts against the relevant party. It would be wrong to think only of the courts when considering the settlement of disputes. Some measures of self-help can be very effective. Recent years have seen the advent of forms of alternative dispute resolution, such as mediation and mini-trials. In other instances the parties may prefer the privacy and convenience of arbitration. Sometimes legislation has provided that a forum other than the courts is more appropriate for the settlement of disputes, for example those cases where any rights must be pursued through specialized tribunals. There are also occasions where a dispute reveals no cause of action capable of being pursued through the courts, and in those cases it may be that intervention by an ombudsman may be more appropriate. The scope of these various procedures is considered separately.

## Self-help

Where a dispute arises between parties, it may be possible to settle matters by negotiation without the need to go to court. For example, in the law of trespass the person in possession of land can ask the trespasser to leave, and if he does not do so, may then use a reasonable amount of force to eject him. What is reasonable will vary with the circumstances of each case. More specialized examples of self-help can be found in the rules relating to set-off and lien. Set-off can occur where one party owes money to another, and is in turn owed money by that other. So if A owes B £10, and B owes A £5, then A may set-off the £5 he is owed by B and thereby reduce the amount he must pay to B.

A lien is the right of a person in possession of goods which belong to someone else to retain them until some demand, usually a demand for payment, has been met. This remedy can be particularly useful to people such as garage owners who have carried out expensive repairs on a vehicle. They can hold on to the vehicle, i.e. exercise their right of lien, until their bill is paid.

Galbraith's Building and Land Management Law for Students. DOI: 10.1016/B978-0-08-096692-2.00003-8

These are legally recognized forms of self-help, but less formal methods may be equally effective. Large numbers of pressure groups now exist which may support individuals who have grievances or who are involved in disputes. The power of the press and the media can sometimes achieve more than the use of formal legal procedures. Moreover, the intervention of an advice agency may be sufficient to prompt the other party to a dispute to resolve matters without resorting to court action.

# Court procedures

If proceedings in court become necessary, there is an elaborate system of civil and criminal courts. The vast majority of civil and criminal matters are disposed of by the inferior courts, respectively the county courts and magistrates' courts (though it should be noted that magistrates' courts do also have limited civil jurisdiction). The more valuable or complex civil cases may be heard in the High Court and the more serious criminal offences may (some of them must) be tried with a jury in the Crown Court. Following the trial of a civil or criminal matter, an appeal or appeals may be possible. The most important appellate courts are the Court of Appeal and, most important of all, the Supreme Court (which has replaced the Judicial Committee of the House of Lords).

# Civil cases

## The county court and the High Court

Civil proceedings may concern disputes between individuals, disputes between individuals and companies, disputes between companies and companies, or even disputes between individuals or companies and public bodies or government departments. Essentially, the party who brings the claim is known as the claimant (he or she was formerly known as the plaintiff) and the party against whom it is brought is known as the defendant. Civil proceedings may have a variety of purposes which may, for example, include obtaining damages as compensation for some civil wrong (for example, where the claimant suffered personal injuries in consequence of the defendant's negligence) or in respect of a breach of contract, enforcing performance of a contract, preventing the continuation of a civil wrong, such as a nuisance, by obtaining an injunction, or recovering land from a squatter. Often a civil dispute can be resolved by agreement between the parties prior to commencing proceedings in the civil court and it is only if such negotiations between the parties fail to result in a settlement that the expense and uncertainty of proceedings in the civil courts becomes necessary.

Civil proceedings may take place in a county court or in the High Court. County courts are essentially local courts which deal with the less valuable or less complex civil matters whereas the High Court deals with the more valuable or more complex civil matters, but since the introduction of the Civil Procedure Rules 1998 (CPR 1998) the distinction between the role of county courts and that of the High Court has become more blurred.

Most of the judges who sit in county courts are district judges, but the most senior judges who sit in the county courts are circuit judges, most of whom also sit in the Crown Court (see criminal procedings, below). Whilst most of the work of the county courts is done by district judges, certain types of claim must be heard by a circuit judge, not by a district judge, and some types of claim must be heard by a circuit judge unless the parties and the judge agree otherwise.

The High Court is based at the Royal Courts of Justice in London but also has district registries around the country and High Court judges do also sit outside London in the various county court districts. Many applications to the High Court, for example applications for injunctions, are dealt with not by High Court judges but, at the Royal Courts of Justice, by Masters or otherwise by district judges. The High Court is divided into three divisions: the Chancery Division, the Family Division and the Queen's Bench Division. The Queen's Bench Division deals, for example, with actions in contract and tort, actions for the recovery of land, applications for judicial review, and encompasses the Divisional Court, the Commercial Court, the Admiralty Court, the Technology and Construction Court, and the Administrative Court. The matters for which the Chancery Division is responsible include, for example, matters concerning dealings in land, mortgages, bankruptcy, administration of estates and intellectual property, and the Chancery Division encompasses the Bankruptcy and Companies Court and the Patents Court. The Family Division is responsible for matters such as divorce, adoption and Children Act 1989 proceedings.

In general, a claimant can choose whether to commence proceedings in the High Court or in a county court, though some claims (e.g. a money claim for £25,000 or less or a personal injuries claim for less than £50,000) should not be commenced in the High Court. Moreover, statute or rules of court may specify that a county court does not have jurisdiction in respect of certain claims, or at least does not have such jurisdiction unless the parties agree otherwise (e.g. libel or slander claims) or, conversely, that certain types of claim (e.g. claims for the recovery of land and mortgage possession proceedings) must be brought in a county court. Certain types of claim (such as applications for judicial review) must always be commenced in the High Court. Where proceedings could be commenced either in the High Court or in a county court, the factors that should govern the claimant's choice are the financial value of the claim, its complexity and whether it is of general public importance.

Essentially, civil proceedings are commenced by the court issuing a claim form (which, amongst other matters, identifies the parties, provides basic details of the claim and specifies the remedy which the claimant seeks), which the claimant or his legal advisor has prepared. The claim form must then be served on the defendant, normally within four months of being issued, and will normally contain or be served with particulars of claim (which, amongst other matters, contains a concise statement of the facts as alleged by the claimant), though the particulars can follow within 14 days if this is within the four-month period following the issuing of the claim form. The particulars of claim will normally be accompanied by the defendant's 'response pack' (containing various forms that the defendant may need), though the response pack will not be required in relation to certain claims that do not raise substantial factual issues (known as 'Part 8 claims').

When the defendant receives the particulars of claim he or she may, within 14 days, either admit the claim by filing (i.e. delivering to the court) a form of admission or contest the claim by filing a defence. Alternatively, if the defendant needs more than 14 days before filing a defence or if they dispute the court's jurisdiction they may, again within 14 days, file an acknowledgement of service. If the defendant does not admit the claim but fails to file either a defence or an acknowledgement of service (other than in the context of a Part 8 claim) the result may be that the claimant can obtain a default judgment, which depending upon the nature of the claim may be obtained with or without a hearing, though the court may set aside a default judgment upon application by the defendant.

If the defendant files a defence, the defence should, amongst other matters, indicate which of the claimant's allegations the defendant admits, which they deny and which they can neither admit nor deny but require the claimant to prove. The defendant may also make a 'Part 20 claim' against the claimant or against some other party (such as a co-defendant or a third party), for example where he or she claims a remedy against the claimant (i.e. a 'counterclaim') or where he or she claims 'contribution' or 'indemnity' from the co-defendant or third party in respect of potential liability to the claimant.

After the defendant has filed a defence or an acknowledgement of service the claimant may apply for summary judgment against the defendant. Equally, the defendant may apply for summary judgment against the claimant. The court may give summary judgment if, respectively, the claimant or the defendant has no real prospect of success and there is no other compelling reason for disposing of the case at trial. Summary judgment may not be given in certain types of proceeding (e.g. proceedings for the possession of residential premises brought against a mortgagor or a tenant who is protected by the Rent Act 1977 or the Housing Act 1988). The most extreme consequences of a summary judgment hearing is that the court may give judgment on the claim or, conversely, may strike out or dismiss the claim. Another possibility is that the court might make a conditional order, which requires a party to take certain teps, and if he fails to do so his statement of case will be dismissed or his claim struck out.

Prior to the trial itself, one or more of the parties to civil proceedings may find it necessary to apply to the court for directions or for interim remedies. The former may relate to matters such as the use of expert evidence (the civil courts now possess broad powers to limit the admissibility of expert evidence and to control the form in which such evidence will be given). The latter may concern remedies, such as an interim injunction, which can either prohibit a party from acting in a certain way or, less commonly, require him or her to perform a particular act. In some circumstances it may be necessary to apply for an interim remedy, such as an injunction, before proceedings have commenced. Normally notice of the making of an interim application must be given to the other parties, but in some circumstances this will not be possible or (where secrecy is required) desirable. Some applications (such as those for interim injunctions) must be supported by evidence. It may be necessary for the court to hold a hearing to deal with an interim application but some interim applications may be dealt with without a hearing. At the end of an interim application the court will normally make a costs order, which will normally, but not always, be made in the successful party's favour.

Apart from the injunction, another important interim remedy is the court's power to strike out part or all of the statement of case (i.e. claim form, particulars of claim, defence, or Part 20 claim) of the claimant or defendant. The court may do so where the statement of claim does not disclose reasonable grounds for the bringing or defending of the claim, where the statement of case abuses the court's process or otherwise obstructs the just disposal of the proceedings (for example, where the claimant could reasonably have brought the claim in the context of earlier proceedings against the defendant) or where the relevant party has failed to comply with a rule of court, a practice direction or a court order.

A further important interim remedy is the court's power to order an interim payment. Essentially, the court may so order if the defendant admits his or her liability to pay damages or money or the claimant has obtained judgment for damages or a sum of money to be assessed against the defendant or where the court is satisfied that if the claim went to trial the claimant would obtain judgment for a substantial sum or the claimant seeks possession of land and the court is satisfied that if the case went to trial the defendant would be held liable to pay the claimant a sum of money. The interim payment must not exceed a reasonable proportion of the likely final amount and in reaching its decision the court must take contributory negligence, set-off and counterclaim into account. Clearly, if the interim payment exceeds the amount finally awarded at the trial, an order for repayment (and for the payment of interest) will be required.

Another matter which may be the subject of interim applications is that of the disclosure and inspection of documents. Where a case is allocated to the fast track or the multi-track (see below) the court will normally order standard disclosure, which essentially requires the parties to disclose documents on which they rely, which adversely affect their case or another party's case or support another party's case. This means that the party must state in a list of documents that the document does or did exist. The other party then has the right to inspect (and to request a copy of) the relevant document unless the party who disclosed it no longer has it in their control or has a right or duty to withhold inspection of it (e.g. because it is privileged) or because he or she considers that it would be disproportionate to permit inspection of it. A party will find it necessary to apply to the court for an order where he or she seeks specific disclosure or specific inspection (i.e. the disclosure or inspection of specific documents or classes of document), where he or she seeks disclosure before the proceedings have started, where he or she seeks disclosure against a person who is not a party or where he or she seeks an order permitting him/her to withhold disclosure or inspection of a document.

Under the CPR 1998 the civil courts possess extensive case management powers and are required to actively manage cases. The courts should exercise these powers so as to give effect to the overriding objective of the CPR 1998 – namely, that of dealing with cases justly. This essentially requires the court, to the extent to which this is practicable: to ensure that the parties are on an equal footing; to save expense; to deal with cases in a way which is proportionate to the money involved, their importance, their complexity and the respective financial positions of the parties; to ensure that cases are dealt with speedily and fairly; and to allot an appropriate share of the court's resources to each case. Active case management includes encouraging the parties to cooperate, identifying the issues as early as possible, deciding quickly which issues need to be resolved at trial, deciding in which order issues should be resolved, encouraging the parties to use alternative dispute resolution where appropriate, helping the parties to settle the case, fixing timetables and controlling the case's progress, considering whether the benefits of a step justify the expense of taking it, dealing with as many issues as possible on the same occasion, dealing with a case without requiring the parties to attend court, using technology and giving directions so as to ensure that trial is fast and efficient. The court possesses the power to give directions concerning the issues in relation to which evidence is required, the nature of the requisite evidence and the method by which it is to be put before the court, and in exercising its powers may both exclude admissible evidence and limit cross-examination. Where a party fails to comply with a rule, a practice direction or a court order, the court can impose a variety of sanctions. These may vary at one end of the spectrum from sanctions concerning costs to, at the other end, the draconian sanction of striking out a party's statement of case. The court may deprive a party of the ability to rely upon certain evidence if, for example, a party failed to disclose a witness statement or an expert's report as directed.

Important case management decisions which the court (normally a district judge or a master) will be required to make normally include both whether the case should be transferred to a different court (e.g. to the county court for the district where the defendant lives or works) and which 'track' the case should be allocated to. A case may be allocated to one of three tracks: the small claims track, the fast track or the multi-track. Essentially, the small claims track is the track for the least valuable or important cases, e.g. for claims the value of which does not exceed £5000 (or in the case of personal injuries claims of £1000) or for claims by residential tenant against landlord requiring the landlord to carry out repairs where the cost of the repairs does not exceed £1000, but claims for harassment or unlawful eviction by a residential tenant against his landlord cannot be allocated to the small claims track. Where the small claims track is not the normal track, the fast track will be the normal track if the financial value of the claim does not exceed £25,000, the length of the trial is not likely to exceed one day, and oral expert evidence at trial will be limited to no more than two expert fields and no more than one expert witness per party per field. Where neither the small claims track nor the fast track are the normal track then the multi-track will be the normal track.

In deciding which track a case should be allocated to, the court will normally take into account 'allocation questionnaires' which have been filed by the parties and may find it necessary to require further information or hold an allocation hearing. The matters that the court will take into account when allocating a claim to a track will include its financial value, the remedy sought, the complexity of the case, the number of parties, the value or complexity of a Part 20 claim, the amount of oral evidence that will be required, the importance of the claim to non-parties, the views of the parties and their circumstances. Essentially, the consequences of allocation to the small claims track are that the pre-trial procedure and the trial itself will be simpler and cheaper, the rules concerning costs are different and the parties' rights of appeal are more limited. Conversely, the more valuable and/or complex cases will be allocated to the multi-track. Cases which are allocated to the multi-track will normally take longer to come to trial (and to try) than fast track cases and will often raise more complex case management issues than fast track cases. Cases allocated to the small claims track will be tried in a county court. Almost all fast track cases will be tried in a county court, not the High Court. It is normally only multi-track cases that will be tried in the High Court, but the vast majority of multi-track cases are tried in county courts and only the most valuable, complex or important multi-track cases will be tried at the Royal Courts of Justice in London.

Civil trials differ from criminal trials in a number of respects. One obvious one is that a civil trial, unlike a criminal trial in the Crown Court, is normally heard by a judge sitting alone; civil trial before jury is now rare and the best known remaining example is that of defamation claims (i.e. libel and slander). Another significant respect in which civil trials now differ from criminal trials is that the evidence in chief of a witness who has been called (i.e. the evidence that the witness gives for the party who calls him or her) now normally takes the form of the witness's written witness statement as opposed to oral testimony, though cross-examination (i.e. examination of the witness on behalf of the party who did not call him or her) will be oral. Further, expert evidence is normally given in civil proceedings by written report, experts now rarely being called to give oral evidence at a civil trial. Moreover a civil court may, in appropriate circumstances, limit expert evidence in relation to a particular issue to the evidence of a single joint expert who is jointly instructed by claimant and defendant. The legal burden of proof in civil proceedings is borne by the party who raises an issue. Thus, in the context of a negligence claim, the claimant would bear the legal burden of proving the defendant's negligence but the defendant, if he or she relied upon the defence, would bear the legal burden of proving the claimant's contributory negligence. The standard of proof is that of proof upon the balance of probabilities, i.e. the party who bears the legal burden of proving an issue must prove that it is more probable than not that the issue is as he or she asserts.

Civil trials normally take place in public but the court may order that a hearing take place in private for a variety of reasons and, for example, the hearing concerns confidential matters, such as personal financial matters, or where a hearing in private would be in the interests of justice. In particular, claims by mortgagers against individuals for possession of land and claims by residential landlords against tenants for repossession of dwelling houses in consequence of non-payment of rent will normally be heard in private.

At the end of the trial when the court gives its judgment the judge may find it necessary to make a variety of orders. Thus, for example, the court may find it necessary to order the sale, mortgage, exchange or partition of land, and consequent upon such an order may find it necessary to order a person to deliver up either or both the possession of land or the rents or profits thereof. In particular, the court will normally make a costs order, normally, though not necessarily, in the successful party's favour. Matters which the court will take into account in the context of making a costs order will include whether a party made an offer to settle or a payment into court, whether part of a party's case was successful and the conduct of the parties in the course of the proceedings.

## Civil appeals

Essentially, in the context of civil proceedings, the appellate structure is that a decision of a county court district judge may be appealed to a circuit judge, a decision of a master, a High Court district judge or a circuit judge may be appealed to a High Court judge, and a decision of a High Court judge may be appealed to the Court of Appeal. Some appeals from a county court go directly to the Court of Appeal, however, i.e. final decisions in claims that were allocated to the multi-track or which were made in specialist proceedings (i.e. admiralty, arbitration, commercial, patents, technology and construction or companies proceedings). Equally, where a county court or High Court decision itself related to an appeal to that court, a further appeal from that decision will go directly to the Court of Appeal. Moreover, where the county court or the High Court would normally hear an appeal but the appeal either raises an important point of principle or practice or there is some other reason why the Court of Appeal should hear it, the court from or to which the appeal is made or the court from which permission to appeal is sought may order the appeal to be transferred to the Court of Appeal.

Permission to appeal is normally required either from the court which made the decision at the time of the hearing or from the appellate court. Where an appeal to the Court of Appeal from the High Court or the county court itself concerns an appeal to the High Court or the county court, permission to appeal to the Court of Appeal will be required from the Court of Appeal. Permission to appeal will only be given either where the appeal would have a real prospect of success or where there is some other compelling reason why the appeal should be heard.

An appeal will not normally take the form of a rehearing of the case (but, rather, will take the form of a review of the lower court's decision) unless it is in the interests of justice to hold a rehearing. The appeal court will not normally receive either oral evidence or evidence that was not before the lower court, though it does possess the power to receive either or both of these types of evidence. An appeal will be allowed either where the decision of the lower court was wrong or where it was unjust in consequence of a serious procedural or other irregularity. The appeal court has a variety of powers (e.g. it could affirm, set aside or vary the lower court's judgment or order a new trial).

The Court of Appeal normally sits at the Royal Courts of Justice in London and the Court normally consists of three Lords Justices or, if the appeal is an interim appeal, of two. (Note: where a party wants to have a decision to make an interim order reconsidered, the party should first apply to the court that made the interim order.) Exceptionally, an appeal may be made from the Court of Appeal to the Supreme Court (which has replaced the Judicial Committee of the House of Lords). A case coming before the Supreme Court will involve points of law of general public importance and the permission of either the Court of Appeal or the Supreme Court will be required in order to bring the appeal. The Supeme Court must comprise at least three Justices of the Supreme Court (formerly, Lords of Appeal in Ordinary), though, in practice, usually five will sit to hear an appeal and at times more Justices will sit, e.g. seven or nine (there must always be an odd number of Justices). Occasionally an appeal may go directly from a High Court judge to the Supreme Court using what is known as the 'leapfrog procedure', but this will only be possible if the trial judge, upon application by a party, certifies that he is satisfied that the case involves a point of law of general public importance, that the point of law concerns the construction of a statute (and has been fully argued in the proceedings and fully considered in the judge's judgment) or a matter in relation to which the judge is bound by the doctrine of precedent, that a sufficient case to justify a leapfrog appeal has been made out and that the parties consent to the grant of a certificate by the judge. Even where the trial judge does grant a certificate the leapfrog appeal can only be brought with the permission of the Supreme Court.

# Criminal cases

## Magistrates' courts and the Crown Court

The scope of the criminal law is much wider than merely dealing with cases, like murder and theft, which are obviously criminal. Rather, the modern criminal law regulates a variety of matters which do not at first glance appear to concern the criminal law (everything from health and safety at work, to cutting down trees without a licence, where one is required, to corporate manslaughter). Thus, today, no business or commercial organization can afford to ignore the criminal law.

Most criminal trials take place in magistrates' courts, such trials being known as 'summary trials'. Some, however, including all of those for the most serious offences, take place in the Crown Court, such trials being known as 'trials on indictment'. (Note: the document containing the charges against the accused to which he or she pleads guilty or not guilty in the Crown Court is called the 'indictment' and, consequently, trial in the Crown Court is known as 'trial on indictment'.) Some criminal offences (summary offences, such as assault and battery) must be tried in a magistrates' court. Other criminal offences (offences triable only on indictment, such as rape) must be tried in the Crown Court. A third category of case (offences triable either way, such as theft) may be tried either in a magistrates' court or in the Crown Court.

The Crown Court sits at numerous centres around England and Wales, the most famous being the central Criminal Court or 'Old Bailey' in London. In the Crown Court a case is tried by a judge sitting with a jury (though, under provisions of the Criminal Justice Act 2003, it is possible for trial on indictment to be conducted by a judge sitting without a jury if there is danger of jury tampering or if jury tampering has taken place). Depending upon the nature of the offence with which the accused is charged, the judge may be a High Court judge, a circuit judge or a recorder (a part time Crown Court judge). The jury consists of 12 persons who are between 18 and 70 years old, are registered as electors, have been ordinarily resident in the UK for a period of at least five years since they were 13 years old, are not mentally disordered and are not disqualified (in consequence of their criminal record or because they are on bail). Persons eligible to serve on a jury who believe that there is a good reason why they should be excused from serving may apply to be excused. The judge governs the trial process, determines questions of law, such as the admissibility of evidence, sums up the case for the jury (both directing them as to the law and summarizing the evidence for them) and determines the accused's sentence, if he or she is convicted. The jury determine questions of fact and, fundamentally, determine whether the accused is guilty or not guilty of the offence(s) with which he or she is charged.

Magistrates' courts also sit in areas all round England and Wales. In a magistrates' court the case will be tried (and sentence imposed if the accused is found to be guilty) either by three unpaid justices of the peace (i.e. magistrates), who are advised by a legally qualified clerk, or by a district judge. Justices of the peace need not be (and most will not be) legally qualified.

Criminal proceedings always commence in a magistrates' court, and this is so even where the accused is charged with an offence which is triable only on indictment. The accused will normally have ended up before the magistrates either because he or she has been arrested by the police, interviewed by them and charged with a criminal offence or because he or she has been summonsed to attend court. Where the accused has been arrested and charged, the decision to charge will normally have been made by the Crown Prosecution Service. In relation to those offences in respect of which there is not normally a power of arrest (e.g. speeding), an 'information' (an allegation that the accused has committed the offence) will be laid (in writing or orally) before a magistrates' court and the magistrates will issue a summons requiring the accused to attend court. Magistrates cannot try a summary offence, however, unless the information was laid within six months of the commission of the offence. Moreover, even where the accused has been charged with a criminal offence, the Crown Prosecution Service may decide not to prosecute, perhaps because there is not a realistic prospect that the accused will be convicted given the weakness of the evidence against him or perhaps because it would not be in the public interest to prosecute him, given his age, poor state of health, the trivial nature of the offence, etc. Further, it should be noted that under provisions of the Criminal Justice Act 2003, the process of laying an information and issuing a summons is being replaced by a new procedure under which a public prosecutor (e.g. the police or the Crown Prosecution Service) issues a written charge and a 'requisition', which requires the accused to attend a magistrates' court to answer the charge. This new procedure is currently being piloted in a number of areas of England.

Whether the accused is eventually to be tried in a magistrates' court or in the Crown Court, a decision which the magistrates will be required to make when the accused first appears before them (unless the accused is attending in response to a summons and has not been remanded in custody or on bail) is whether the accused should be remanded in custody or remanded on bail. Essentially, subject to certain exceptions, the accused has a right to bail, but bail can be refused on a variety of grounds. For example, where the offence is imprisonable, bail may be refused: if there are substantial grounds for believing that the accused might fail to surrender to custody, might commit an offence whilst on bail or might interfere with witnesses; if the defendant should be kept in custody for his own protection; if the defendant is already serving a custodial sentence; if there has not been sufficient time to gather sufficient information to make a decision regarding bail; or if the defendant has already absconded in relation to the instant offence. If the accused is granted bail this may be subject to conditions, such as a curfew, a requirement that he resides in a bail hostel or the taking of sureties (persons who guarantee that the accused will attend court and who will potentially lose a specified sum of money if the accused fails to attend). The decision of whether to remand in custody or on bail is one which a magistrates' court or the Crown Court may be required to make again, perhaps several times, if the accused comes before them on subsequent occasions.

In relation to 'either way' offences the magistrates will be required (at a 'mode of trial hearing') to determine whether to hear the case themselves or whether to commit the accused for trial in the Crown Court. If the accused indicates that he will plead guilty, the magistrates must sentence the accused unless they decide that their sentencing powers are not sufficient, in which case they must commit the accused to the Crown Court for sentence. If the accused indicates that he will plead not guilty or does not indicate his plea, the magistrates must decide whether to 'accept jurisdiction'. If the magistrates decide to 'accept jurisdiction' (i.e. to hear the case themselves), the accused can elect to be tried in the Crown Court. If the magistrates decide to commit the accused for trial in the Crown Court, however, the accused cannot elect to be tried by the magistrates. If the accused is to be tried in the Crown Court a committal hearing will first take place in the magistrates' court to determine whether there is a prima facie case against him. If the committal takes the form of an 'old style committal', at which the court considers the prosecution evidence on paper (no witnesses being called), the magistrates will discharge the accused if the prosecution evidence does not disclose a case for him to answer. If the prosecution evidence does disclose a case for the accused to answer or the committal is a 'new style committal' at which no evidence is considered, the case will be committed to the Crown Court for trial. The committal must take the form of an old style committal if either the prosecution or the accused so require. In practice, since it is difficult to challenge the prosecution evidence, there will commonly be no good reason for requesting an old style committal. Indeed, when the relevant provisions of the Criminal Justice Act 2003 are in force, committal proceedings will be abolished and will be replaced by a process under which, if the magistrates regard the either way offence as more suitable for trial on indictment or if the accused does not consent to be tried summarily in respect of the either way offence, the magistrates will send the accused to the Crown Court for trial. Another major change, when the relevant provisions of the 2003 Act are in force, will be that if the magistrates (when making what will be known as the decision as to allocation) decide that the offence is suitable for summary trial, the accused will be entitled to request an indication of whether he would be likely to receive a custodial or non-custodial sentence if he pleads guilty. If the accused makes such a request, the court gives such an indication (which it will not be required to do) and the accused then decides to plead guilty, the magistrates will proceed to summary trial of the accused on his guilty plea and the court may not impose a custodial sentence unless they had indicated a custodial sentence.

Prior to a criminal trial, the prosecution is required to disclose to the defence both the material that they intend to rely on at the trial and 'unused material' (i.e. material they are not relying on which might undermine their case). The defence then may (must in the case of Crown Court trial) serve a defence statement which, amongst other matters, sets out the nature of the accused's defence, indicates the matters in relation to which the accused takes issue with the prosecution and indicates why he does so. This may then draw the attention of the prosecution to other unused material, which they will be under a duty to disclose, the prosecution being under a duty to keep under review whether there is any unused material which might reasonably be considered capable of undermining the prosecution case or of assisting the defence case. If the defence fails to serve a defence statement where there is a duty to do so, the court may be entitled to draw an inference in respect of this failure.

In the context of Crown Court trial there will be a Plea and Case Management Hearing before the trial for the purpose of determining the accused's plea and resolving a variety of procedural, legal and evidential issues pre-trial. There is no plea and case management hearing in the context of summary trial, though there may be a 'pre-trial hearing' to deal with the admissibility of evidence or questions of law. Moreover, it should be noted that in relation to some summary offences it is possible for the accused to plead guilty by post.

Following a not guilty plea, the trial process itself is similar whether trial takes place in the Crown Court or in a magistrates' court. It should be noted, however, both that there is provision in the context of summary trial for the trial to take place in the accused's absence where he fails to attend and that whilst in the context of trial on indictment prosecution and defence counsel normally both make opening and closing speeches, normal practice in a magistrates' court is that the prosecution only make an opening speech and the defence only make a closing speech. Essentially, the trial process is that the prosecution call their witnesses first, the witnesses being examined in chief by counsel for the prosecution, cross-examined by defence counsel and then may be re-examined by prosecution counsel. The defence may then, if appropriate, submit that the accused has no case to answer. If the judge or magistrates accept this submission the accused will be acquitted. If the submission fails then the defence will call their witnesses who they will examine in chief, the prosecution will cross-examine and the defence may re-examine. Essentially, the purpose of examination in chief is, if possible, to obtain the evidence from the witness upon which the party calling the witness intends to rely in support of his case. The purposes of cross-examination may include obtaining evidence from the witness which supports the cross-examining party's case, contradicting the witness's evidence in chief or to discrediting the witness (for example, by showing that he has previous convictions or is biased against the accused). The purpose of re-examination is to deal with matters arising from cross-examination, not to permit the re-examining party to adduce new evidence that he failed to adduce during examination in chief. When all of the prosecution and defence witnesses have been examined and following any closing speeches (and following the judge's summing up to the jury in the context of trial on indictment), the magistrates, district judge or jury determine the guilt or innocence of the accused.

If the accused, having pleaded not guilty, is convicted, or if the accused pleads guilty, the judge or the magistrates will determine the sentence. They may do so immediately or may adjourn to await the production of a pre-sentence report (which will normally be prepared by a probation officer). Before imposing a sentence the judge or the magistrates will have the facts of the case summarized for them by the prosecution and will be informed by the prosecution of the accused's antecedents. In the context of a guilty plea, unless the court is prepared to accept the accused's version of the facts, the court will find it necessary to hear the evidence of prosecution witnesses at a '*Newton* hearing' if the accused's version of the facts differs from that asserted by the prosecution. Where a Newton hearing takes place in the Crown Court the judge sits in the absence of a jury. Whether or not a Newton hearing is required, the accused will be entitled to a defence mitigation speech, in which defence counsel has an opportunity to put before the court factors which may potentially persuade the court to impose a lesser sentence.

The maximum sentence that the court is entitled to impose in relation to an offence may vary from a small fine at one end of the spectrum, to life imprisonment at the other. Examples of other possible sentences which the court may be entitled to impose, apart from fines and custodial sentences, include an absolute or conditional discharge, a suspended sentence or a community sentence (which will impose one or more of: an unpaid work requirement; an activity requirement; a programme requirement; a prohibited activity requirement; a curfew requirement; an exclusion requirement; a residence requirement; a mental health treatment requirement; a drug rehabilitation requirement; an alcohol treatment requirement; a supervision requirement; or an attendance centre requirement).

The maximum sentence that a magistrates' court is entitled to impose for a single offence is six months' imprisonment, though magistrates may impose a sentence of up to 12 months where the accused is guilty of two or more either way offences. Moreover, where the magistrates convict the accused of an either way offence and decide that their sentencing powers are inadequate, they may commit the accused for sentence to the Crown Court (though when changes to be brought in under the Criminal Justice Act 2003 come into force, magistrates who accept jurisdiction in respect of an either way offence will lose the power to commit for sentence to the Crown Court).

## Criminal appeals

Where the accused is tried summarily (i.e. in a magistrates' court), he or she has a right of appeal to the Crown Court against conviction and/or sentence, the appeal taking the form of a rehearing of the case. The accused must send notice of appeal to the magistrates' court within 21 days of being sentenced. The appeal is heard by a judge and two magistrates, not by a jury. A further right of appeal by way of case stated (see below) lies from the Crown Court to the Divisional Court of the High Court's Queen's Bench Division.

Alternatively, where the accused is tried summarily, he (or the prosecution) may appeal to the Divisional Court of the Queen's Bench Division of the High Court 'by way of case stated', but only where it is asserted either that the magistrates made an error of law or that they exceeded their jurisdiction (i.e. that they did something that they were not empowered to do). The application to the magistrates to state a case must be made within 21 days of conviction, acquittal or sentencing. The Divisional Court consists of at least two High Court judges. The appeal does not take the form of a rehearing but, rather, the Divisional Court hears legal argument and considers a written 'case', prepared by the magistrates' clerk in consultation with the magistrates, who will have taken into account representations made in relation to a draft 'case' by the parties before agreeing and signing the final version of the 'case' which will be considered by the Divisional Court. The 'case' sets out the charges, the findings of fact, the arguments raised before the court, the authorities upon which the parties relied, the decision reached by the magistrates and the question of law of jurisdiction which the appeal raises. It should be noted that if the accused first appeals by way of case stated he cannot subsequently appeal to the Crown Court. Thus, if the accused wishes to challenge both the findings of fact which the magistrates raised and to raise issues of law or jurisdiction he should appeal the magistrates' decision to the Crown Court rather than follow the case stated procedure as the possibility of an appeal by way of case stated from the decision of the Crown Court will, if required, remain available.

A third alternative which is open to an accused (and to the prosecution) where the accused has been tried summarily is to bring an application for judicial review, which again, if the High Court gives leave to bring the application, will be considered by the Queen's Bench Division's Administrative Court. The nature of an application for judicial review is considered in Chapter 4 but, for present purposes, it should be noted that whilst the accused will generally be better advised to appeal by way of case stated rather than to bring an application for judicial review (and should appeal by way of case stated where he asserts that the magistrates made an error of law within their jurisdiction), judicial review is more appropriate when the assertion is that the magistrates made a procedural error.

In the context of criminal proceedings, a decision made by a Divisional Court of the Queen's Bench Division may be appealed to the Supreme Court (which has replaced the Judicial Committee of the House of Lords) provided that the Divisional Court certifies that the case involves a point of law of general public importance and that either the Divisional Court or the Supreme Court grants leave to appeal.

Where the accused is tried on indictment (i.e. in the Crown Court) he or she may appeal to the Criminal Division of the Court of Appeal. The accused can appeal against conviction or sentence if the trial judge certifies that the case is fit for appeal or the Court of Appeal grants leave to appeal. The accused must serve notice of appeal within 28 days of conviction or, in the case of an appeal against sentence, within 28 days of sentence. In relation to appeals against conviction, the Court of Appeal consists of at least three judges, normally Lords Justices of Appeal or High Court judges, though circuit judges may also sit. In the case of appeals against sentence, the minimum number of judges is two. The Court of Appeal does not normally hear the examination of witnesses or receive new evidence, though the Court does possess discretion to do so. The Court of Appeal will allow an appeal against conviction if they think that the conviction is unsafe. In the context of appeals against sentence, the Court of Appeal may quash the sentence and impose a new sentence but cannot deal with the accused more severely than the Crown Court did.

An appeal lies from the Criminal Division of the Court of Appeal to the Supreme Court if the Court of Appeal certifies that the case involves a point of law of general public importance and either the Court of Appeal or the Supreme Court grants leave to appeal.

Where the accused is acquitted in the Crown Court the Attorney-General may refer a point of law which arose in the case to the Court of Appeal in order to obtain its opinion. The procedure is known as an 'Attorney-General's reference'. The accused's acquittal will be unaffected but the obtaining of the Court of Appeal's opinion should ensure that the error of law made by the trial judge, if indeed he did err in law, is not repeated by other judges. Having given their opinion on a point of law, the Court of Appeal may then refer the point of law to the House of Lords.

Another form of Attorney-General's reference relates to unduly lenient sentencing in the Crown Court. The Attorney-General requires the leave of the Court of Appeal in order to refer to them a sentence which he believes was unduly lenient. In the context of a reference of this type the Court of

Appeal may replace the accused's sentence with a more severe sentence. Following the Court of Appeal's review of the case the Attorney-General or the person whose sentence the reference concerned may refer a point of law to the House of Lords, though the Court of Appeal must certify that the point of law is of general public importance and the leave of either the Court of Appeal or of the House of Lords is required.

A final route by which a matter may be referred to the Court of Appeal (or, in the context of summary trial, to the Crown Court) is by the Criminal Cases Review Commission (CCRC). Where the CCRC investigates a case and decides that there are grounds for referring it, the CCRC will refer the case to the Court of Appeal or (if the offence was tried summarily) to the Crown Court, the effect being, respectively, as if the accused had appealed from the Crown Court to the Court of Appeal or from a magistrates' court to the Crown Court.

It should also be noted that, as was seen in Chapter 1, the fact that the European Convention on Human Rights now forms part of UK law does not prevent a person who believes that his or her Convention rights have been violated in the context of the criminal process from making an application to the European Court of Human Rights.

Finally, under Part 9 of the Criminal Justice Act 2003, the prosecution, in the context of trial on indictment, has a right to appeal to the Court of Appeal against rulings by the trial judge that terminate the trial. The result of such an appeal may be that the accused is acquitted, that fresh proceedings are commenced or that the existing proceedings, having been adjourned, are resumed. Under provisions of the 2003 Act that are not yet in force, the prosecution will also have the right to appeal against evidentiary rulings made by the trial judge.

## Tribunals

A significant feature of dispute-solving since 1945 has been the rise in the number of tribunals created by Act of Parliament to deal with specific questions. Examples of such tribunals included Employment Tribunals, the Lands Tribunal, Valuation Tribunals and Agricultural Land Tribunals. The reasons for this trend were that:

- The volume of work could not be given to the ordinary courts as the system would have become overloaded;
- The questions to be resolved by tribunals were frequently specialized, and the expertise of specialists could be used in the decision making;
- Tribunals could dispose of cases quickly, cheaply and informally.

Each type of tribunal had its constitution fixed by the statute creating it. For example, under the Lands Tribunal Act 1949, the Lands Tribunal (the jurisdiction of which included determining issues concerning compensation in respect of compulsory purchase of land and hearing appeals from Valuation Tribunals) consisted of a legally qualified president, a number of legally qualified members and a number of members who were surveyors. The President selected one or more members to deal with a particular matter. Appeals on a point of law from the Lands Tribunal were heard by the Court of Appeal, if the court gave permission to appeal.

Recently, a new tribunal structure has been created, under the Tribunals, Courts and Enforcement Act 2007. This consists of the First-tier Tribunal and the Upper Tribunal.

The First-tier Tribunal is currently divided into five chambers, namely the General Regulatory Chamber, the Social Entitlement Chamber, the Health, Education and Social Care Chamber, the War Pensions and Armed Forces Compensation Chamber and the Tax Chamber. For example, appeals under the Estate Agents Act 1979 are dealt with by the General Regulatory Chamber. The former tribunal chairmen are now tribunal judges, appointed via the Judicial Appointments Commission.

The Upper Tribunal hears appeals from First-tier Tribunal decisions and also decides some cases itself that are not dealt with by the First-tier Tribunal (e.g. matters that were formerly dealt with by the Lands Tribunal are now dealt with by the Upper Tribunal). The Upper Tribunal is divided into three chambers, namely the Administrative Appeals Chamber, the Tax and Chancery Chamber and the Lands Chamber.

Decisions of the Upper Tribunal are either made by judges or members (e.g. former surveyor members of the Lands Tribunal are now members of the Upper Tribunal). The decision will normally be made by a judge or a member sitting alone.

Where no right of appeal is available, the Upper Tribunal possesses the power to hear applications for judicial review. Currently, applications for judicial review will be heard by the Upper Tribunal, rather than the High Court, where they relate either to decisions of the First-tier Tribunal on appeals against review decisions of the Criminal Injuries Compensation Authority and where they relate to decisions of the First-tier Tribunal under its new Procedure Rules in circumstances in which there is no right of appeal to the Upper Tribunal.

# Arbitration

Going to court can be an expensive, public and time-consuming activity for the parties to a dispute. It may be better to agree to arbitrate, when the parties themselves can exercise some control over the choice of arbitrator. This process is particularly useful where a dispute involves points of a technical nature. In some types of contract (e.g. building contracts) it is common for the parties to agree not to refer disputes arising from those contracts to the court until they have submitted to arbitration. Then, if a dispute does arise, it will be a breach of contract if one of the parties tries to take the case to court instead of using the arbitration agreement. Although the parties must finance the arbitration themselves, the process has obvious appeal because of its speed and privacy. Once the arbitrator gives his decision (award), it can be enforced like a court judgment.

The law governing arbitration is now to be found in the Arbitration Act 1996 and the general law of contract. The 1996 Act is intended to restate and improve the law relating to arbitration and is founded on the following principles:

(a)   the object of arbitration is to obtain the fair resolution of disputes by an impartial tribunal without unnecessary delay or expense;

(b)   the parties should be free to agree how their disputes are resolved, subject only to such safeguards as are necessary in the public interest; and

(c)   in matters governed by Part I of the Act, the court should not intervene except as provided by Part I.

For the Act to apply to an agreement to submit to arbitration in present or future disputes it must be in writing, although this is defined very widely. The written agreement need not name the arbitrator or indeed specify the number of arbitrators. In the latter situation the Act states that in the absence of agreement as to numbers there shall be one arbitrator.

On normal contractual principles, parties are, in general, free to fix the terms of their own agreement and can decide which disputes the arbitration agreement will cover. The House of Lords (in *Fiona Trust and Holdoing Corp v. Privalov*) held that where the issue of whether a dispute is covered by an arbitration clause arises, unless the arbitration clause makes clear provision to the contrary, the assumption is that the parties intended it to be determined by arbitration.

Whilst parties cannot oust the jurisdiction of the courts, it is possible for them to exclude by agreement an appeal against the arbitrator's award. In any event, it is only possible to appeal on a point of law, lack of jurisdiction or procedural irregularity, and not on findings of fact. The courts have power to supervise arbitration, and may remove an arbitrator, for example, for lack of impartiality, or for a failure 'to use all reasonable dispatch in conducting the proceedings or making an award and substantial injustice has been or will be caused to the applicant'.

A number of terms are implied into an arbitration agreement by the Arbitration Act 1996, unless the parties have agreed otherwise. These include that the parties to the agreement must be prepared to be examined on oath by the arbitrator; that they must produce all documents required by the arbitrator; that the arbitration will be final and binding on the parties; and that the arbitrator can award costs.

Once the arbitrator makes his decision, 'the award', it can be enforced just like a court order or judgment. The parties are more likely to find the decision of an arbitrator acceptable but they pay a high price for this alternative to court proceedings as they are responsible for financing the arbitration proceedings.

The distinction between arbitration and valuation has always been regarded as problematic. In arbitration the aim is to settle an existing dispute. A valuation seeks to prevent a dispute arising, by allowing a third party to fix the value or price. The distinction is not always clear and the third party may be anxious to be sure of his status, as a valuer can be sued for negligence but an arbitrator cannot be sued for acts or omissions unless done in bad faith (see s.29 of the 1996 Act in this respect). This was an important point in issue in the case of *Sutcliffe v. Thackrah* in 1974, where an architect issued an interim certificate on the basis of which his employer paid the contractor. The employer subsequently sacked the contractor and wanted to recover damages from him for proven shoddy work. The contractor went into liquidation, so the employer then sued his architect for his negligence in certifying the poor work. In issuing the certificate, does the architect act as a valuer or an arbitrator? Once the architect issues a certificate, the employer is then obliged to pay. But the architect is *not* at that point determining a dispute between the employer and the contractor, although his duty when certifying is to act impartially between them. In consequence the architect in this case could be held liable for his negligent certification.

Whilst arbitration does have some advantages over litigation (for example, it is private and the arbitrator may have expertise in the relevant technical field that a judge may not possess), arbitration can sometimes take just as much time and be just as expensive as litigation. Indeed, since the Civil Procedure Rules were introduced in 1998, judges have been given significant powers to reduce the length and expense of civil proceedings (for example, by directing that expert evidence be given by a single joint expert). It is interesting to note that in the JCT 05 Standard Building Contract, if the parties wish arbitration to be the method of dispute resolution, this must be expressly stated in the contract. If this is not expressly stated, the default method of dispute resolution will be litigation. In contrast, under the earlier JCT 98 Standard Form of Building Contract, arbitration was, formerly, the default method of dispute resolution.

# Alternative dispute resolution (ADR)

The last two decades have seen the increased use of alternative forms of dispute resolution as an alternative to litigation in the civil courts. Indeed, under the Civil Procedure Rules 1998, a civil court must encourage the parties to use ADR, where this is appropriate. Alternative dispute resolution refers to alternatives to going to court and therefore, strictly speaking, includes arbitration. However, as arbitration, at least prior to the Arbitration Act 1996, was widely seen as very court-like, ADR may also be viewed as an alternative to arbitration. ADR has arisen because of concerns over the time and expense associated with litigation and, indeed, arbitration. Additionally, litigation and arbitration are adversarial processes that tend to destroy existing relationships. Hence the cost of litigating a dispute will often include the breakdown of long-term commercial dealings. The purpose behind ADR is to 'blunt adversarial attitudes' and to promote settlement of disputes, speedily and with the minimum of expense. In order for ADR to succeed it depends upon the parties exhibiting a spirit of cooperation and ultimately compromise. There are several forms of ADR – for example, mediation, non-binding arbitration, the mini-trial and adjudication.

Mediation involves the use of a third party, a mediator, to promote communication between disputing parties. After a joint meeting, the mediator will then listen to each party's side of the dispute in private, but must not divulge such, without permission, to the other party. The process will end with another joint meeting. No view as to the dispute will be given by the mediator. However, by exploring each party's case it is hoped that common ground between the parties will emerge which may form the basis for settlement of the dispute by the parties themselves. The drawback in this process is that, unlike with litigation or arbitration, there is no guarantee that the dispute will be resolved. If the parties cannot settle the dispute after mediation, then the cost of mediation, which is borne by the parties, is additional to the cost of going to court or arbitration. This drawback is common to all forms of ADR.

Non-binding arbitration is similar to mediation, but if the parties fail to settle the dispute the mediator will give a view as to what will be the likely outcome of a trial of the dispute. This may have the effect of encouraging the parties to settle.

A mini-trial is a form of structured settlement. Senior representatives of the disputing parties sit with a neutral chairman and listen to arguments for each side presented by advocates and experts. The panel are then able to assess the strengths and weaknesses of each side's case. Again the object is to encourage the senior representatives to settle the dispute.

Adjudication has in the Housing Grants, Construction and Regeneration Act 1996 been put on a statutory footing in relation to 'construction contracts'. Sections 104–107 provide a definition of this term. Clearly building work is covered, but the term 'construction contracts' also includes agreements to do architectural work, design, or surveying work and to give advice on building, engineering, interior or exterior decoration or on the laying out of landscape, in relation to construction operations. Certain types of contract are excluded from the definition. A party to a construction contract has a right to refer a dispute arising under the contract to adjudication in accordance with the Act. Note there is no obligation upon the parties to submit a dispute to adjudication, but if a party wishes to do so then such dispute must go to adjudication. An adjudicator's decision is binding until the dispute is finally determined by legal proceedings, arbitration or agreement. The decision is intended to allow the disputing parties to continue construction work, but it does not finally dispose of the dispute; the parties are free to reopen the issue. However, the parties may accept the adjudicator's decision as finally determining the dispute.

Should a construction contract not contain an adjudication provision or if such provision fails to comply with the terms of the 1996 Act then the law will impose an adjudication clause. This will be done through a piece of delegated legislation called the Scheme for Construction Contracts (England and Wales) Regulations 1998.

## Administrative control

Where a dispute involves allegations of maladministration by a local or central government department or agency it may be possible to refer the dispute to the appropriate ombudsman. This, for example, would be the Parliamentary and Health Service Ombudsman in relation to complaints concerning government departments or the National Health Service in England or would be the Local Government Ombudsman in relation to complaints concerning local authorities. Where the Parliamentary and Health Service Ombudsman accepts a complaint, the ombudsman investigates the complaint and sends a report both to the government department or health authority and to the person who made the complaint. If the ombudsman finds that the complaint is justified, the report will make recommendations but the ombudsman does not possess the power to require the government department or health authority to comply with the recommendations. Similarly, a local authority is not obliged to act in compliance with recommendations made by the Local Government Ombudsman, though, in practice, local authorities normally do comply with such reccomendations.

Examples of other ombudsmen include the Legal Services Ombudsman, who deals with complaints about lawyers, the Housing Ombudsman, who deals with complaints about certain landlords (e.g. housing associations), the Surveyors Ombudsman Service, which deals with complaints about surveyors and estate agents who are members of the Surveyors Ombudsman Service, and the Property Ombudsman, who deals with complaints about estate agents who are members of the Property Ombudsman Scheme.

## Affording the law

Where the ordinary man or woman in the street is a party to a dispute, many of the procedures outlined above may seem inhibiting to him or her, either because of fears of what it may cost to be involved with 'the law', or because of doubts over ability to cope with complex rules and procedures. If a person wants to engage a lawyer for advice, what will that cost? To what extent can someone be helped financially out of public funds to afford the services of a lawyer? The answer lies in the Community Legal Service and the Criminal Defence Service.

## Civil proceedings

Public funding in the context of civil proceedings may be available via the Community Legal Service (CLS). The CLS is one of two schemes for which the Legal Services Commission is responsible, both of which provide public funding for legal services. The other scheme, considered below, is the Criminal Defence Service. At the time of writing the government had indicated that it intends to change the Legal Services Commission from a non-departmental public body into an executive agency of the Ministry of Justice. This will enable the government to exercise more direct control over the legal aid budget.

Essentially, the CLS encompasses the following types of funding. First, 'Legal Help' can only be provided by persons or bodies (e.g. firms of solicitors) who have contracts with the Legal Services Commission and covers initial advice and assistance. Secondly, 'Help at Court' provides assistance and advocacy for the purposes of a particular hearing, but does not provide the same level of service that a client would be entitled to from a legal representative, and again can only be provided by persons or bodies who have contracts with the Legal Services Commission. Thirdly, 'Legal Representation', which can encompass the litigation and advocacy work that a legal representative would provide for a client, may be provided if a certificate is granted by the Legal Services Commission. Fourthly, 'Support Funding', which is like 'Legal Representation' except that the case is only partially funded by public funds, the remainder being privately funded. The other two types of funding, 'Family Help' and 'Family Mediation', concern family disputes and fall outside the scope of this book.

It should be noted that the CLS does not fund all types of legal work. Examples of matters that are not funded include conveyancing, boundary disputes, company law and partnership law. Moreover, CLS funding is only available to individuals, not to companies. It should further be noted that CLS funding is dependent upon financial eligibility (i.e. depending upon a person's income and capital the person may be required to pay a contribution or may not be entitled to CLS funding at all). Moreover, even where a person is financially eligible, funding will only be provided if a number of conditions (criteria) are satisfied. The criteria vary depending upon the type of funding concerned but relate to matters such as the benefit to the client if the proceedings are successful, the availability of other sources of funding and the likelihood of success.

Other potential sources of funding civil litigation might include a conditional fee agreement (under which, for example, the client may be required to pay no fee if the case is lost but an increased fee if it is won), legal expenses insurance (if the client has an appropriate insurance policy already or takes one out following the commencement of proceedings) or the client funding the civil litigation out of his or her own pocket.

## Criminal proceedings

Public funding in the context of criminal proceedings may be available via the Criminal Defence Service (CDS). Like the Community Legal Service (see above), the CDS is the responsibility of the Legal Services Commission. Essentially, publicly funded legal services in the criminal context are provided by private firms of solicitors who are contracted as part of the Criminal Defence Service (the 'General Criminal Contract'). The Legal Services Commission also employs its own 'Public Defenders' (at present only available in a few parts of England and Wales), the Public Defender Service providing an alternative to instructing a solicitor from a private firm.

The suspect at the police station is entitled to free legal advice. The Defence Solicitor Call Centre will contact the suspect's own solicitor or provide a CDS-accredited 'duty solicitor' (or a Public Defender) or (in relation to less serious matters) put the suspect in touch with Criminal Defence Service Direct (which provides advice by telephone).

The accused who is required to appear before a magistrates' court may decide to pay his or her own legal adviser or may rely upon the duty solicitor (who is available free of charge) when first appearing before the magistrates. Alternatively it may be that public funding for legal services is available via a CDS contracted firm of solicitors. Three types of CDS funding are potentially available. First, 'Advice and Assistance', which essentially funds the cost of a lawyer giving legal advice and providing assistance with various preliminary matters. Secondly, 'Advocacy Assistance', which funds preparation of the accused's case and a limited amount of courtroom advocacy by a lawyer. Finally, a 'Representation Order', which funds preparation for and representation at criminal proceedings by a lawyer. The accused's income and capital will affect his or her eligibility for 'Advice and Assistance', for 'Advocacy Assistance' and for a 'Representation Order' in a magistrates' court but not his or her eligibility for a 'Representation Order' in the Crown Court.

In relation to 'Representation Orders', the accused makes an application to the court (to the magistrates' court in relation to proceedings before magistrates and to the Crown Court in relation to proceedings in the Crown Court). The court will make a 'representation order' if the making of such an order is in the interests of justice. In deciding whether this is so the court will be required to consider the risk to the accused's liberty or reputation, whether a substantial question of law is involved, the accused's ability to understand proceedings or to put his case forward, whether tracing, interviewing or expert cross-examination of witnesses will be required and whether the accused being legally represented is in the interests of a person other than the accused. As was indicated above, eligibility for a representation order in a magistrates' court is also subject to a means test, relating to the accused's income and capital.

Whilst the making of a representation order for representation in the Crown Court is not means tested, where this is reasonable, the Crown Court may make a 'Recovery of Defence Costs Order', which requires the accused to pay part or all of the costs of representation. In determining whether the making of such an order is reasonable, one factor that the court will take into account is the accused's means. It should be noted, however, that a Recovery of Defence Costs Order may not be made where the accused is tried in a magistrates' court and will not normally be made if the accused is found not guilty in the Crown Court.

## The legal profession

Practising lawyers are divided into solicitors and barristers. The two professions are separate, with their own entrance requirements and examinations, though it is possible to convert from one profession to the other. Solicitors are sometimes compared to general medical practitioners with barristers being seen as the equivalent of consultants. This can be rather misleading, however, as a solicitor, more usually in a large firm, may have an extremely specialized practice, while a young barrister seeking to make his name as an advocate may have to be prepared to be a 'jack of all trades'.

Most solicitors work in private firms. These vary from very small high street firms of a few partners to large city firms. Some solicitors work for other organizations, for example for companies, for local authorities or for the Crown Prosecution Service. Some firms specialize in very narrow areas of legal work whereas others offer a wide range of legal services.

Solicitors possess rights of audience in county courts and magistrates' courts but most solicitors do not possess rights of audience in the Crown Court, the High Court, the Court of Appeal or the Supreme Court. Solicitors can, however, obtain 'higher rights' and, thus, some solicitors do appear as advocates in the higher courts. Many solicitors, however, never appear in court at all but, rather, specialize in non-contentious work, such as conveyancing. When litigation arises the work may be passed on to their firm's litigation department and/or a barrister may be instructed to represent the client in court.

Barristers in private practice are self-employed. Groups of self-employed barristers operate in sets of 'chambers', which provide barristers with the administrative infrastructure that they need in order to perform their functions. They cannot be instructed by a lay client directly but, rather, the 'lay client' must instruct a person or body such as a solicitor, patent agent or licensed conveyancer who is entitled to instruct a barrister. Like solicitors, some barristers specialize in narrow specialist areas of law whereas others have a much more general practice. Unlike most solicitors, barristers do have higher rights of audience. This does not mean, however, that every barrister will spend the majority of his or her time in court. Many barristers will spend much of their time writing opinions for solicitors (i.e. advising solicitors in relation to the application of specialist areas of law to the facts of specific cases), engaged in negotiations or conferences on behalf of their clients or drafting documents on their clients' behalf. Moreover, a considerable number of barristers are employed by other organizations (e.g. companies, local authorities or the Crown Prosecution Service). Indeed, some barristers are now employed by firms of solicitors. Some employed barristers (e.g. those employed by the Crown Prosecution Service) will spend much of their time in court whereas others (e.g. those employed by companies) may never go near a courtroom.

In future, in consequence of changes to be brought in under the Legal Service Act 2007, lawyers and non-lawyers will be able to adopt 'Alternative Business Structures'. For example, solicitors and estate agents will be able to form Alternative Business Structures and, thus, will be able to integrate the legal and other professional services that they can offer to their clients. Firms who wish to adopt an 'Alternative Business Structure' will be required to apply to a licensing authority for a licence.

# 4
# Central and local government

## The structure of central government

When society develops to the point where the activities of its members need to be directed and controlled, it will require a government. The tasks of government are: to formulate and carry out policies, which is its executive function; to frame laws, which is its legislative function; and to enforce those laws, which is its judicial function. At the same time as it grants power to a government to fulfil these functions, society wants to see control exercised over the government so that it does not become too powerful or dictatorial. Such control is exercised through the rules of the constitution, rather as a club or association is controlled by means of its rules. A striking feature of the British Constitution is that it is unwritten. Unlike countries such as France or the USA, we cannot point to one document embodying our guaranteed rights. Nevertheless, important constitutional laws are contained in a number of historic Acts of Parliament. These Acts are not invested with any special protection and could be revoked or altered by the same processes as any other Act of Parliament. Thus, whilst the European Convention on Human Rights is now part of UK Law and the UK is a member of the European Community, the Acts of Parliament which give effect to Human Rights Law and European Community Law in the UK could be repealed by the UK Parliament (though, certainly in the case of the European Community, not without causing considerable political problems). By contrast, in certain foreign constitutions there are entrenched provisions which can only be changed by special procedures.

The leader of the political party which commands a majority in the House of Commons will be invited by the monarch to form a government. The House of Commons is one part of Parliament, which consists of the monarch, the House of Lords and the House of Commons. The Queen is a constitutional monarch who exercises her powers only on the advice of her ministers. The House of Lords, which is the upper chamber of Parliament, is a non-elected assembly. Its members include hereditary peers, life peers and the Lords Spiritual. The Lords Spiritual are not peers. They comprise the Archbishops of Canterbury and York, the Bishops of London, Durham and Winchester, and the next 21 most senior bishops of the Church of England. Inevitably the major criticism of this element of representation in the Lords

Galbraith's Building and Land Management Law for Students. DOI: 10.1016/B978-0-08-096692-2.00004-1
Copyright © 2010 by Elsevier Ltd

is that it is exclusive to the Church of England. Indeed, over the years, the make-up of the House of Lords has been subjected to regular criticism. This mainly centred on the fact that the House was largely aristocratic and non-elected, with an inevitable inbuilt permanent Conservative majority. In 1999, however, the number of hereditary peers entitled to sit and to vote in the House of Lords was reduced from 759 to 92. Thus, the vast majority of the members of the House of Lords (about 594 members out of a total of about 706 who are currently qualified to sit in the House of Lords) are now life peers. Members of the House of Lords do not receive a salary, but are entitled to claim expenses.

Until October 2009 the House of Lords also possessed a judicial function, via the Lords of Appeal in Ordinary (a maximum of 12 judges who were specifically appointed to the House of Lords to hear appeal cases from the civil and criminal courts but who were also entitled to take part in all business of the House of Lords). In October 2009, however, the judicial function of the House of Lords was transferred to the new Supreme Court, the former Law Lords becoming Justices of the Supreme Court, Lord Phillips of Worth Matravers becoming the first President of the Supreme Court.

It is often thought that the House of Lords has little power today since the Parliament Acts of 1911 and 1949 restricted its role to the delaying of legislation rather than its out and out rejection. In answer to the criticisms of the House of Lords, however, it is worth remembering that the quality of membership of the Lords overall is high, reflecting a wide range of backgrounds and experience, with life peers being selected from trade unions, commerce and industry, public life and the armed forces. Standards of debate are generally accepted to be high. The part which the House of Lords can play in improving and refining legislation sent from the Commons is regarded as an important and significant part of its functions.

The House of Commons is the elected chamber of Parliament with 646 Members of Parliament. Each MP represents a constituency and elections must be held at least once every five years, at a time chosen by the Prime Minister when he or she believes it will be politically opportune.

Once the result of the general election is known, the leader of the political party which commands a majority of the House of Commons will be invited to form a government. Many of the MPs will be hoping to hold office in the government. The new Prime Minister will then select the Cabinet team and ministers for the various government departments. The Cabinet represents the most important departments but is of no fixed size and the selection of ministers to be in the Cabinet may reflect the government's policies and priorities. Traditionally the Cabinet will always include the Chancellor of the Exchequer, the Home Secretary, the Foreign Secretary and the Lord Chancellor (the Lord Chancellor formarly acted as the speaker of the House of Lords and sat as one of the Lords of Appeal in Ordinary but the House of Lords now has a Lord Speaker and has lost its judicial function and the current Lord Chancellor, who is also the Secretary of State for Justice, sits in the House of Commons). To be of a reasonable working size, the Cabinet is likely to consist of about 24 ministers (there are currently 27 ministers in the cabinet, examples of other key cabinet posts including the Secretaries of State for Defence, Health, Environment Food and Rural Affairs, Business Enterprise and Regulatory Reform, Work and Pensions, Transport, Communities and Local Government, Children Schools and Families, and Innovation Universities and Skills). Outside the Cabinet, however, there are other Ministers of State and Parliamentary Under Secretaries of State (for example, the Minister of State for Housing and Planning is in the Department for Communities and Local Government) and in total the government is likely to have over 100 members. Some ministers will be members of the House of Lords but it would be an unpopular move to appoint too many from the Lords because they cannot be questioned in the House of Commons, a corrective and control feature much prized by MPs.

The government, once formed, is Her Majesty's Government, and the ministers are Her Majesty's Ministers. The new session of Parliament will be formally opened by the Queen, reading the Queen's Speech (which outlines the government's legislative programme) from the Throne in the House of Lords. Originally all executive power was vested in the monarch, but a series of historical events reduced the power and produced the modern figurehead monarchy of today. All acts of government are done in the name of the monarch but not necessarily with the monarch's personal participation. The Queen does, however, preside at meetings of the Privy Council. All bills going through the two Houses of Parliament require the Royal Assent to become Acts of Parliament. In theory the Queen could refuse her Assent but in this, as in most of her activities, she is bound by constitutional conventions. The Assent has not been refused since 1708. The principal convention is that the Queen exercises her formal legal powers only on the advice of her ministers. It is said that the Queen has 'the right to be consulted, the right to encourage and the right to warn'.

There is very little formal law governing our constitution and, therefore, many of the 'rules' under which central government and the Crown operate are simply conventions. A constitutional convention is a rule of political conduct, not a strict law in the sense that there is no sanction to enforce such a convention. Conventions can change imperceptibly over a period of time and it can therefore be argued that they keep the constitution flexible. Conventions have been described as 'the flesh which clothes the dry bones of the law, they make the legal constitution work, they keep in touch with the growth of ideas'.

It has been seen that the government of the day will emerge largely from members of the House of Commons. The House of Commons itself has three main functions: to make laws; to control national expenditure and taxation; and to criticize policy. Its law-making function has already been considered. In order to carry on its business the first task of the newly elected members of the Commons following a general election is to elect their Speaker, who is an impartial chairman of the proceedings in the House. Usually this will mean the re-election of the previous Speaker if he or she is prepared to stand. Proceedings in the House of Commons take place in the Chamber with the government side sitting to the right of the Speaker's chair and the opposition to the left. The Chamber is not large enough to accommodate all the members at once, but usually they only want to be present in force for a limited number of great debates, or when there is going to be a division (i.e. a vote). MPs are paid salaries and receive allowances regarding the expense of running an office, provision of accommodation in London and in their constituencies, and travelling between their constituencies and Parliament. MPs who are Government Ministers, or who hold certain other jobs, receive additional salaries.

There is no formally recognized career path towards becoming an MP. The 646 MPs in the House of Commons represent a wide range of backgrounds. At the time of writing (i.e. prior to the 2010 general election), a significant number of MPs were lawyers, teachers, lecturers or journalists, or came from a business background, though more than 10% were manual workers. After the 2005 general election the average age of an MP was 50 years. At the time of writing, 126 MPs are female.

# Procedure of the House of Commons

There is a 'bible' of procedure in the House of Commons, a book called *Parliamentary Practice* by Erskine May. It started off as a small handbook, but as the procedure of the House of Commons has become more complex, the book has grown and grown. The procedure is complex because of the range of activities and interests with which the House must deal. Inevitably rules are important where the Speaker needs to control the activities of 646 volatile and articulate MPs of strong conviction, many of whom hold violently opposing views. If there are disputed points of procedure, the final word lies with the Speaker.

After numerous assorted items of business including question time, private notice questions, ministerial statements, introductions of new members and requests for emergency debates, the House eventually gets down to the main business of the day, which might be a major debate or the second reading of an important bill. When an issue is put to the vote in the House of Commons, the Speaker puts the question and he must then weigh up whether the 'ayes' or the 'nos' have succeeded. If his assessment is challenged, he orders the lobbies to be cleared and the division bells will be rung, tellers are appointed and members must file through the appropriate lobby to record their votes. The tellers give the results to the Speaker, who announces the outcome to the House.

## Parliamentary questions

Question time takes place in the House of Commons on Mondays to Thursdays and lasts for one hour. On Wednesdays the Prime Minister answers questions. Other ministers take it in turn to be questioned on a rota basis.

There are a number of reasons why an MP may wish to ask a question. He or she may wish to elicit factual information. He may well be asked by a minister to table a question to give the minister an opportunity to make a public statement in the House. The question may seek redress of a grievance for a constituent. Very often in such a case the question will only be asked if initial approaches to the minister or department concerned have drawn a blank. Questions in Parliament are not the only option open to an MP acting for a constituent. He or she can also refer the matter to the Parliamentary Commissioner for Administration. The question may be designed to embarrass a minister. This can be achieved particularly successfully in the follow-up to the oral question in what are called supplementaries. Although the minister will have notice of the question tabled, a cunning supplementary might find him or her unprepared. It is likely, however, that he or she will be well briefed by his/her civil servants with information to anticipate any supplementaries. Again, the question may be designed to enhance the reputation of the MP concerned, as opportunities to speak in the House are extremely limited.

The value of the parliamentary question can be considerable, especially if it ensures that ministers give their personal attention and consideration to issues or problems which would otherwise be handled by officials in their department. Often decisions will have been taken at quite a low level within the department. On a closer look at the file, ministers may have occasion to reconsider their decision or redefine the policy within their department.

Question time can be regarded as an important time for the ordinary Member of Parliament, the backbencher, particularly when it is remembered that this is part of the day when the parliamentary whips are not in control. Moreover, starred questions for oral reply are not usually asked by leading frontbenchers (members of the Government or Shadow Cabinet) although supplementaries may be put by them. The publicity potential is very great as question time takes place at a fixed time each day at the start of the day's business when the chamber is likely to be at its fullest and the press are in attendance, as there is always the prospect that a minister may be caught out or shown up in some way. A leading writer on constitutional law, De Smith, has said of question time:

> Parliamentary reputations have been made and ruined in the rapid cut and thrust of question time. For a few minutes the House comes to life audibly and visibly; wit, feigned or genuine outrage, cheers and jeers intrude upon the solemnity of the proceedings; government and opposition are briefly locked in verbal combat; the Prime Minister and the Leader of the Opposition may gain or lose a point or two in the public opinion polls; a backbencher shows his ministerial potential, and the House wonders how much longer the Minister of Cosmology can last.

## Private member's bills

If the opportunities to ask questions are restricted, then the opportunities to introduce a private member's bill are even more limited. The number of private member's bills passed in each session is always small, often not running into double figures. As a general rule private members will seek to introduce bills which do not require public expenditure, because if this were necessary the bill would face further hurdles. The government would then need to be persuaded to put forward a financial resolution. If a member wishes to introduce a private member's bill there are three possible ways of doing so: by winning a high place in the annual ballot; under the 10-minute rule; or under Standing Order No. 37. The safest of these three courses is winning a high place in the ballot for promoting a private member's bill. Successful MPs may have some personal preference for the measure which they wish to introduce, or an MP may be solicited by various pressure groups anxious to promote some particular measure. Occasionally the government itself has small measures that it wishes to promote. Where it has no time in the government programme, it may approach an MP who has been successful in the ballot asking him or her to take on a particular bill. Once a member knows what the subject matter of a bill is to be, he or she must then see to its drafting. This can be a costly process, especially if the bill is lengthy. The first 10 placeholders in the ballot do, however, receive a small allowance towards drafting costs.

# The committee system of the House of Commons

Whenever there is a very large organization with significant amounts of business to transact, it is common to find delegation of some of that business to specialized committees. In the House of Commons, delegation can be to a Committee of the Whole House, when the committee consists of all of the MPs. But, usually, two main types of committee are used – the general committee (formerly known as the standing committee) and the select committee. The most common form of general committee is the public bill committee. A public bill committee is specifically set up to consider the committee stage of a particular bill, after which the committee will be suspended. Select committees, on the other hand, will usually operate for the duration of a Parliament.

## Public bill committees

These consist of 16 to 50 members who are chosen to reflect the political strength of each party in the House. A government with a large majority will, therefore, enjoy a large majority in these committees. The actual members for each committee are chosen by the Committee on Selection, which is a select committee. The chairman of the standing committee will be chosen by the Speaker from a panel of members which is selected at the start of the parliamentary session. The main task of the standing committee is to consider amendments to individual clauses of the bill before them. This is the first time that the specific terms of the bill will be given detailed clause-by-clause consideration. Because of the number of votes on amendments, many of them of critical importance to the government if the substance of its policy is to be actively carried through, the whip system operates fully over the activities of public bill committees.

### Select committees

Some of the select committees are permanent features of the life of the House of Commons but with extremely varying functions. One of the most important select committees is the Committee on Standards and Privileges which, amongst other matters, oversees the work of the Parliamentary Commissioner for Standards and considers complaints relating to the declaring of interests or breaches of the Code of Conduct by MPs. There are a number of departmental select committees, such as the Science and Technology Committee and the Home Affairs Committee, which scrutinize the work of the government departments. Select Committees have a permanent membership for the duration of a Parliament. The committees are small, many only having 11 members, and the chairmanships are divided between the political parties, although the government would tend to retain the chairmanship of the more important committees like Home Affairs. They are free to determine for themselves which areas they will investigate.

Once a committee decides upon an area of investigation it will usually appoint its own expert advisers. It can call before it witnesses from industry and commerce, from the government and the civil service. Persons who are called to give evidence before it take an oath and, if they lie, may be prosecuted for perjury. It publishes its findings in a report to the House, to which the government must respond. Because of lack of parliamentary time, only a very small proportion of these reports is ever debated in the House, but there is, nonetheless, considerable value in the work of these committees. Their work is a means by which scrutiny and control of the executive is provided. They produce a considerable amount of information with which MPs may arm themselves for debate and questions. That information can be useful to add to the armoury of material used by pressure groups. Because issues are considered in some detail, MPs can develop considerable expertise. Membership of these committees is highly prized amongst MPs and they foster a cooperative sense of cross-party unity.

# Financial procedures of the House of Commons

The Treasury, under the Chancellor of the Exchequer, is the government department with responsibility for the management of the economy. Inevitably, as Treasury approval is required for all large-scale expenditure, it is a significant department, but its influence may wax and wane depending on the forcefulness of individual Chancellors.

In order to run the essential services of the country, money must be raised and collected, and some priorities have to be established about how it should be paid out. There must also be mechanisms for controlling how that money is used. Whilst the government does obtain revenue from other sources (e.g. from Crown lands and loans), significant amounts have to be raised by taxes, which may be annual taxes like income tax, or longer term taxes like stamp duty land tax.

The major financial statement of the year in the House of Commons is the Budget speech delivered by the Chancellor of the Exchequer. The proposals contained in the Chancellor's speech will ultimately be implemented in the Finance Act, but it may be several months before the Act receives Royal Assent. In the meantime, authority to collect annual taxes like income tax needs to be renewed, and the gap between the Budget and the Finance Act will be bridged using the authority contained in the Provisional Collection of Taxes Act. The appropriate resolutions will be passed as soon as the Chancellor has finished delivering his Budget speech.

In former times, it was common to raise taxes specifically for particular items, but now all moneys raised are lumped together into one fund, the Consolidated Fund, which is effectively just an account at the Bank of England.

There is a Treasury Committee, a select committee which monitors the Treasury, the Revenue and Customs, the Bank of England and the Financial Services Authority as well as other bodies. There is also a Public Accounts Committee, a select committee which examines reports concerning the economy, efficiency and effectiveness demonstrated by government departments and other bodies in using the resources allocated to them so as to further their objectives.

There are many ancient conventions with regard to financial matters in Parliament. Most significant amongst these are that the granting of public money and the imposing of taxes is the function of Parliament, not the government; that the granting of public money and imposing of taxes must begin in the House of Commons and be finally determined by the Commons; and that the redress of grievances must precede any grant of public money. There must be no taxation without representation.

## The Civil Service

Initially, when a minister takes over a new department or ministry, it might seem a daunting task to have to quickly learn all that is necessary for him or her to be able to make appropriate decisions. However, the role played by the Civil Service should not be forgotten. Civil servants are the permanent officials employed in the ministries. Unlike the American system where all officials change when the President changes, our Civil Service is permanent.

The civil servants work for the government of the day whatever its political complexion. The civil servants in most senior positions are in close touch with their ministers, and play a part in the formation of policy, watching over bills on their way through Parliament and keeping their ministers supplied with information to enable them to answer questions in Parliament. They will frequently make decisions by applying rules and policy to individual cases, and when this happens, such a decision is regarded as having been made on behalf of the minister, who, according to a Constitutional Convention, is responsible for what has been done.

## The structure of local government

There has been a long tradition of local administration in England, dating back to 1066. With poor networks of roads and the dangers of travel, the advantages of control and decision making resting with local organizations were obvious. But as those justifications have gradually disappeared, it may be questioned whether it is necessary in the twentieth century to divide responsibilities between central and local government. Two of the main advantages are:

1    The variations necessary for different local conditions can be introduced.
2    Local people will be involved, and those living in the area will see some benefit from the effort which they contribute.

Basically, the structure of local government in England is based upon the division of the country into counties, which in turn are divided into districts, the two basic types of local authority being county councils and district councils. The type of council will determine the range of functions, although some functions may be shared. County councils have responsibility for matters such as highways, fire, police and social services. District councils have responsibility for housing, public health, rates and refuse collection. In some parts of England there is a third tier of local government, in the form of parish councils and town councils, which are responsible for matters such as footpaths, allotments, burial grounds and bus shelters. Planning is an example of a responsibility shared between county and district councils. Equally, some local authorities share responsibility for services such as police and fire services.

In certain parts of the country the county/district council dichotomy is replaced by unitary authorities, which exercise the functions of county and district councils. Moreover, London, whilst divided into a considerable number of boroughs, each a unitary authority governed by its own council, now also has its own Assembly and elected Mayor, with responsibility for or involvement in matters such as transport policy, development, policing, fire and emergency planning, town planning, environmental issues, culture, media and sport.

Local authorities are examples of bodies enjoying corporate personality, i.e. the local authority is a 'person' recognized by the law. It is run by a council of elected members within the powers granted to the local authority by the Act of Parliament which created it. The members of the council may change at each election, but the council has permanent officials, called its officers, who fulfil a role rather like that of civil servants. The business of a local authority is traditionally transacted at council meetings, sometimes meetings of the whole council, but more often meetings of smaller committees set up for specific purposes, and with powers delegated to them, e.g. the planning committee. These days many decisions are made by the local authority's execuctive, the decisions of the executive being subject to scrutiny by groups of elected councillors sitting in scrutiny committees or scrutiny panels. A few local authorities have elected Mayors, in which case the council's executive will either take the form of the elected Mayor plus a cabinet of some of the council's elected members or an elected Mayor plus a council manager, but in most local authorities the executive takes the form of a council leader and cabinet comprising some of the council's elected members. Some local authorities have ceremonial Mayors, the functions of the ceremonial Mayor (e.g. chairing council meetings) being performed in councils which do not have a ceremonial Mayor by the council's chairman. Council officers may exercise powers delegated to them by the council (e.g. the power to determine straightforward planning applications may be delegated to planning officers by the planning committee).

Council meetings are open to the public, unless the council resolves to meet in private, usually because publicity about a particular matter would be prejudicial to the public interest.

Council members may represent a particular political viewpoint and it is quite often the case at local government elections that the votes of the electorate will swing against the political party in power in central government. This can produce conflicts between local and central government, e.g. on issues such as the sale of council houses or the size of rate increases. This naturally raises the question: to what extent is a local authority truly independent? The answer to this lies in the amount of control which can be exercised over the local authority.

## Control of local authorities

Every local authority must act within the scope of the powers granted to it. A local authority is a statutory corporation, which can only do such things as are authorized expressly or impliedly by the Act of Parliament which created it. If the local authority acts in excess of its powers, its actions are said to be *ultra vires* (beyond its powers) and void (of no legal effect). Obviously, it would require immensely detailed legislation to spell out everything a local authority could do. Consequently, by s.111 of the Local Government Act 1972, authorities are empowered to do anything which is 'calculated to facilitate, or is conducive or incidental to the discharge of any of their functions'.

Where a local authority has acted in a way which causes someone to be aggrieved, it will be necessary to check whether the authority acted under some express power, some implied power, or in some way reasonably incidental to an existing power. Not only must the local authority show that it had power to carry out a particular act, it must also be able to show that, procedurally, it carried out the act in accordance with the prescribed rules. A typical example of how things can go wrong occurred in 1967, in *Bradbury v. Enfield London Borough Council*. This case involved the reorganization of schools and the introduction of the comprehensive school system. Where the local authority intended to 'cease to maintain' a school, it was obliged to submit its proposals to the minister and give public notice of its intentions, to allow affected persons to make objections. The local authority took the view that what it was proposing to do in its reorganization plans did not amount to ceasing to maintain any of the schools involved. An aggrieved person applied to the court to stop the authority going ahead with its plans. An injunction was granted, because the behaviour of the local authority was *ultra vires* as it had failed to observe a mandatory procedural requirement.

When the help of the courts has to be sought, it is on the basis of judicial review of the local authority's decision or actions by the High Court. The High Court has a number of remedies available, including: a quashing order, which is used to quash an *ultra vires* decision; a prohibiting order, which is used to prevent *ultra vires* action which is about to take place; a mandatory order, which is used to compel the performance of a public duty; and a declaration, which simply makes clear what the court determines the legal position to be. A more detailed treatment of these remedies is provided later in the present chapter.

One possible way to give greater scope for action to a local authority, without rendering it constantly liable to claims of *ultra vires*, is to give the authority discretion as to whether, or how, it will act. Although this may appear at first sight to be a welcome prospect, the exercise of discretion is fraught with problems. Complaints may arise from the way in which the discretion has been exercised and in making its decisions the local authority will usually be obliged to exercise its discretion reasonably, to exercise it proportionately when breaches of European Union Law or European Human Rights Law are potentially involved, and to observe the rules of natural justice. These rules are usually stated to be *audi alterem partem* (hear both sides) and *nemo judex in causa sua* (no one should be a judge in their own case). In practice, these rules often require a local authority to give prior notice of its intentions to persons who may be affected by their decisions, so that such persons have an opportunity to put forward their own case, and a local authority should exclude from the decision-making process anyone who has some pecuniary interest in the decision, or who has some other bias. A more detailed treatment of the rules of natural justice is provided later in the present chapter.

Where it is suggested that an application for judicial review to the High Court may be needed to control the activities of local authorities, many individuals would be deterred by the expense involved or by the formality of the procedures. As was seen in Chapter 3, however, an alternative to litigation may be to have recourse to a Local Government Ombudsman. A Local Government Ombudsman can investigate complaints of maladministration. The word 'maladministration' is not defined, but had been said to cover instances of bias, neglect, inattention, delay, incompetence, ineptitude, perversity and arbitrariness. If the Commissioner finds that there has been maladministration resulting in injustice, the local authority concerned is then under a duty to consider the Commissioner's report, and notify the Commissioner of the action it proposes to take. If the Commissioner is dissatisfied with the response to this report by the authority, he or she may issue a second report.

## Control by central government

The main type of control which central government can exercise is financial. The control can consist of withdrawing grants, refusing loans or capping expenditure. Without financial support from central government, and without the ability to raise loans, no local authority could function for long. Admittedly, there are other forms of income available to local authorities, especially the rates levied on business property within the area (the sums raised by business rates are pooled nationally and are then redistributed to local authorities) and the sums raised by council tax, which local authorities levy on domestic properties. There are also less significant amounts, such as rents from council houses, profits from undertakings run by the local authority (e.g. swimming pools, markets, buses) and fees from the granting of various licences. However, the inadequacy of their income is apparent when it is realized that their funds have to be supplemented annually by the central government revenue support grant.

When considering the role of central government in controlling local authorities, it is important to emphasize that much of the control is informal, exercised in the process of consultation which takes place with various government departments. Circulars, codes of practice and memoranda issued by central government departments form the basis of much decision making at local authority level. Central government does also possess more formal powers to control the activities of local authorities, however. For example, the Secretary of State (under s.21 of the Planning and Compulsory Purchase Act 2004) possesses the power to call in all or part of a development plan document at any time before it is adopted and to direct that modifications are made to the document by the local planning authority.

The range of services and amenities provided by local authorities is vast. Not surprisingly, there will be some people living in the area covered by a council who will be dissatisfied with the council's efforts. It should be remembered that ultimate control lies with the electorate, whether at the time of the elections, or by the strength of opinion they express to their councillors, or through the media. It is clear from the low level of votes cast at local elections that some of these controls are not fully used.

## Bye-laws

Within its own area, a local authority will have statutory power to regulate activities by means of bye-laws, for example the power to make bye-laws for 'the good rule and government of' the district, under s.235 of the Local Government Act 1972. When making bye-laws, a local authority is usually governed by the procedures of s.236 of the Local Government Act 1972. A bye-law must be made under the seal of the council and must be confirmed by the appropriate authority, which will usually be the relevant Secretary of State. Before confirmation, the local authority must publish notice of its intention to submit the bye-law. Bye-laws are only valid if they are *intra vires* (i.e. the local authority possesses the power to make them). Moreover, in order to be valid they must be reasonable, they must be certain, and they must be consistent with the general law. If valid and effective, a bye-law may impose penalties for breach (e.g. a fine) which may be imposed upon conviction by a magistrates' court.

## Non-departmental public bodies

A non-departmental public body (NDPB) is not a government department and does not form part of a government department but, rather, operates at arm's length from the government. The two main types of NDPBs are Executive NDPBs and Advisory NDPBs.

Executive NDPBs (such as the Health and Safety Executive, the Environment Agency, English Heritage, and the Judicial Appointments Commission) perform executive, regulatory, administrative or commercial functions. Advisory NDPBs (such as the Low Pay Commission, the National Housing and Planning Advice Unit, and the Commission for Integrated Transport) provide Ministers with independent, expert, advice.

The minister who is responsible for an NDPB is accountable to Parliament for the NDPB.

## Challenging decisions made by public bodies in the courts

A person who is unhappy with a decision made by a public body, such as a local authority or a government department, may be able to challenge the decision in court. There are three ways in which it may be possible to raise an issue of public law before the courts. These are by appeal, by application for judicial review, or in the course of an ordinary 'private law' action.

## Appeal

There is no automatic right to appeal to the courts against a decision made by a public body. Such an appeal is only possible where an Act of Parliament has created the right to appeal against decisions made by a particular type of public body. For example, the Town and Country Planning Act 1990 permits persons aggrieved by planning decisions made by local planning authorities to appeal to the Secretary of State. If the aggrieved person is not then satisfied with the Secretary of State's decision, the Act allows him or her to make an application to the High Court within six weeks of the making of the Secretary of State's decision.

Where an Act of Parliament creates a right of appeal, it will also lay down the grounds of appeal, i.e. it states the reasons which, if established, justify the courts in substituting their decision for that of the original decision-maker. An appeal against a decision will not succeed simply because the appellant can show that he or she has good reason to be unhappy with the decision. In order to bring a successful appeal, it is necessary to establish one or more of the specified grounds of appeal. An Act of Parliament which creates a right of appeal also specifies the nature of those persons who are entitled to appeal.

## Application for judicial review

Where there is no right to appeal against a decision made by a public body it may still be possible to challenge the decision by making an application for judicial review to the Administrative Court. The Administrative Court is part of the Queen's Bench Division of the High Court.

### 1   Remedies of judicial review

An applicant for judicial review may apply for one or more of the following remedies. All of the remedies of judicial review are discretionary remedies. Thus, even where the claimant has established the grounds on which the application is based, the court may still refuse to award the remedy if, for example, the claimant failed to pursue a right of appeal or if his or her behaviour does not justify the award of a remedy.

(a)   *Quashing orders.* A quashing order can be used to quash an unlawful decision made by a public body. It may be awarded if the decision can be shown to be illegal or unauthorized. Following the making of the order the decision ceases to have legal effect though, if appropriate, the court can require the public body to reconsider the relevant matter and make a new decision in accordance with the law.

An example of the award of a quashing order is provided by *R. v. Medical Appeal Tribunal, ex parte Gilmore.* The tribunal misinterpreted regulations made under an Act of Parliament with the result that the award of compensation which they made to an injured workman was less than that to which he was legally entitled (an error of law). A quashing order was awarded to quash the tribunal's decision. The court recognized that the workman's case would then go back to the tribunal where a valid decision would be made.

(b) *Prohibiting orders.* A prohibiting order can be used to prevent a public body from exceeding or abusing its powers. The effect of the order will be to prohibit the public body from acting or continuing to act in excess or abuse of its powers.

(c) *Mandatory orders.* A mandatory order can be used to force a public body to do something which it is legally required to do. It may be awarded if it can be established that a public body has failed to perform its public duties. A mandatory order cannot be awarded against the Crown, but can be awarded against an officer of the Crown, including a government minister.

In *R. v. Shoreditch Assessment Committee, ex parte Morgan*, the committee refused to initiate a process whereby the rateable value of a house could be reduced to take account of a fall in its value. Since the committee had a duty to initiate this process, a mandatory order was awarded in order to oblige them to do so.

(d) *Declaration.* This is a court order which declares what the legal rights of the parties to an application for judicial review are.

(e) *Injunction.* An injunction is a court order which may prohibit an act from taking place or, more unusually, may require an act to take place.

(f) *Damages.* Where one or more of the five remedies referred to above is applied for by way of judicial review, the applicant may also claim damages in respect of any matter to which the application relates. The court may award damages if an ordinary civil action for damages (e.g. a contract or tort action) would have succeeded or if the party is entitled to claim damages under the Human Rights Act 1998 or under European Union Law.

## 2 Applications for judicial review and 'private law' matters

The application for a judicial review process is only available in relation to public law matters, not in relation to private law matters. Issues of private law should normally be raised in the form of an ordinary private law action, such as an action for damages in contract or tort, and not by application for judicial review. The distinction between public and private law matters may be difficult to draw in certain cases. The following guidelines may be helpful.

1 Where the legal issue involved is one of contract, such as employment law, the issue will normally be one of private law even if the employer is a public body (such as the BBC).

2 Where power is given to a body by an Act of Parliament, decisions made in relation to the exercise of the power will normally be public law matters (e.g. the exercise of planning powers possessed by local authorities or by the Secretary of State).

3       Where power is given to a body by a contract or by consent, decisions
        made in relation to the exercise of the power will normally be private
        law matters (e.g. the powers of the National Greyhound Racing Club).
4       Where a power stems from the Royal Prerogative (ancient common
        law powers of the Sovereign) decisions made in relation to the
        exercise of the power will normally be public law matters (e.g. refusal
        or withdrawal of a passport).
5       When determining whether the exercise of a power by a body is a public
        law matter, the courts take into account not just the source of the power
        (i.e. Act of Parliament, contract or Royal Prerogative) but also consider
        whether the function performed by the body is of a public or private law
        nature. For example, decisions of the Panel on Takeovers and Mergers, a
        body appointed by City of London Institutions such as the Stock Exchange,
        which devises and polices the code on takeovers and mergers, have been
        held to be subject to the application for judicial review process.

## 3   The process of application for judicial review

(a)     *The pre-action protocol.* The steps that parties should follow before
        making a claim for judicial review are set out in the pre-action
        protocol. A letter before claim sent by the claimant to the defendant
        in compliance with the pre-action protocol to which the defendant
        responds may result in the out-of-court settlement of the claim, thus
        avoiding litigation.
(b)     *Permission to apply.* Before an application for judicial review can be made,
        permission to apply must be obtained. Normally, an application for
        permission will be determined by the judge on the papers without a
        hearing. If permission is refused, it is possible to request a hearing to have
        the decision reconsidered. Basically, permission will be granted if the
        applicant has an arguable case. If permission is granted, the application
        for judicial review will be heard in open court, usually by one judge.
(c)     *Delay.* An application for review must be made promptly and, in any
        event, within three months from the date when the grounds for the
        application arose, though the court does possess discretion to extend
        the time period. Even if made within three months, the court may
        refuse an application if it has not been made sufficiently promptly.
        The issue of delay may be considered both by the court when it is
        deciding whether to grant the claimant permission to apply and, if
        permission is granted, by the court when it hears the application for
        judicial review itself and is deciding whether to grant the remedy for
        which the applicant has applied.
(d)     *Standing.* An applicant for judicial review must have 'standing'. This
        means that he or she must have a sufficient interest in the matter
        which the application concerns. Basically the test is whether the

applicant has been or may be affected by the act or decision which he or she wishes to question, as opposed to busybodies who are meddling in matters of no concern to them. Whether a pressure group has standing to bring an application for judicial review will depend upon the facts of the particular case before the court. The issue of standing will be considered by the court both when it is deciding whether to grant the claimant permission to apply and, if permission is granted, at the hearing of the application for judicial review itself. It should be noted that where it is asserted that a public authority has acted unlawfully within the meaning of s.6 of the Human Rights Act 1998, the effect of s.7 of the 1998 Act is that only the 'victim' of the relevant act may bring proceeedings against the public authority under the 1998 Act (see Chapter 1).

## Ordinary 'private law' actions

Basically, a public law matter should be raised by appeal or application for judicial review and not by bringing an ordinary private law action (e.g. an action in contract or tort). It may, however, be permissible to raise a public law matter in the course of an ordinary private law action as a defence to that action. Further, it may be appropriate to bring an ordinary private law action where a civil claim raises both private law and public law issues.

## Privative clauses

Privative clauses (or ouster clauses) are clauses in Acts of Parliament which remove or restrict the courts' power to hear matters. The normal approach of the courts is that such clauses will be interpreted very narrowly, so that if there is any doubt as to their effect, the courts' powers will not be reduced. Where there is no such doubt, however, the effect of the clause will be to restrict or remove the courts' power to hear the matter in question.

An example is provided by *R. v. Secretary of State for the Environment, ex parte Ostler*. Ostler discovered that a trader had withdrawn his objection to a compulsory purchase order after the Department of Transport had secretly promised him that a road would be widened. The road widening affected Ostler but he had not been told about it so he had not objected to the compulsory purchase order. Ostler claimed that the order was invalid, as it was made in breach of the fair hearings rule of natural justice (see below). The compulsory purchase order was protected by a statutory clause which provided that it could only be challenged in the courts within six weeks of notice of its making being published. Ostler had not discovered the secret promise, and consequently had not challenged the order until more than a year after the order was made. Thus the court could not interfere with the order.

## Application for judicial review: the grounds of challenge

Upon application for judicial review of a decision, the High Court will not award a remedy simply because it disagrees with the decision. It is not for the court to substitute its decision for that of the decision maker. Rather, a remedy will only be awarded if the decision is unlawful or unauthorized. Traditionally, the High Court may be prepared to award a remedy upon proof of procedural impropriety, illegality or irrationality. More recently, the courts have also demonstrated an increasing willingness to intervene in circumstances in which the conduct of a public body has given rise to a legitimate expectation that it will act in a particular way and the public body subsequently acts in a diffent way without a lawful reason for failing to give effect to the legitimate expectation. It is also important to note that where rights that are guaranteed by the European Convention on Human Rights are in issue or, indeed, where fundamental rights in general are involved, the principle of proportionality (that was referred to in Chapter 1) may come into play. These grounds will be considered in more detail.

### 1   Procedural impropriety

(a)   *Failure to observe procedural requirements laid down by Parliament.* Procedural requirements specify the process which a public body must adopt in acting or making decisions. If, in the course of making a decision, a public body fails to observe procedural requirements laid down either by Parliament or under statutory authority, this may render its decision invalid but will not necessarily do so. The traditional approach of the courts is that breach of a mandatory procedural requirement normally makes a decision invalid. Breach of a merely directory procedural requirement will not usually have this effect.

Mandatory procedural requirements are those which serve an important purpose. Directory procedural requirements are those which are not so important.

An example is provided by *Howard v. Secretary of State for the Environment,* which concerned a requirement of the Town and Country Planning Act 1968. Notice of appeal against an enforcement notice had to be served on the Secretary of State within 42 days of service of the enforcement notice. (The nature and purpose of enforcement notices is considered in Chapter 14.) The Court of Appeal classified this requirement as mandatory. They did so because without a time limit the enforcement provisions of the Act would not have worked effectively, so the time limit was of great practical importance. The Act also required that the notice of appeal contain certain specified information relating to the grounds of appeal and the facts of the appeal. The Court of Appeal classified this requirement as directory. They did so because once the notice of appeal had been served, the appellate mechanism began to operate. Failure to provide the specified information could easily be remedied at a later date and thus it was not essential that the information be contained in the notice. On the facts of the case, notice of appeal had been served within 42 days but a letter containing the required information had not arrived until more than 42 days had passed. The Court held that the procedural defect did not render the appeal invalid.

At times the courts may regard the breach of a minor procedural requirement, which appears to be merely directory, as rendering the decision to which it relates invalid. This may be the case, for example, where the court believes that the requirement has been deliberately ignored. Equally, the courts may at times regard the breach of an important procedural requirement, which appears to be mandatory, as not rendering the decision to which it relates invalid. This may be the case where, for example, the court believes that the breach has not caused any harm.

In *R. v. Soneji*, Lord Steyn suggested that rather than drawing rigid distinctions between mandatory and directory requirements, the courts should place the emphasis on the consequences of not complying with a requirement and should consider whether Parliament could fairly have intended the consequence of non-compliance to the total invalidity of the decision. Lord Carswell, however, whilst recognizing that the distinction between mandatory and directory requirements has gone out of fashion in recent years, suggested that the principles that underlie this distinction are still of value.

(b) *Breach of the rules of natural justice (or the requirements of fairness).* The rules of natural justice are common law rules of procedure which the courts impose both on themselves and on other public bodies. The rules do not apply to all public bodies and sometimes apply differently to different bodies. It is possible, however, to draw some general conclusions as to what the rules are and how they work. Basically there are two rules. First, the rule that a decision-maker must not be biased and, secondly, the rule that a person whose rights, interests or legitimate expectations are affected by a decision has a right to a fair hearing before the decision is made. It is also important to remember that apart from the common law rules of natural justice, Article 6 of the European Convention on Human Rights (which was considered in Chapter 1) guarantees that when a person's civil rights and obligations or a criminal charge against a person are determined, they will be determined via a fair hearing by an independent and impartial tribunal.

(i) *The bias rule.* The bias rule itself has two elements. The first requires that the decisions of public bodies must not be tainted by pecuniary bias. The second requires that the decisions of public bodies must not be tainted by non-pecuniary bias.

*Pecuniary bias.* The decision of a public body is vitiated by pecuniary bias if a member of the body possesses a personal financial interest in the subject matter of the decision. This rule is applied strictly in the sense that even if the court is satisfied that the person who had the interest was not influenced by it, the decision will still be quashed.

An example is provided by *R. v. Hendon RDC, ex parte Chorley*. The case concerned a councillor who was acting as estate agent in relation to a land transaction. The transaction would only go ahead if planning permission was granted in relation to a development on the land. The councillor took part in the meeting which decided to grant planning permission for the development and voted in favour of granting permission. The court quashed the grant of planning permission.

*Non-pecuniary bias.* The decision of a public body is vitiated by non-pecuniary bias if a member of the body has a non-financial interest in it. This could occur, for example, where the decision-maker is a close relative or professional colleague of a person who is affected by the decision. Even where a decision-maker does not have a direct personal interest in the outcome of a decision and even though the court accepts that there was no actual bias, the existence of an indirect interest will still infringe the bias rule if, in the circumstances, there was a real possibility of bias.

An example of a case in which non-pecuniary bias arose is provided by *R. v. Barnsley Council, ex parte Hook*. A market trader urinated in a side street. The market manager reported this to the council who revoked the trader's licence. The market manager was present whilst the relevant council committee made its decision and thus acted both as prosecutor and judge. The decision was held to be vitiated by non-pecuniary bias and was quashed.

(ii) *The requirement of a fair hearing.* Basically, fairness requires that a person whose rights, interests or legitimate expectations are affected by a decision is entitled to a fair hearing before that decision is made. (A legitimate expectation of a fair hearing may arise where, for example, a promise made by or a policy adopted by a decision-maker leads a person to believe that he or she will be given a fair hearing before a decision is made.) The rule has three basic requirements, though exactly what it requires will vary depending upon the facts of the case (e.g. depending upon the nature of the decision-making body which the case concerns).

*Firstly,* the person so affected by the decision must be made aware of the charges, complaints, grounds, arguments or evidence against him or her. In *Annamunthudo v. Oilfields Workers Union,* A was accused of breaking four union rules, none of which could result in expulsion from the union. He did not attend the meeting at which the charges were to be considered. At the meeting, a new and more serious charge was raised and he was expelled.

The expulsion was invalid because, since A had not been made aware of the charge, he had not been given a fair hearing.

*Secondly*, the person so affected by the decision must be provided with an opportunity to prepare his or her own defence or arguments. In *R. v. Thames Magistrates' Court, ex parte Polemis*, a ship's captain was charged with discharging oil into navigable waters. Hs summons was served at 10.30 and he was to appear in court at 14.00. This did not provide time for chemical analysis which might have shown that the oil did not come from his ship. The magistrates refused to allow an adjournment to allow the analysis to be carried out as the ship was sailing that day. They convicted him and fined him £5000. Thus the magistrates, in failing to allow the captain time to prepare his defence, did not give him a fair hearing.

*Thirdly*, the person so affected by the decision must be provided with an opportunity to present the defence or arguments which he or she has prepared. This does not necessarily mean that he or she must be allowed to appear in person before the decision-maker and it does not necessarily mean that he or she will be entitled to be legally represented. What is required in order for the hearing to be fair will depend upon the circumstances of the case. It may be sufficient to allow him or her to hand in written arguments, provided that both sides to a dispute are treated in the same way.

In *Errington v. Minister of Health,* following a public inquiry, the Secretary of State received further information from a local authority without allowing objectors to the local authority's scheme to comment upon it. Since the objectors had not been given an opportunity to put their arguments forward, they had not been given a fair hearing.

The rules of natural justice do not generally require the giving of reasons, but may do so in appropriate circumstances (*ex parte Murray*).

## 2   Illegality

A public body may only do those things which it possesses the power to do. (Normally such power will be given by or under an Act of Parliament or, more unusually, by the common law.) If a public body does not possess power to make a decision then, in making such a decision, it exceeds its powers: its decision is thus *ultra vires* and invalid.

In *White and Collins v. Minister of Health*, a local authority made a compulsory purchase order in respect of land which formed part of a park. The authority possessed power to make compulsory purchase orders but did not possess power to make such orders in respect of land which formed part of a park. Consequently, the order was quashed.

*R. v. Willesden JJ, ex parte Utley*, concerned an offence of failing to stop a motor vehicle when requested to do so by a police officer. The maximum penalty for the offence was a fine of £5. The magistrates imposed a fine of £15, which was quashed.

*Incidental matters.* Where a public body is expressly empowered to do something, it may be implied that the public body can also do things incidental to that which it is expressly empowered to do. In *Loweth v. Minister of Housing and Local Government*, a council was empowered to make a compulsory purchase order to acquire land on which to build a civic centre. The empowering Act did not specify whether the order could also include land on which to construct an access road to the civic centre. The court held that it could, as this was a matter incidental to development of the civic centre.

*Unlawful delegation.* Power should only be exercised by the body to which it is given. Only where a decision-maker is given the power to delegate can it lawfully delegate its powers to another person or body.

In *Barnard v. National Dock Labour Board*, a local dock labour board, without the power so to do, attempted to delegate disciplinary powers to a port manager. The port manager suspended some dock workers. The suspensions were *ultra vires*.

A common example of lawful delegation is encountered in relation to local authorities. Section 101 of the Local Government Act 1972 empowers a local authority to delegate its powers to a committee, a subcommittee or an officer or to another local authority. Delegation to an individual councillor is not authorized, however.

## 3   Irrationality

Where a public body exercises a power which it does possess, it does not exceed its powers but it may abuse them. Power is abused when its exercise is not in accordance with law. Thus, where in the course of making a decision, a decision-maker makes an error of law, he abuses his power.

Basically, an error of law is an error of legal reasoning. Such an error may take a variety of forms. In particular it may consist of:

(a)   the making of a decision which is so unreasonable that no reasonable person properly understanding the law would make it;

(b)   a failure to take legally relevant matters into account or the taking into account of legally irrelevant matters;

(c)   the use of a power for a purpose which is legally improper.

(a)   *Unreasonableness.* The courts may be prepared to quash a decision on the basis of unreasonableness if it is so unreasonable that no reasonable decision-maker would ever have made it. The fact that the court does not agree with a decision does not necessarily mean that it is unreasonable because, sometimes, two reasonable people can come to different conclusions on the same set of facts. Consequently, not every mistake is necessarily an unreasonable mistake. Thus, for a decision to be classified as unreasonable, it should be overwhelmingly clear that no reasonable decision-maker would have regarded it as one which he or she was legally entitled to make in the circumstances.

In *Hall v. Shoreham-by-Sea UDC*, a council granted conditional planning permission to an applicant. The condition imposed was that the applicant construct a road on his land and dedicate it to the public. No compensation was to be paid by the council to the applicant for the loss of his land. The court held that the condition was invalid. It was clearly unreasonable of the council to seek to force the applicant to give up his property without compensation by making the grant of planning permission dependent upon this.

(b) *Failure to take legally relevant matters into account/taking legally irrelevant matters into account.* The courts may be prepared to quash a decision if satisfied that, in reaching it, the decision-maker failed to take a legally relevant matter into account. Equally, they may be prepared to quash a decision if satisfied that, in reaching it, the decision-maker took a legally irrelevant matter into account.

Legally relevant matters are matters which, according to law, the decision-maker should take into account in reaching his or her decision. Basically, these include all matters which a reasonable decision-maker would take into account. They also include all matters which an Act of Parliament, expressly or by necessary implication, indicates should be taken into account.

Legally irrelevant matters are matters which, according to law, the decision-maker should not take into account in reaching his decision. Basically, these include all matters which a reasonable decision-maker would not take into account. They also include all matters which an Act of Parliament, expressly or by necessary implication, indicates should not be taken into account.

*Roberts v. Hopwood* concerned the validity of the wage rates paid by a council to its employees. The council was empowered to pay its employees 'such wages as it thought fit'. The council fixed wage levels far higher than those paid either in the public sector or by other councils. The council fixed these wage levels because it sought to become a model employer. In so doing, it failed to take into account ordinary economic considerations, such as the market price for labour. Thus, the council based its decision on a legally irrelevant matter, its desire to become a model employer, and failed to take a legally relevant matter into account, namely the market price of labour. The importance of the latter matter becomes obvious when it is realized that the effect of paying extremely high wages to council workers was that workers doing the same or similar jobs in the private sector, and being paid the lower market rate, would additionally be required to face higher rate bills in order to meet the increased public sector wage bill.

Sometimes, a public body adopts a policy by which to exercise its powers. If such a policy is rigid, in the sense that exceptions to it will not be considered, the public body is said to 'fetter its discretion'. This means that by rigidly applying the policy it fails to take into account legally relevant considerations, namely the individual facts of each separate case that comes before it. An example is provided by *Lavender v. Minister of Housing and Local Government*, in which the Minister of Housing adopted the policy of refusing to allow gravel working on agricultural land if the Minister of Agriculture objected.

Where a policy is adopted but it is made clear that exceptions will be considered and, consequently, the individual facts of each separate case are examined in order to determine whether they are exceptional, discretion is not fettered.

(c)   *Improper purposes*. The courts may be prepared to quash an exercise of power if they are satisfied that the power has been exercised for a purpose other than that for which it can lawfully be exercised. In order to decide whether power has been exercised for an improper purpose the courts must first determine the purpose for which Parliament intends the power to be exercised. Having determined this they must then decide whether power has been exercised for this or for a different purpose.

In *Webb v. Minister of Housing and Local Government*, a local council possessed a power to acquire land by compulsory purchase for the purpose of building sea walls for coastal protection. They used this power to acquire land to build a promenade along the sea front. Their decision was invalid because the purpose for which the power had been used was improper.

Sometimes a power may be exercised for more than one purpose. As long as the main purpose for which it is exercised is proper, it does not matter that a subsidiary purpose is also achieved.

In *Westminster Corporation v. LNWR*, a council was empowered to construct public conveniences on or under any road. It constructed conveniences under a road with access from both sides of the road. It was asserted that the decision to build the conveniences achieved an unlawful purpose, namely the building of a subway. The House of Lords held, however, that the council's primary object was to build conveniences and thus the fact that the development also achieved the secondary objective of constructing a subway did not make the development unauthorized.

## 4   Proportionality

Where rights guaranteed by the European Convention on Human Rights are in issue or, indeed, where other fundamental rights are involved, the principle of proportionality (which was referred to in Chapter 1) comes into play. As Lord Steyn recognized in *Secretary of State for the Home Department, ex parte Dayley,* in such circumstances the intensity of judicial review is greater than would be the case under the traditional grounds of judicial review. This, Lord Steyn indicated, may be so both because the court may be required to consider the balance that the decision-maker struck between the aims that it sought to achieve and the means that it adopted in order to achieve those aims, and because the court may be required to consider the relative weight that the decision-maker accorded to various interests and considerations. For example, in *R, on the application of Baker v. First Secretary of State,* a decision to compulsorily purchase property which was unfit for human habitation and the condition of which was having an adverse effect on adjoining residential occupiers was held to be necessary and proportionate.

## 5   Legitimate expectations

Where the conduct of a public body has given rise to a legitimate expectation that it will act in a particular way, the public body should not act in a different way without properly taking the legitimate expectation into account and should give effect to the legitimate expectation unless there are lawful reasons for not so doing. For example, in *R v. The London Borough of Newham, ex parte Bibi,* where the council promised to provide tenants with suitable accomodation with security of tenure but failed to do so, the Court of Appeal made a declaration that the Council was under a duty to consider the housing applications on the basis that the applicants had a legitimate expectation that they would be provided with suitable accomodation on secure tenancies.

# 5
# Business organizations

## Partnership

When persons want to set up in business there is nothing to prevent them simply trading on their own account, with only their own money at risk. As such, the law would describe such a person as a sole trader. Trading in this way, however, may involve a number of drawbacks:

- There may be lack of adequate funds to develop the business.
- There is no one to share responsibility for running the business.
- If the business gets into financial difficulties, the sole trader will have unlimited liability for its debts. Creditors will not be restricted to seizing the assets connected with the business. Even the person's home may have to be sold.

The sole trader may be able to overcome the first two of these disadvantages by forming a partnership. This relationship is described by the Partnership Act 1890 as 'existing between persons who carry on business in common with a view to profit'. The arrangements for a partnership can be very informal, but it is advisable to have a properly drawn up agreement, in case disputes arise between the parties.

Partnership is a very suitable form of business association for professional people, e.g. accountants, architects, quantity surveyors and consulting engineers. Under the 1890 Act all partners are legally entitled to take part in the running of the business, but in practice the parties can make whatever arrangements they like. All partners are agents of each other for the purposes of partnership business (e.g. making contracts, paying money and engaging staff on behalf of the firm). For this reason, it may be sensible to limit the number of partners. A partnership is often referred to as a firm, e.g. a firm of solicitors, but legally that term has no significance. The law is concerned to know: is this a partnership or a company? The word 'firm' might be equally appropriate to a firm of builder's merchants trading as a limited company, but the legal consequences of dealing with a partnership and a company can be entirely different.

The relationship among partners is a very special one, as potentially each could expose his or her fellow partners to unlimited liability. The law determines that it is a relationship of utmost good faith. This means that

Galbraith's Building and Land Management Law for Students. DOI: 10.1016/B978-0-08-096692-2.00005-1

intending partners should make a full disclosure of any relevant facts or circumstances. Wherever possible, potential disputes should be avoided by a carefully drafted partnership agreement. Unless such an agreement is drawn up, the rules of the Partnership Act 1890 will apply. The Act provides that partners are entitled to share equally in the capital and profits of the business. That certainly may not accord with the wishes of partners who have contributed capital in differing proportions. A formal agreement allows for detailed rules about capital and profit sharing, and matters like the running of the firm, engaging in other business activities, retirement of partners and dissolution of the partnership, which under the 1890 Act is automatic on the death or bankruptcy of a partner.

Forming a partnership introduces others who will share liability for debts of the business, but that liability continues to be unlimited. All the partners are fully liable for all the debts, and again creditors are not restricted to assets connected with the business. Even the advantages of a partnership (e.g. privacy with regard to the conduct of business affairs and accounting) may not outweigh the drawback of unlimited liability. Alternatives which can avoid this are the formation of a company or a limited liability partnership for the running of the business.

## Company incorporation

The process of forming a company is referred to as incorporation. This is a very significant legal step, because the business then becomes known in law as a corporation. This is to distinguish it from groups or bodies which are not incorporated, and which are usually referred to as unincorporated associations. These latter groups have no separate legal existence or personality distinct from their members. So, the Mid-Tyne Angling Club is simply an identifying name for a group of people who have come together with a common interest. There is no legal person called Mid-Tyne Angling Club, i.e. it is not a separate legal entity or personality.

The phrase 'legal personality' means a person or organization which is recognized by the law as having legal rights and owing legal duties. All human beings in this country have a legal personality, although the law may impose restrictions on their capacity to act in certain ways (e.g. any child under 18 years of age has legal personality but lacks capacity to make a will, or own land). The law can also confer legal personality on an organization which has become incorporated.

Incorporation can take place by three methods:

1    *By Royal Charter* – this was the method used to create the original trading companies (East India Trading Company and Hudson Bay Trading Company) but it was more important as the means of creating many royal boroughs and institutes.

2    *By Act of Parliament* – many of the public corporations and local authorities have been established in this way, e.g. British Airways.

3    *By virtue of the provisions of the Companies Act 2006* – under the rules of the Act, a number of different types of company can be created. For trading and general business purposes, a company limited by shares is by far the most common.

The fact that a company has a separate legal personality, quite distinct from its members (i.e. its shareholders) confers a number of advantages. These include a power to sue and be sued in its own name, the right to own and transfer property, the possibility of perpetual succession unless the company is wound up and the possibility for its members to enjoy limited liability.

If a company gets into financial difficulties, creditors will be paid out of the assets of the company. If those assets are insufficient, the individual shareholders do not become personally liable. The extent of the liability of shareholders is the amount they owe for their shares. Once they have paid for their shares they risk losing that amount if the company founders, but no more. Limited liability is the most appealing feature of forming a company. However, a price must be paid for this advantage; limited liability companies are under statutory duties to publish certain financial details in the interests of both their creditors and the general public.

## Creating a company

The actual creation or formation of a company is achieved by drawing up appropriate documents (including a Memorandum of Association and Articles of Association) which are submitted to the Registrar of Companies. The Memorandum was historically the company's most important document as it set out the objects for which it was formed and thereby determined or limited the powers of the company. The Companies Act 2006 has changed the status of the Memorandum of Association to a historical document simply reflecting that the subscribers to it wish to form a company under the 2006 Act and agree to take at least one share in it. Much of the information which used to be required to be included in the memorandum is now found in form IN01 (application to register a company).

The company will have a registered name but it may choose to trade under a different name, its business name. Business names are regulated by statute (Companies Act 2006). The name chosen for the company must not be the same as that of an existing registered company, and if it is too similar to that of an existing company the new company may be directed to change its name. Permission from relevant bodies has to be sought to use certain words in a company's name, such as bank or charity (the Financial Services Authority and the Charity Commission respectively must be asked for permission). In the case of a limited company, the word 'limited' must usually be the last word of the company's name. Where the company is a public company, it must include the words 'public limited company' or the abbreviation 'plc' in its name.

The company must have a registered office. Both the situation of this office and its address must be recorded on form IN01. The location of the registered office establishes the 'domicile' of the company, which may be important in choosing which system of law should apply. For example, Scots law applies to companies registered in Scotland. The company must also file a separate note of the address of its registered office.

Historically, all companies had an objects clause. This clause sets out exactly what business the company can engage in. Traditionally, these clauses have been very long, with the draftsman attempting to cover every possible activity which a company may wish to undertake. This was justified because of the *ultra vires* principle, whereby any activity undertaken by the company outside the scope of its objects clause was *ultra vires* (beyond its power) and void. So far as the *ultra vires* rules are concerned, for any third party who deals with a company, successive Companies Acts (currently section 40(1) of the 2006 Act) have set out that the validity of acts done by the company shall not be called into question on grounds of lack of capacity by the company. Of course, the directors of the company which thus acted beyond the scope of its objects clause would be in breach of duty, and a shareholder could seek to bring proceedings to restrain any proposed activity which would be beyond the scope of the company's capacity. New companies formed under the 2006 Act have unrestricted objects, unless restrictions appear in the company's Articles of Association. The contents of existing company's memoranda, including any restriction on their objects, are now deemed to be included in the company's Articles of Association.

A statement of the company's initial issued (allotted) share capital appears on form IN01 and includes its division into shares of a fixed amount. For private companies there is no legal maximum or minimum initial share capital fixed, but a public company must have allotted shares with a minimum nominal value of £50,000. Each share in a company has a nominal value – this represents the minimum amount that the company must receive in return for allotting the share. So, if a company has shares with a nominal value of £1.00 each, the company must be paid at least £1.00 per share by anyone subscribing for shares in the company. Once shares have been issued by the company to individuals subscribing for them, those shares can be transferred (sold or given away) by the shareholder, subject to any restrictions on transfer of shares in the company's Articles.

The Articles of Association set out the internal rules and regulations for the running of the company, covering matters such as meetings, rights of shareholders, voting, appointment of directors, powers and responsibilites of directors, and payment of dividends. There is a model set of Articles for private companies provided by the 2006 Act, called the Model Articles for

Private Companies Limited by Shares (the predecessor to these articles was called Table A). It is common for companies to adopt model articles with suitable amendments, although an entirely bespoke set of articles can be used. Which form of articles are being adopted is indicated on form IN01.

The Articles are particularly significant for the members, as they are effectively the terms of the contract between the member and the company. Any violation of the rules contained in the Articles is a breach of contract. Members can sue to enforce their rights in ordinary breach of contract claims.

## Public and private companies

It is common to think of private companies as small family firms and public companies as the large business giants, and in practice this is often the case. A company which starts as a private company may expand its activities to such an extent that it decides to 'go public' and seek the change of status to a public company. There are a number of simple differences between public and private companies. For example:

1   A private company can be formed with one member and one director (a single member company) whereas a public company needs two members and at least two directors.

2   A private company need not obtain from the Registrar a certificate entitling it to do business whereas a public company must have such a certificate, and commits offences by trading without it.

3   A public company must satisfy the minimum capital requirement (i.e. its allotted share capital must be at least £50,000). This rule does not apply to private companies.

4   A private company may exercise its borrowing powers immediately upon incorporation. A public company may not do so until it has allotted shares up to its authorized minimum capital of £50,000.

5   Private companies may still restrict the transferability of their shares, if they choose. They would usually choose to do so in cases where they want to preserve control of the company within a family. Restrictions normally operate by giving the directors the power to refuse to register transfers. Where such a discretion is given by the Articles, it places immense power in the hands of the directors, but they must act throughout in good faith. The new Model Articles referred to above contain provisions allowing the directors to refuse to register a transfer of shares, although the Companies Act 2006 does oblige the directors to give reasons for any such refusal. By contrast, public companies may not restrict the transferability of their shares if they want to be quoted on the Stock Exchange (listed shares). There is a model set of articles for public companies which do not give directors a discretion to refuse to register a share transfer of fully paid shares.

6  Private companies may be freed from some of the rigours of the rules about publication of accounts, depending on their size. For these purposes size is determined by factors like turnover, assets and numbers of employees.

7  Public companies must have a suitably qualified company secretary. Under provisions in the Companies Act 2006, private companies are no longer required to have a company secretary.

8  Public companies are subject to rules prohibiting their giving financial assistance for the purchase of their shares except in very restricted circumstances. Financial assistance would include, for example, a company lending money to a person which that person then used to buy shares in the company. Private companies are no longer subject to this restriction.

9  Private companies are no longer required to hold annual general meetings.

## Management of a company

Companies are run by a board of directors. The company's first directors are appointed on form IN01 on incorporation. Thereafter, directors can be appointed by the board or the shareholders in general meeting. With private companies it is a common requirement that directors must have a certain holding of shares in the company. Where it is a small family business this is usually not difficult as directors will often also be the members of the company. The company, being an artificial person, must act through human agents, and that normally means through the board of directors. The board will often delegate duties to paid employees, who may also be directors. Once appointed, a director is under duties to show good faith to the company and to take reasonable care. This means, for example, that a director is obliged to reveal any personal interest in a contract made by the company, particularly if he or she stands to profit by the contract. The duties owed by directors to the company that have been established at common law are now codified and extended under sections 171–177 of the Companies Act 2006. These duties cover: acting within powers; promoting the success of the company; exercising independent judgment; exercising reasonable care, skill and diligence; avoiding conflicts of interest; not accepting benefits from third parties; declaring an interest in proposed transactions or arrangements with the company.

The directors, or those acting under delegated powers, must act within the powers granted to them by the documents creating the company. The Articles should provide mechanisms for the appointment and retirement of directors. If shareholders are dissatisfied with one of the directors, they may seek his or her removal from office by passing an ordinary resolution,requiring a simple majority in favour of the decision. However, so dramatic a step is surrounded by numerous safeguards. The member proposing the resolution must give 28 days' special notice to the company before the relevant meeting; the company must then give other shareholders 21 clear days' notice of it; the company must notify the director concerned and he or she has the right to circulate a written statement setting out his or her point of view prior to the meeting and speak at the meeting personally. It should also be remembered that a company is not obliged to call a general meeting just because a shareholder has served a special notice proposing the removal of a director on the company. The board normally decides whether or not to call general meetings, although shareholders with 5% of the compay's share capital can compel the board to call a general meeting through a members' requisition. Although the Companies Act 2006 makes clear that a company's articles cannot require anything more than an ordinary resolution to remove a director, case law has established that giving directors enhanced voting rights on any resolution to remove them in the company's articles is legally effective. The director concerned may therefore have enhanced voting rights which allow him or her to defeat any such resolution. If a resolution to remove a director is successful, it should be remembered that he or she may also have a contract of service with the company, and substantial compensation may need to be paid.

A board of directors is likely to have a Chairman, who will preside at directors' meetings. The board may also decide to appoint a Managing Director, who will often have a contract of service with the company, and to whom much of the day-to-day management of the company is delegated.

How a company divides its management between directors and shareholders is largely a question to be determined by the Articles. In practice, the rights of members to participate in management, particularly in large public companies, is extremely limited. A member does have some say by virtue of voting power at meetings, but he or she may be one of a very large number of shareholders, holding very few shares, and his or her opinion may carry no weight. The wishes of the majority should normally take effect but not where the impact of that is to perpetrate a fraud on the minority.

The law has been alert to the problem of protecting minority shareholders and provides a number of means which can help to secure some protection. One example is shareholders' agreements, whereby the members enter into a separate contract with each other regulating how certain aspects of the company will be run. A statutory example of protection is the right of a minority shareholder to apply to the court if the company's affairs are being carried on in a way which is unfairly prejudicial to the members or some part of them. The Companies Act 2006 has also codified the common law rules which allowed a shareholder to bring a claim on behalf of the company (a derivative action) where some wrong was done to the company and the wrongdoers controlled the company and therefore refused to take action against themselves. The 2006 Act allows a member to bring a derivative claim in respect of a cause of action arising from an actual or proposed act or omission involving negligence, default, breach of duty or breach of trust by a director of the company, although there are substantial hurdles to overcome before a court will give the member permission to continue the claim. Alternatively, a member could petition the court to wind up the company on the grounds that such a winding up would be just and equitable. These rules could be a particularly effective sanction in small private companies with, say, only half a dozen members, where the rights of one of them are being seriously prejudiced.

It may be that what concerns members most is the income that their shares will produce. A declaration of a dividend is usually made in general meeting (GM) or annual general meeting (AGM). Private companies are no longer required to hold an annual general meeting although public companies must continue to do so. Any kind of business may be transacted at an AGM as long as appropriate notices have been served and appropriate resolutions are passed. If business needs to be transacted before another AGM is likely to be held, or if a private company had dispensed with holding AGMs, a GM may be convened, either by the directors or by members who together hold 5% of the company's paid-up share capital (members' requisition). Resolutions may be ordinary or special, the latter requiring 75% of the shareholders in favour. A private company can pass a written resolution, avoiding the need to hold a meeting and requiring the same percentage of the members in favour as would have been required had the resolution been put to a meeting.

The Chairman of the board will usually chair company meetings. He or she must process the business in accordance with the rules contained in the Articles, and in accordance with the general rules of the common law. Voting on resolutions can be on a show of hands but sometimes the more formal procedure of a poll may be demanded. The Articles usually provide for situations when members can insist on a poll. The advantage of a poll is that it reflects the voting power that an individual shareholder has in the company – instead of having one vote per person as on a show of hands, voting reflects the number of shares held.

## Company finance

In order to provide adequate resources to allow the company to develop and flourish it may be necessary for a company to borrow money or look at other methods of funding such as government grants, as well as taking account of the capital raised by the issue of shares. The advantages of share capital are that it does not have to be repaid (unless redeemable shares have been issued) and therefore has a permanence about it, and dividends are only payable if the company makes a profit and declares a dividend.

The size of the company's share capital may well determine how much it can borrow, especially in the early stages of a newly formed and developing company. Lenders will be reluctant to advance more than the members of the business have themselves risked. Where a company needs sizeable amounts of money it may be possible to attract funds under government-sponsored schemes.

A company can sometimes improve its financial position by siting itself where it may benefit from central and local government or EU grants and loans. The schemes will vary from time to time, reflecting the political stance of the government of the day. Examples of schemes have included Enterprise Zones and Regional Development areas. Sometimes, the incentive to set up in a particular place will consist of very favourable rents and rate-free periods.

Where a company needs to borrow money it may turn first to its own shareholders, who can provide loans called debentures. Like all forms of borrowing the issue of debentures is only possible if the directors are empowered to borrow by the Articles. It is possible to give debenture holders greater security by creating a charge over the assets of the company.

All lenders of money worry about how they will be repaid if the company gets into difficulties. A lender may be prepared to make the loan only on condition that he or she is given some security in the event of the company being unable to pay. If that happens the lender can then realize this security, i.e. sell it, to repay what is owed.

One type of security is the fixed charge, which is a mortgage over a specific thing like a building. Inevitably the company cannot dispose of, or realize, the property which is subject to the charge without the consent of the lender. This could be particularly restricting in a developing business, so the company may prefer to borrow against the security of a floating charge. This is a mortgage of assets of the company, such as the stock in trade, which changes from time to time. It is in the contemplation of both lender and borrower that the company should be free to carry on its normal business, dealing as usual with those assets. The company continues to have the flexible use of whatever is comprised in the charge. However, once the borrower defaults the floating charge is said to crystallize. This means that it becomes a charge attached to the assets available at the moment of crystallization and the company is henceforth prevented from dealing with those assets. The parties may have specified a number of circumstances which will give rise to crystallization.

Floating charges are a device by which a company can raise money without affecting its ability to trade. However, they could have the effect of misleading persons who are thinking of dealing with the company, because the company might look in a more affluent and healthy financial position than is truly the case. The problem is overcome to some extent by requiring company charges to be registered with the Registrar of Companies within 21 days of being made. The company is legally obliged to effect the registration, but it would be sensible for the lender to do so too, because a charge which is not registered in time is void. Once the charge is registered it is part of the information which is available for public inspection. Any person about to deal with a company, and needing information about its financial viability, could make a search at Companies House. That would reveal any charges registered, the date the charges were created, the person in whose favour the charge is made and the property over which the charge operates.

Raising money by the device of the floating charge is a procedure only available to companies and limited liability partnerships. This gives a company and a limited liability partnership an advantage over a trading partnership, which could borrow only against the security of fixed assets.

## Winding up of companies

A company is an artificial body and theoretically it could continue forever with a constantly changing membership as shares are bought and sold (transfer of shares). If a company is to be dissolved the process is referred to as 'winding up' or 'liquidation'. The directors may decide on this course, in which case the dissolution is voluntary, or it may occur by order of the court.

If the winding up is brought about because a company has financial problems the rules of liquidation are contained in the Insolvency Act 1986.

When a company is in financial difficulties a full liquidation process may be unnecessary. It may be possible to save the company by using voluntary arrangements. The proposal for such a voluntary arrangement can be made by a director. The proposal will be for a composition in satisfaction of the company's debts or a scheme of arrangement of the company's affairs. An insolvency practitioner will be appointed to act as a supervisor, implementing the scheme. The value of such a voluntary arrangement is that it can be organized with the minimum of involvement of the court, as it will be approved at meetings of the creditors and the company.

Where a company is likely to be unable to pay its debts another alternative procedure is an Administration Order. The court will make such an order if it believes that by doing so it can secure the survival of the company or that it would give time for a voluntary arrangement to come into being. There will be a court hearing. If an order is made it directs that, for the time being, the affairs, business and property of the company are to be managed by an administrator. While the order is in force there can be no order for the liquidation of the company. The administrator has three months to draw up a statement of his or her proposals, which are put to meetings of the creditors and the company.

The Enterprise Act 2002 amended the Insolvency Act 1986 to provide a procedure for the out-of-court appointment of an administrator by the holder of a floating charge, the company itself or its directors. Notice of such an appointment must be filed with the court but no court hearing or order is required to effect the appointment. The purposes of administration are to rescue the company as a going concern or to achieve a better result for its creditors than would be likely on a winding up or to realize property to pay one or more secured or preferential creditors.

Where liquidation of the company is necessary, this may be: a members' voluntary liquidation, which can occur where the company is solvent; a creditors' voluntary liquidation, which occurs where the company is insolvent and where the wishes and interests of the creditors will prevail; or a compulsory liquidation, which arises by order of the court, following the presentation of a petition seeking such an order.

For a compulsory liquidation, a winding-up petition may be based on a number of grounds, including the ground that the company is unable to pay its debts. This is deemed to be the case where a creditor who is owed more than £750 has made a statutory demand for this money that has remained unsatisfied for three weeks. Other evidence accepted by the court that the company is unable to pay its debts may be proof that the company's assets are less than its liabilities, or proof that a creditor entitled to execution has been unsuccessful. This means that a creditor was granted a warrant of execution to levy distress against the goods of the company, but this warrant has proved ineffective because there are insufficient goods to satisfy the demands.

Proceedings for compulsory liquidation are commenced in the county court or the High Court, depending on the amount of the company's share capital. Once a petition has been presented any disposition of company property is void unless approved by the court. This discretion is important because it means that the company can continue to trade, but it allows control over activities which may be detrimental to the interests of some of the creditors.

If the court decides to make a winding-up order, certain immediate consequences follow:

- A liquidator is appointed.
- All actions against the company must be stayed.
- A first meeting of creditors may be convened.
- The directors' duties and powers cease.
- The business of the company can carry on only to the extent that it is beneficial to the liquidation.

The liquidator must take control of all the company's property. He or she must call in all the assets, realize them, and distribute the proceeds to the appropriate creditors. Some property of the company may not be available to the liquidator, e.g. where the contract by which the company was acquiring the property provides that the company's interest in the property will end in the event of liquidation, or where a company has purchased goods under a contract containing a valid retention of title clause. The liquidator will also be concerned to discover whether transactions immediately prior to the liquidation were undertaken to defeat the rights of creditors, or some groups of them. If so, the transaction may be void. Examples include cases where a company transfers property for no consideration, or for a consideration significantly less than the real value (an undervalue transaction), or a transaction involves a preference, where a company puts one of its creditors into a more favourable position than the creditor would be under the liquidation.

Once the liquidator has realized all the assets he or she then pays the costs and expenses of the liquidation. He or she must next pay off the creditors, who must have submitted formal proof of their debt. Secured creditors will usually realize their security to settle what is owed to them but they may prove any outstanding balance not covered by the security. Fixed chargeholders take separately, but those holding floating charges rank after preferential creditors. This tends to emphasize the superior security provided by a fixed charge. The creditors rank in the following order:

1   *Preferential creditors* – this group ranks equally among themselves after the expenses of the winding up. The categories of preferential debts have been significantly reduced by the Enterprise Act 2002. Examples of preferential debts include:
    (a)   Contributions to occupational pension schemes.
    (b)   Employees' remuneration due for the previous four months, not exceeding £800 (this figure can be adjusted from time to time by the Secretary of State).
2   *Amounts secured by floating charges* – the Enterprise Act 2002 provides that, where there is a floating charge over the property of a company, a percentage share of the company's net assets must be made available to unsecured creditors. This provision improves the position of unsecured creditors but means that there is less money available to distribute to the holder of the floating charge. There are some circumstances in which these provisions will not apply, for example if the company's net assets are less than a prescribed minimum threshold.
3   *Unsecured creditors* – all other creditors rank equally amongst themselves.

If there are surplus funds left over after distributing to the creditors these are distributed to the members of the company according to their rights and interests.

Where a company has been wound up this may reflect seriously on the competence and integrity of the directors. In that case, the courts have extensive powers to disqualify directors for specific periods. In some cases the disqualification may be based on activities of the director which could involve criminal liability but there is no overall requirement that the behaviour must infringe specific criminal law rules. One example where a director could be disqualified is for 'unfitness' to be concerned with the management of a company.

Once the process of winding up has finished, the company is removed from the Register and is dissolved three months later.

# Limited liability partnership

Since 6 April 2001 it has been possible to incorporate a limited liability partnership. The Limited Liability Partnerships Act 2000 created this new form of organization for businesses, which is a hybrid between the partnership and the limited liability company.

In the same way as a company, a limited liability partnership (LLP) has a separate legal identity to that of its members. It is the LLP that enters into contracts, owns property etc., not the individual members. This gives members of the LLP the benefit of limited liability for the debts of the partnership. There are formalities associated with setting up and running an LLP, although these are less onerous than those associated with setting up and running a company.

## Creating a limited liability partnership

To form an LLP, the Limited Liability Partnerships Act 2000 requires that two or more persons carrying on business with a view to profit subscribe their names to an incorporation document. These individuals are known as designated members and have responsibility for sending any necessary information to Companies House on formation and during the life of the LLP. All the members of the LLP can be designated members, although the legal requirement is that there must be two.

The incorporation document contains the name of the LLP, the situation and address of its registered office, the name, address and date of birth of each member of the LLP, and which of the members are to be designated members. This document, the appropriate fee and a statement that the requirements of the Limited Liability Partnerships Act 2000 have been complied with must be sent to the Registrar of Companies. On registration, a certificate of incorporation is issued.

## Managing a limited liability partnership

An LLP has far more in common with an unlimited liability partnership than with a limited company in terms of managing its affairs. Unlike a company, an LLP is not required to have formal constitutional documents on incorporation. It is, however, sensible for the members of an LLP to enter into a formal agreement. This agreement should deal with the sort of issues that would be covered in a partnership agreement for an unlimited liability partnership. In the absence of a partnership agreement, the Limited Liability Partnerships Regulations 2001 will apply. Under these Regulations, profits, losses and capital are shared equally and all members have an equal right to manage the LLP. Unlike a company, the members of an LLP are not obliged to make decisions in formal, minuted meetings and hold meetings of all the members for certain matters. The members' agreement should cover which members make decisions of day-to-day management and any matters requiring the agreement of the members.

# 6
# The law of contract

## The nature of a contract

In business and commercial life the law of contract underpins a huge range of activities as diverse as engaging staff, buying supplies, arranging insurance, leasing premises, raising loans and buying shares. These transactions all involve making a contract which must comply with the general principles governing all contracts. The modern law of contract has largely evolved from case law, although major pieces of legislation such as the Unfair Contract Terms Act 1977 have radically altered some of the rules. There is no complete code of rules governing the making of a contract. The courts have generally taken a *laissez-faire* approach to the making of contracts, indicating that the parties are free to make the contract of their choice, with which the courts will rarely interfere. Although this is a guiding principle the law will intervene in certain circumstances, as will be explained below.

Where a dispute arises between the parties, litigation, arbitration or a form of alternative dispute resolution may be necessary to determine the rights of the parties. However, the role of the courts ought not to be overemphasized. Of all the millions of contracts entered into daily only a very small proportion ever involve litigation before the courts. On the whole the parties carry out their obligations and by that means discharge their contract. Even where problems do arise some compromise or settlement may be reached between the parties without the need to take action in the courts.

Contracts are based on the idea of a bargain. Each side must put something into the bargain. A contract may be defined as 'an agreement which is binding on the parties'. Within that definition lies the first problem. The ultimate bargain arrived at between the parties may not have been 'agreed' in the commonly understood sense of that word. There are a number of reasons why this is so. First, English law adopts an objective approach to decide whether there is an agreement between the parties. It is pointless to ask an individual who is now in dispute whether he or she intended to make a contract. Rather, the courts ask the question: 'Would a reasonable person think that there was an agreement between the parties?' Second, 'agreement' may not seem to be present where one party has little or no choice in the terms he or she is obliged to accept. There are many contracting situations where, because of the superior bargaining strength of one of the

Galbraith's Building and Land Management Law for Students. DOI: 10.1016/B978-0-08-096692-2.00006-3
Copyright © 2010 by Elsevier Ltd

parties, the other has little or no choice but to accept the terms. Third, there are many cases where little or no negotiation takes place between the parties (e.g. sales in a self-service store) but the 'agreement' between the parties is largely created by implied terms (e.g. terms implied into contracts for the sale of goods by virtue of the Sale of Goods Act 1979).

To return to our definition of a contract (an agreement which is binding on the parties) it is always possible for an injured party who is suffering from some breach of contract to sue to enforce it. Enforcing the contract means seeking an appropriate remedy from the court. In most cases that remedy will be an award of damages. It is only in very limited types of case that a court would order one party specifically to perform their obligations. If the court does grant such an order it is called specific performance. This is an equitable remedy and as such is discretionary, i.e. the court may or may not award the remedy.

## Formation of the contract – agreement

It has become customary in English law to regard agreement as consisting of offer and acceptance. This is only a method of analysing the bargaining process. It is not necessary for the parties to use these words. An offer exists when one party effectively declares a readiness, or intention, to be bound by a set of terms without any further negotiation. In other words, the offeror (the person making the offer) has reached a stage where he or she is content for the other party to say he or she accepts, and at that point a valid, binding contract will come into existence. If the parties are not at such an advanced stage in their negotiations, no offer may yet exist. Often, what appears to be an offer may only be some kind of enticement to bargain, which the law calls an invitation to treat.

The courts have had to consider numerous instances where the argument turned on whether a definite offer was made. In the case of goods on display in a shop window, or goods on supermarket shelves, the court would normally find that such a display was an invitation to treat – the shopper must come forward and make an offer to buy. Similarly in auction sales, when an auctioneer calls for bids, he or she is not 'offering' to sell the goods, but rather inviting potential buyers to put forward their offers. Equally it may be said that a website advertising goods (it may be viewed as a virtual shop) will usually constitute an invitation to treat in the absence of a clear intention to be bound by certain terms.

The problem regarding offers can also arise in relation to price lists or catalogues which are sent to customers. These are usually construed to be a mere indication of the goods available and their likely selling price and, in consequence, an invitation to treat. Any other interpretation could give rise to serious practical difficulties. Consider a case where a firm sends out a price list of second-hand earthmoving plant which it has available for sale.

The circular may go to dozens of companies who could be interested to buy. If the circular was considered to be an offer and there is only one model of a particular type on the list and several purchasers come forward saying that they wish to buy it the supplier could find itself in breach of contract over and over again as it is unable to supply them all with the equipment. It is better to construe the price list as a mere invitation to bargain, a 'display' of what the firm has to sell, and think of the approaches by potential buyers as the offers, which the seller can accept or reject as it chooses.

Where a case arises which causes doubt the court approaches the issue of whether any offer has been made by using an objective test of the intention of the person concerned. Would a reasonable person, knowing all the circumstances of the case, think that an offer has been made? The court takes into account the words used. In one case, where a letter said 'I may be prepared to sell ...', the writer could not yet be said to be ready to be bound in contract without further negotiation. Even if the word 'offer' is used, that is not conclusive in law that an offer exists.

If it is clear that the intention is to make a firm offer it may be made orally or in writing. It comes as something of a surprise to many people to discover that English law has very few rules about formality or written evidence in contracts. Apart from exceptional cases such as contracts for the sale of land, transactions worth hundreds of thousands of pounds can be binding even when made orally. Of course, there may be sound reasons for making a contract in a more formal way, as subsequent disputes may be capable of being resolved simply by reference to a written agreement.

An offer can be made to an individual, when only that person can accept. The offer may be made to a group and, in certain cases, it may be made to the world at large, for example in cases offering a reward to a finder. The offer is only effective when it has been communicated to the other party, the offeree.

## Counter offers

The offeree has a number of options. He or she may wish to accept the offer, in which case he or she must exactly accept the precise offer which was made. If he or she seeks to bargain further or to introduce new terms he or she is not accepting, rather he or she is making a counter offer. The law provides that the effect of a counter offer is to terminate the original offer. So if the counter offer is not acceptable the person who made it cannot try to revive the original offer. He or she must make a fresh offer. This situation can be of vital importance where the person making the final offer may be fixing the terms of the deal. The offer–counter offer situation arises frequently in business where each side in the negotiations may be trying to make the contract on the basis of their own 'pro forma' standard documents. This gives rise to what is known as the battle of forms. A classic example can be seen in *Butler v. Ex Cello Machine Tool Co.*, where the seller offered to sell a machine tool for £75,000, which would be delivered in 10 months' time.

The seller's offer was made on a printed form containing various conditions, one of which was a price variation clause allowing for an increase in certain circumstances. The buyers placed an order, using their own standard printed order form. On the bottom of the buyers' form was a tear-off slip, which the seller was asked to sign and return to the buyers. The seller duly signed and returned the slip. It contained the words 'we accept your order on the terms and conditions thereon'. When the tool was ultimately delivered, the seller sent a bill for over £78,000. Its claim for the increased amount was unsuccessful. Its original quotation was an offer. The buyers had made a counter offer which did not contain a price variation clause. The seller had accepted the counter offer by returning the tear-off slip.

Of course, not every communication from an offeree will necessarily amount to a counter offer. Once an offer has been made, the offeree may simply wish to seek further explanation of its terms to satisfy himself that he is getting the best deal possible. So there may be a fine distinction between the cases. For example, X offers to sell Y his car for £3000. If Y replies 'Will you take £2700?' he is still negotiating and making a counter offer. But if Y replies 'Is that cash or is there any possibility of HP?', he may simply be seeking to clarify the terms of the offer. A mere inquiry does not terminate an offer.

## Acceptance

Acceptance is defined as a final, definite and clear assent to the terms of an offer. Not only must the offeree accept the exact and precise offer made, he or she must also communicate acceptance to the other party. Normally, the offeror must actually receive the acceptance. This tends not to create problems where the parties are face to face, or using some form of instantaneous communication such as the telephone or a telex machine. In all of these cases, the contract is made when the acceptance is received by the offeror.

The situation is somewhat different if the parties are negotiating at arm's length, i.e. not in each other's presence. This will usually be a case where they have been dealing with each other by post. The offeror can lay down rules about how the offeree has to accept but in most cases will not have done so. Then it may be appropriate to use what is called the 'postal rule'. When this applies, an acceptance made by letter is effective, and the contract comes into existence when the letter is posted. Inevitably this means that a legally binding contract exists for some time before the offeror actually learns that he or she is legally bound. That can sometimes be inconvenient, but an offeror could overcome the problem by stipulating as part of the offer that he or she will only be bound on *receipt* of the acceptance. In *Holwell Securities v. Hughes*, the offeree wanted to accept an offer under the terms of which he had to accept by a notice in writing by a fixed date. He posted a letter of acceptance but it did not arrive until after the fixed date. The court held that the phrase 'notice in writing' indicated that the offeror must actually receive the acceptance by that date. Merely posting a letter was no use in those circumstances. The 'postal rule' is merely a rule of convenience and the courts would not apply it in cases where it was manifestly absurd to use the post (e.g. a case where the offeror had indicated that he or she needed a very speedy acceptance). Nor will the rule be applied if the offeror has prescribed some other method of acceptance.

The offeror can lay down quite elaborate rules for acceptance. The courts may then be faced with the problem that an offeree has purported to accept but has not followed the precise method stipulated. Generally the courts will allow an acceptance by any method which is equally as effective as the one which was prescribed, so long as the offeror has not stipulated that acceptance *must* be made by the method prescribed. It is a question of construing whether the offeror was making a suggestion, or whether that method and no other must be used. For instance, if the offeror says 'Drop me a line to tell me if you want to buy', does that mean he or she is insisting on acceptance only by letter?

It is unclear whether or not the postal rule applies to e-mailed acceptances. As an e-mail may be seen as an instantaneous form of communication (like a telex) it is suggested that there must be actual communication to complete acceptance.

The rule that acceptance must be communicated means that silence cannot usually amount to acceptance. This is important because otherwise an offeror could impose obligations on the offeree: 'I offer to sell you my car for £3000. If I do not hear from you before 9 p.m. I will assume that you have accepted.' If this were permitted, it would force the offeree into a bargain.

## Termination of offers

An offer can be said to be terminated by its acceptance because at that moment there is a legally binding contract. There are, however, a number of ways in which an offer may be brought to an end. The offer itself may have been made subject to a time limit. If there is no acceptance before that time, the offer will automatically lapse. Where there is no fixed time limit the offer will remain open for acceptance for a reasonable time. What is reasonable will vary with the circumstances and the nature of the subject matter.

Once an offer has been made, the offeror may have a change of mind and wish to withdraw it. Revocation is always possible before an offer has been accepted. To be effective, the fact of revocation must be communicated to the offeree; in postal cases this means the letter must actually be received and not merely posted; in other words, the postal rule does not apply. This can cause conflict where the parties are negotiating at arm's length using the post where the postal rule of acceptance is relevant. The letter revoking the offer may be posted first, but if it arrives *after* a letter of acceptance has been posted it will be too late to withdraw.

Where an offer has been made for a fixed time the question may arise whether it can be revoked during that time. In reliance on the offer remaining open, the offeree could be going to considerable trouble to put him- or herself into a position where he or she can accept, e.g. by raising finance. If the offeror revokes before the fixed time is up the offeree may have no legal basis for complaint. Effectively this fixed offer situation is really two distinct offers: 'I offer to sell you my car for £3000, and I offer to keep the offer open for seven days.' If the offeree wishes to enjoy the benefit of the seven-day time limit he must accept the second offer and 'buy' it by giving some consideration for it. This is what happens when a person buys an option but in most everyday situations it would not occur to people to offer to 'buy' time in this way. The rules of consideration are considered later.

## Certainty of the agreement

Even though the parties think they have reached agreement the courts may find that there is no legally binding contract because of lack of certainty. Naturally the courts do not want to disappoint people's expectations so they approach any question of vagueness or uncertainty with the idea in mind that 'a thing is certain if it is capable of being made certain'. An agreement which appears vague at first sight may be capable of being given a sensible meaning, either because the parties have dealt together before or because there is some custom of the trade or business. In some cases the parties themselves may be aware that their agreement is vague or incomplete and they may provide for future resolution of these aspects by including an arbitration clause. Sometimes it is possible to fill in missing terms by reference to a statute. For example, if the parties have failed to agree on a price and goods have now been supplied under the contract, s.8 of the Sale of Goods Act 1979 provides that where no price has been fixed and no method has been agreed for determining the price, the buyer must pay a reasonable price.

## The agreement

Once the processes of offer and acceptance have culminated in the formation of the agreement the terms have become fixed between the parties. That has been the object of their negotiations – to determine the extent of their obligations to each other. Once agreement is reached it cannot be altered by one of the parties without the consent of the other. If any alterations or variations to the agreement are to take place then the relevant rules of consideration should also be borne in mind.

There can be great difficulty in establishing what has become a term of contract. A distinction has to be drawn between statements made which go to the core and essence of the transaction and those statements which merely influence the decision to enter into the contract. The former are terms, where a breach gives rise to a right to claim for damages for breach of contract. The latter are representations, where rights may be available if the statement made was untrue and in some way induced the contract.

# Consideration

English contract law is based on the idea of a bargain. There must be some exchange between the parties. One party must be doing something or giving something in consideration of what the other is doing or giving. Consideration can be defined simply as the 'price' each party pays for the right to enforce the other party's promise. It is important to remember that consideration will not always be money although it is quite usual for money to be the consideration provided by one of the parties. If I sell you my car for £3000 my consideration is giving you the car and your consideration is giving me the £3000. If we make that agreement and provide for the exchange to take place at the end of the month then our agreement is still a binding contract *now* because our exchange *now* is an exchange of promises. In consideration of you promising to pay me £3000 at the end of the month I promise to hand over my car to you then. As you can see there is no need for a physical exchange of car for money to create a contract, the exchange of promises is sufficient. This allows parties to plan for the future and is particularly important in the commercial and business world.

This type of consideration, based on promises which are to be performed in the future, is called executory consideration. It is surprising how often people fail to realize that there is a legally binding obligation in such a case. Take the example of a shopper who orders a chair in a furniture store, delivery to be in six weeks' time, when the customer will be invoiced for the chair and will then send payment. Later the same day, having placed the order, the customer wants to back out. That would amount to breach of contract. The customer might say that the furniture store had not even had time to send the order off, but the store has lost the profit on the sale and no doubt has been put to some trouble and expense in dealing with the customer.

In order to provide a valid consideration, the law insists that it must not be past. *Re McArdle* illustrates the principle. Members of a family were all entitled to a share in a house on the death of their mother. While the mother was still alive one of the sons and his wife went to live with her. The son's wife made several improvements and alterations to the property for her own comfort and convenience. When the mother died the other members of the family promised to pay the son's wife for all the work that had been done. When they failed to pay she could not enforce their promise as she had given no valid consideration for it. The work she had done was in the past and had not been done in consideration of their promise to pay. The promise and the act were independent of each other. To be a valid consideration there has to be an interdependence of promise and act.

There are some situations which look like past consideration but which are not true instances. Take the case of a person requesting someone to perform services for him, in circumstances where he must be expecting that he will have to pay, for example where he asks an accountant to deal with his tax affairs. Once the service is performed any subsequent promise to pay in such a case is enforceable because by asking for the service there is an implied promise to pay a reasonable amount.

Although contracts are based on the idea of a bargain the law does not insist that the two sides of the bargain must be equal. This is often expressed as a rule that consideration need not be adequate. It is up to the parties to make the contract of their choice. One person, for good reasons of his own, may be prepared to pay far more for goods than they are really worth. Generally, contract law allows great freedom to the parties in making their bargain and each is supposed to look after their own interests and strike the keenest bargain possible. One of the maxims of the law is *caveat emptor*, which means 'let the buyer beware'. Where the two sides of the bargain are seriously unequal that may be evidence of fraud, duress, mistake or incapacity, in which case there may be some basis on which the court could set aside the contract. However, if the person complaining has simply made a bad bargain the law will not intervene.

While not insisting that the two sides of the bargain must be equal the courts will insist that each side of the bargain must have some value in the eye of the law. This may be so even where the consideration consists of something intrinsically worthless. This point is illustrated in *Chappell & Co. v. Nestlé & Co.*, where the chocolate company was running a promotional scheme. In return for three chocolate wrappers and a postal order for 1/6d (71/2 p) the chocolate company would send the customer a pop record. Chappell's were entitled to receive royalties on the selling price of all sales of the record. The question arose here whether the record was sold for 1/6d or whether the wrappers formed part of the consideration. Even though the chocolate company simply threw the wrappers away the House of Lords found that they did form part of the consideration.

Consideration must move from the promisee, meaning that the person who receives the promise must provide consideration. However, it does not mean that the consideration must move to the promisor. In *Tweddle v. Atkinson*, a young couple were about to be married and their respective fathers promised *each other* that each would pay a sum of money to the son. Indeed, their agreement said that the son could sue on the agreement should either fail to pay. One father did not pay. The son attempted to sue him but failed as he was not a party to the contract and, additionally, had provided no consideration for the promise. Note that the consideration given by the fathers did not move to the other but to the son, a third party. Consideration moved from the promisee but not to the promisor, i.e. it was for the benefit of a third person.

It should be noted that the above difficulty of contracts not being enforceable by third parties is connected to the notion of *privity of contract*. The rule that contracts cannot be enforced by a person who is not a party to a contract has been subject to qualification by the Contracts (Rights of Third Parties) Act 1999. Section 1(1) provides: Subject to the provisions of this Act, a person who is not a party to a contract (a 'third party') may in his own right enforce a term of the contract if – (a) the contract expressly provides that he may; or (b) subject to subsection (2) …, the term purports to confer a benefit on him. If the facts of *Tweddle v. Atkinson* were to be repeated today, then the son would be able to enforce the promise under s.1(1)(a) of the 1999 Act. The 1999 Act may entitle a third party to rely on an exemption clause in a contract.

## Duties already owed

It seems obvious that you are not providing any consideration where you simply behave in a way in which you were legally obliged to act. This view is accepted by the courts in relation to both public and contractual duties owed. Take the case of *Collins v. Godefroy*, where a witness was under a public duty to attend and give evidence because he had been ordered to do so by the issue of a subpoena. The witness was promised money by the defendant in the case if he came and gave evidence. Once he had given his evidence was it possible to say he had done so in consideration of the defendant's promise to pay? The court held that the witness gave no consideration for the promise as he was already under a duty to give evidence. He was doing nothing more than his duty.

In those cases where it can be shown that a person does more than his public duty requires, then he or she may have provided consideration for a promise to pay. This can be seen to be the basis of the decision in *Glassbrook Bros v. Glamorgan CC*, where a mine owner requested police to provide a 24-hour guard over his property. Although the police were under a duty to prevent crime and protect property they were not required to mount a continuous guard. The mine owner promised to pay for the service and it was held that the local authority could recover the money from him as the police were doing more than their public duty. They had, therefore, given consideration for his promise to pay.

The same sort of situation can arise where a duty is owed under a contract. Suppose you have asked me to supply 100 sausage rolls for your village fair and you have paid me for them. Under the terms of our agreement I am due to deliver the sausage rolls at 12 noon on 1 June. I indicate that I am going to struggle to arrive at 12 noon and sound reluctant to make a delivery. You then promise to pay an extra £10 and, in consequence, I make the delivery. What consideration do I give for your promise to pay £10? I was already contractually bound to deliver the goods so I am doing nothing more than performing an existing duty. In an example such as this, the courts have traditionally found that I could not enforce your promise to pay £10 as I had given no further consideration for it. An old case, *Stilk v. Myrick*, has long represented that view of the court. There, sailors were promised extra pay for sailing a ship back to port when some of the crew had deserted. Later, the shipowner refused to pay, as the sailors were already obliged by their contract to sail the ship back, and had therefore simply performed the duty they already owed under the contract.

Such a decision may be entirely correct as a pure application of the law, and no doubt it prevents situations of near blackmail arising where a person could apply pressure to the other contracting party to pay more, simply by threatening not to perform. But inevitably that 'pure' approach may fail to recognize the true pressures of business life and the marketplace. In a more modern shipping case, the *Atlantic Baron*, the court was able to enforce a promise of extra payment for the building of a ship by finding that the shipyard did more than they originally promised, and had therefore given consideration for the extra payment.

Some judges have gone so far as to say that any promise to perform a pre-existing duty should be good consideration, because it is a benefit to the person receiving performance. When the argument is carried to these lengths, in our sausage roll example, I could recover the extra £10. The benefit to you would be that you receive performance with none of the aggravation of needing to sue for breach. The detriment to me is that I give up the possibility of being in breach of contract – and there is no doubt that this can be a very valuable right. Having agreed to sell sausage rolls to you for a fixed price, I may now find that I can sell the same goods so advantageously elsewhere that it actually pays me to be in breach with you and suffer the consequences.

Much of this thinking has been further developed by the Court of Appeal decision in *Williams v. Roffey*. The facts there were that building contractors had subcontracted the carpentry work on a block of 27 flats to the claimant, for a price of £20,000. The claimant was entitled to interim payments under the contract. After he had completed nine of the flats, and done preliminary work on the others, he had already received over £16,000. He then realized that his price for the work was far too low. The building contractors began to get anxious, as they would become liable under penalty clauses in their main contract if the work was not completed on time. They knew that the claimant was having difficulty because he had underpriced the job. They suggested a meeting with him, at which they agreed to pay him £10,000 more, on condition that the work was completed on time. In view of the extra payment, the claimant went on with the work. He completed eight further flats but then stopped work. When he tried to recover what he alleged was due to him under the contract, the building contractors refused to pay any part of the extra £10,000, on the grounds that the carpenter had given no extra consideration for it. In the Court of Appeal, the carpenter was held to be entitled to the extra money, because the building contractor had obtained practical benefits in having the work completed and in avoiding the penalty under another contract.

Of course, an important feature of the case was that there was no suggestion of 'blackmail' by the carpenter. Indeed, he had been approached in the first instance by the builder who had made the suggestion to pay extra. Moreover, if the court had been determined to look for something in the carpenter's behaviour which could amount to performing more than his mere contractual duty, no doubt they could have 'found' it. For example, it could be said that the carpenter's agreement to undertake the work in a different sequence could be a detriment to him and a benefit to the builders and consideration.

It is difficult to make a judgment yet as to the ultimate effect of the Williams case on the development of this area of law, but it is clear that the courts have found it easier to discover a valid consideration where performance of a duty owed to a third party is involved. Consider the following example. A is contractually bound to supply the engines for a ship being built by B. The engines are due to be delivered by 1 January. C has a contract with B to charter the ship from 1 June, immediately on its completion. It begins to look as if A is going to be late in making delivery of the engines. C is extremely anxious that the ship should be finished on time as he stands to lose an enormous sum of money if his charter of the ship is delayed. C approaches A and promises him £10,000 if he will get the engines to B by 1 January. If A does deliver the engines to B on time, can he then claim the £10,000 from C? The courts have been inclined to take a more relaxed view in this type of three-party situation, and in a number of cases (e.g. *New Zealand Shipping Co. v. A. M. Satterthwaite & Co.*) they have allowed A to sue for the sum promised by C. The reasoning seems to be that there is a detriment or burden to A because, by binding himself in a contract to C as well as to B, he lays himself open to two actions for breach of contract if he fails to deliver the engines. Moreover, it might have been more to A's liking to be in breach of contract with B and face whatever consequences that might have brought. Now he foregoes that possibility because of his agreement with C. C also gets what C wants and this is a benefit to C.

In such cases it would be important to check that there was nothing in the transaction which was contrary to public policy. Latterly the courts have shown themselves willing to control these situations by developing rules of economic duress (see page 135).

## Part payment of a debt

Problems can arise in finding valid consideration where one party seeks to perform a contract by doing something less than contractually obliged. Take the example of A, who owes £100 under a contract to B which is due for payment now. A may be in financial difficulties, so much so that B agrees to accept £75 in full settlement. What consideration has A given for B's promise to forego £25? The courts would say that the £25 was still due and B could sue to recover it despite this promise. Thus, part payment of a debt does not discharge the full debt; this rule was approved by the House of Lords in *Foakes v. Beer*. It would be quite different if, at B's request, A had paid £75 at a date earlier than the debt was due. That would be a detriment to A and a benefit to B, and A would then be providing consideration for B's forbearance.

In *Re Selectmove Ltd* it was considered whether the approach adopted in *Williams v. Roffey* could be extended to part payment of a debt. It was argued that by paying less than the sum owed this could constitute consideration provided there are practical benefits to the creditor. For example, the practical benefit could be that the creditor gets part of the sum owed in circumstances where they might otherwise receive no payment. However, the Court of Appeal refused to follow *Williams v. Roffey* as to do so would be inconsistent with *Foakes v. Beer*, a House of Lords case. At present, therefore, the rule as to part payment of a debt, as seen in *Foakes v. Beer*, remains unaffected. It is fair to say that *Williams v. Roffey* has created inconsistency in the law.

The difficulty with the proposition advanced above is that, in practice, people dealing together do frequently make concessions to each other within the framework of their legally binding agreement. These concessions or variations are usually informal and may work perfectly well so long as the parties remain on good terms. If their relationship becomes strained one of them may wish to revert to the strict terms of their original agreement. The other will want to know if the variation can be regarded as binding. Traditionally, in order to enforce the variation, it has to be 'paid for' by some extra consideration. There is sometimes a way round this problem, however, by the application of the equitable principle of promissory estoppel.

Equitable principles operate to abate the rigours of the operation of the common law. Undoubtedly, there can be grave hardship where one contracting party has relied on some concession by the other, only to see it abruptly withdrawn. That was the state of affairs in *Central London Property Trust v. High Trees House*. The contract in that case related to a lease of a block of flats in London where the rent was to be £2500 per annum. Because of the war it became extremely difficult to find tenants for the flats, so the landlord agreed to reduce the rent by half. When the war ended the landlord wanted to revert to the original rent and was also keen to know whether he could recover the arrears of rent. Effectively, what the landlord had done was to make a promise, within the framework of an existing contract, where the tenant gave no consideration for the promise, but where obviously the tenant had relied on it. On the authority of earlier cases the High Court found a principle of equity whereby if A leads B to suppose that A will not enforce strict rights under an existing contract, then A cannot go back on a promise where it would be inequitable. Applying this principle in the High Trees case, the court would not allow the landlord to recover the arrears of rent. But the court also found that the end of the war acted as notice that the promise (made only because of the war) was no longer operative.

At first sight this case may seem to strike at the very core of the doctrine of consideration. Further examination shows that not to be the case. A number of points must be borne in mind. It was made clear in cases subsequent to the High Trees case that this equitable principle of promissory estoppel can only be pleaded by way of defence. In the colourful language of judges it is a shield and not a sword; estoppel of this type cannot be used to create a contract. There must be circumstances which show it would be inequitable for the promisor to go back on his or her word, at least without giving notice of the intention to revert to the original terms of the contract. Of course, there may be cases where once a concession has been made it is impossible to revert to the original terms, e.g. if the concession related to an extension of time for delivery. This raises important questions about the scope of promissory estoppel, as it seems that the effect of the promise or concession should be intended only to suspend legal rights and not to extinguish them completely. This fits well with the idea that the promisor can revert to the original terms by giving reasonable notice. Certainly if the scope of the rule is seen to be thus limited it gives rise to fewer legal difficulties.

It must be stressed that the development of promissory estoppel is based on equity. 'He who seeks equity must do equity' and 'he who comes to equity must come with clean hands'. It follows from these maxims that a person seeking to take advantage of promissory estoppel must have behaved properly. This aspect was critical to the decision in *D & C Builders v. Rees*, where two jobbing builders were owed money by Mr and Mrs Rees. Mrs Rees knew they were in a grievous financial state and she persuaded them to accept less than they were owed in full settlement. So desperate was their plight that they accepted the smaller sum but determined to see what they could do later to recover the rest. Mrs Rees claimed that they had promised to accept the smaller sum, that she had acted in reliance on their promise and that it would be inequitable to allow them to go back on their word. The court had no hesitation in finding that her own behaviour was inequitable as it verged on intimidation so they refused to allow her to take advantage of the equitable principle.

Although promissory estoppel can be seen as an important safeguard against some of the harsher aspects of the common law rules it must be remembered that consideration is still essential for the creation of a valid contract, although there are limited circumstances where it may not be necessary to support some modification within the contract.

## Intention to create legal relations

If both parties to a contract provide valid consideration, that may be taken to indicate that they intend to make a legally binding arrangement. However, in English contract law, the courts look for a separate element of intention to create legal relations. Traditionally the courts have viewed the cases as falling into two categories: commercial arrangements where the parties are presumed to have intended to create a contract; and family, domestic and social arrangements where the parties are presumed not to intend to create a contract. The presumptions are sensible in most cases. Take the example of a parent who gives pocket money to a child or two friends who arrange to meet for supper. In either case, one party would never expect to be sued for breach of contract if such arrangements collapsed.

The problem facing the courts arises where one party to an arrangement is seeking to rebut the normal presumption. In family arrangements a person may want to prove that a legally binding agreement was indeed intended. He or she must then satisfy the court that this was so. Many of these cases turn on problems between husband and wife on the breakdown of marriage. In one case, *Merritt v. Merritt*, relations between the parties were so strained that they met in the neutral territory of a car park to discuss the disposal of their matrimonial home. Once they had arrived at a decision the wife insisted that the husband should jot down the main points on a piece of paper. The court found here that the parties did intend to create a legally binding contract, influenced no doubt by the rather formal steps that the wife had insisted on, and also by the fact that the relationship between the parties was breaking down, so that they were no longer closely bound in a family or domestic situation.

Where one party seeks to rebut the presumption, the court may also be influenced by the degree of certainty and detail with which the parties made their arrangements. In *Jones v. Padavatton*, the details of the arrangements with regard to a house were so sketchy that it was impossible to believe that the parties could have intended to make a legally binding agreement. In other cases it seems that as the family relationships become less close, so the strength of the presumption diminishes. In *Parker v. Clark*, where the parties were uncle and nephew, there was found to be an intention to create a legally binding agreement, but no doubt the amount at stake, together with the fact that the terms of the agreement obliged the nephew to give up his own home to come to look after the uncle, also influenced the court. There seemed to be little difficulty in establishing the appropriate intention in *Simpkins v. Pays*, where three people living together regularly participated in a competition. When one of them was ultimately the winner, she was obliged to share her winnings with the other two.

In a commercial situation it is considerably more difficult to rebut the presumption that a contract was intended. The courts will require convincing evidence that the parties intended otherwise. In *Rose and Frank Co. v. Crompton Bros* the parties indicated in a written document that their arrangement was a 'gentleman's agreement, binding in honour only'. In the face of such convincing evidence the court was obliged to find that there was no contract.

Sometimes the parties want to lift their transaction outside the framework of contract law because they do not want the trouble and expense of litigation. This can frequently be seen as a condition in competitions and is one explanation of the football pools cases, where a punter who believes he has a winning line cannot sue to enforce his win. Provided the parties have made their intentions sufficiently clear, the court will accept their decision.

## Form of the contract

As a general rule, English contract law does not require contracts to be made in any particular form. In practice many contracts are oral, which often leads people to assume that it would be difficult to sue to enforce such a contract. Certainly there may be problems of proof but in the end it may be a matter of judging which party the court can more reliably believe. Where a contract is put into writing to provide the necessary proof, the writing can in turn prove to be problematical. There may be a dispute about its precise meaning, or it may fail to spell out the entire contract between the parties. A good example of written contracts is to be found in the construction industry's use of standard form contracts (see Chapter 8).

Exceptionally the law sometimes requires contracts to be made in writing (e.g. consumer credit agreements, such as hire purchase contracts). In these cases it is usually necessary to look at the appropriate statute to see the precise form the writing must take and also to determine what will happen if the parties have not complied with the rules. Rules about the precise form of the contract can be a useful device for protecting consumers. In the consumer credit legislation certain information contained in HP agreements needs to be displayed in prominent boxes, by which means it is hoped to draw those aspects more particularly to the attention of the consumer.

Distance selling, e.g. contracts made in reponse to catalogues or press advertising with an order form or teleshopping via the medium of a television, is subject to the Consumer Protection (Distance Selling) Regulations 2000. By the Regulations certain information must be given by the supplier to a consumer before a contract is formed, e.g. description of subject matter of the contract, price, delivery costs and the existence of a right to cancel the contract. The Regulations do not apply to certain contracts, including contracts for the sale or other disposition of an interest in land or for the construction of a building where the contract also provides for a sale or other disposition of an interest in land on which the building is constructed.

Despite the freedom about the form of the contract there has long been special treatment for contracts for the sale of land. As early as 1677, rules were laid down that contracts for the sale of land must be evidenced by a note or memorandum in writing, otherwise the contract would be unenforceable. These rules were re-enacted in more or less the same form by s.40 of the Law of Property Act 1925. Evidence in writing was meant as a safeguard against fraud but ironically the need for such evidence has sometimes been the very means by which people could perpetrate a fraud. The rules were regarded as unsatisfactory and, when the Law Commission took a detailed look at the workings of s.40, it reported: 'As a result of judicial attempts to prevent the statute being used as an instrument of fraud, it is virtually impossible to discover with acceptable certainty, prior to proceedings, whether a contract will be found to be enforceable … s.40 would appear ripe for reform.'

That reform took place with the repeal of s.40 by the Law of Property (Miscellaneous Provisions) Act 1989. Section 2 of the new Act provides that contracts for the sale or other disposition of interests in land can only be made in writing, and must incorporate all the terms expressly agreed between the parties. The contract must be signed by both parties, or their agents. Incorporation of the express terms may be by reference to another document, and the rules also make provision for the methods employed by solicitors engaged on behalf of the parties, where it is common to 'exchange contracts'. Sales of land by public auction are excluded from these rules, as are short leases.

Although these new rules seem designed to promote greater certainty, in that it should be easier to determine when a legally enforceable contract for the sale of land exists, one major problem remains to be tackled. It is still the case that people who are unaware of these rules may 'shake hands' on an oral agreement for the sale of land, and the purchaser may then incur expenditure on the land concerned, only to find that the vendor refuses to complete the transaction. Take the example of an oral agreement for the sale of a house, where the parties agree that the prospective purchaser can have a key for the property, and may undertake work on the property in advance of the legal transfer. If the purchaser were to install central heating and undertake extensive decorating work, it would be disastrous if the vendor then refused to transfer the house. In such a case, the vendor would seem to receive improper protection from the new rules. The aggrieved purchaser would be unable to sue for breach of contract because the s.2 rules state that a contract for the sale of land *can only be made in writing*. Lack of writing, therefore, means that there is no contract.

Problems of the sort outlined above were previously solved by the use of discretionary equitable relief, in the form of the equitable doctrine of part performance. In the absence of the necessary note or memorandum in writing, this equitable principle used to permit specific performance of the contract if the behaviour of the parties was in itself sufficient to indicate the existence of a contract. Unfortunately, there is no scope now for the application of this discretion, as it depends for its efficacy on the fact that lack of writing under the old s.40 rules merely rendered a valid contract unenforceable. It did not deny the very existence of a contract, which is the result achieved under the s.2 provisions of the new Act. Clearly, the equitable doctrine of part performance has been repealed along with s.40.

The Law Commission did consider the difficulties that the new rules would create, and in particular they examined the plight of a would-be purchaser who had expended money and effort on the property. They considered that the present law contained a sufficient armoury of weapons by which to assist such a person, such as an action for restitution, or suing on a *quantum meruit* if work had been undertaken on a property. It will take time for cases to come before the courts to show whether the Law Commission's faith in these solutions is justified.

Section 2 has generated a significant case law on its scope and meaning. In *Spiro v. Glencrown Properties Ltd*, the vendor and prospective purchaser had both signed an agreement creating an option to purchase land, which conformed in all respects with s.2. Later, when the would-be purchaser served a signed notice to exercise the option, the vendor attempted to argue that the documents already signed merely amounted to an irrevocable offer and that it was the notice exercising the option which brought about a contract for the sale of an interest in land. As this notice was signed only by the purchaser, it did not satisfy the rules of s.2. The court dismissed that argument, preferring instead to construe the original agreement as the contract, conditional upon the exercise of the option. When the purchaser served the signed notice, he was thereby fulfilling the condition. It was the original agreement which needed to satisfy the s.2 rules, and as it clearly did, the vendor was bound.

In *Record v. Bell*, before contracts for the sale of a property were exchanged, the purchaser asked to see entries in the Land Register to check the ownership of the property. There was a delay in the production of this information. Nonetheless, the exchange of contracts went ahead, but the continued existence of the contract was made conditional on the entries confirming what the vendor had previously stated. The condition was not contained in the contracts exchanged, but in letters between the parties. The question arose when the purchaser sought to escape from the agreement, was there compliance with s.2? It was held by the court that the condition formed the basis of a separate contract, collateral to the contract for the sale of the property. As the collateral contract was not one for the sale of land it did not have to satisfy the requirements of s.2 and therefore as the condition had been fulfilled the contract was valid and enforceable.

The courts adopted a different approach to the problem of terms missing from a written agreement for sale of land in *Wright v. Robert Leonard (Developments) Ltd*. Here the contract was for a flat and furnishings. The written contract made no mention of the furnishings and before completion these were removed by the vendor. The Court of Appeal held there was one contract which had to satisfy the requirements of s.2, but the written agreement did not reflect the agreement reached by the parties so rectification of the contract was granted. This meant that the written agreement included a promise as to the furnishings and the purchasers were entitled to damages for the vendor's breach.

Where parties enter into a written contract for the sale of land and later seek to vary a material term of that contract, then such variation must comply with s.2. This was held in *McCausland v. Duncan Lawrie Ltd*, where the parties altered the original completion date stated in a written contract in their subsequent correspondence. The completion date was a material term of the contract and the variation was not in a written document or documents signed by each of the parties; the variation of the contract did not comply with s.2.

Section 2 provides that a written contract must be signed by both parties. The signature of the parties may be contained in the same document or, where documents are exchanged, each party signs a document then sends it to the other. *Firstpost Homes Ltd v. Johnson* explored the question of what was meant by signature. In this case the parties orally agreed on a sale of land. The purchaser prepared a typed letter for the vendor to send to the purchaser containing: the names and addresses of the parties; the price; and reference to an enclosed plan identifying the land. This letter was signed by the vendor, but not by the purchaser. The plan was signed by both parties. The Court of Appeal held that there were two documents and as the letter referred to the plan it was necessary for the letter to be signed by both parties to comply with s.2(3). It was argued that the typed name and address of the purchaser on the letter served as a signature, but this was rejected by the court. The problem mentioned above concerning part performance or reliance on a 'contract' for the sale of land was partly addressed by s.2(5), which states that 'nothing in this section affects the creation or operation of resulting, implied or constructive trusts'. Additionally, it has been suggested that proprietary estoppel may have a role to play in rendering a contract for the sale of land, not made in writing, enforceable. The cases are not clear on this latter point but the better view seems to be that as s.2 does not refer to proprietary estoppel and the use of such estoppel has the potential to undermine the purpose of s.2, it should not be applicable here.

### Subject to contract

It is common in dealings relating to land for the parties to be anxious to take expert advice before finally committing themselves to a binding contract. Negotiations are often carried on 'subject to contract'. Those words will usually be interpreted by the courts to mean that the parties have not yet concluded a contract. The very words deny the existence of a present contract. Of course, much depends on the precise words used, as the role of the court is to seek to construe what the parties must have intended. In one case the parties declared their agreement to be a 'provisional agreement', but the effect of that phrase was quite different and the court held that their agreement was binding from the start (*Branca v. Cobarro*). It seems to follow from the cases that where the parties use the phrase 'subject to contract' in a written document, that document is not a contract, even if it complies in all other respects with s.2.

Agreements which are subject to contract may be contrasted with conditional contracts. For example, a contract for the sale of land may be subject to the condition that planning permission is obtained by a particular date. If the condition is not satisfied, then the parties are not bound by the contract.

# The contents of the contract

Once it is clear that a contract has been agreed, it may then be necessary to establish precisely the obligations which each party has undertaken. Of course, this analysis will not be necessary if everything goes well between the parties and each performs the contract to the entire satisfaction of the other. However, the rules are important not only in problem or dispute solving, but also in contract planning, particularly in a business situation.

The more the parties understand the rules, the better they are able to use them to ensure that subsequent problems do not occur, or where they do, that they can be solved with a minimum of expense, delay and irritation.

Where a contract consists of buying a cup of coffee from a machine, or paying the fare on a bus, it may be difficult to imagine what the contents of such a contract are. But all contracts have terms. These may be express or implied. The express terms will be those specifically negotiated between the parties, either orally or in writing. Remember that the parties are largely free to fix their own terms. Contract law still adheres generally to the principle of freedom of contract. The parties must strike their own bargain. In effect, this imposes a serious restriction on the extent to which terms can be implied in a contract.

Implied terms may be incorporated from an Act of Parliament. One of the commonest examples is the set of implied terms incorporated into sale of goods transactions by the Sale of Goods Act 1979. Without any need for specific negotiation by the parties, terms are implied relating to fitness for purpose of the goods, and the requirement that goods be of satisfactory quality. The courts have a limited role to play in judicially implying terms, but great care is needed to ensure that judges are not seen to be making the agreement for the parties. It is usually said that terms can only be implied by the judges if it is necessary to give 'business efficacy' to the arrangements made by the parties. This may occur where the parties have already undertaken a substantial part of the contract and they come up against some problem or difficulty which they did not provide for. In such cases a judge can imply a term where it is clear that, had the parties put their minds to the problem at the appropriate time, both would have agreed to the term proposed.

Terms are the obligations under a contract but inevitably they will not be of equal importance. Take, for example, a term about the time of a delivery of goods. This may be of paramount importance where goods are perishable or where it is clear that they will be of little commercial use unless delivered on time. In other cases the time for delivery may be far less significant. The law has to find some way to grade or classify terms, because it must seek to offer remedies appropriate to the significance of the term which has been broken. Traditionally, the law classified terms as major terms, which it called conditions, and minor terms, which it called warranties. The terminology is unfortunate, as both of those words have sundry other meanings and uses in the law of contract. Conditions are said to be those obligations which are of the essence, which go to the very core and heart of the contract. Breach of a condition always gives the injured party the right to sue for damages. Over and above that, the injured party may also have the right to repudiate the contract. This means that he or she is entitled to regard him- or herself as no longer bound by the contract. A warranty is of more peripheral importance, often said to be merely collateral to the main purpose of the contract. An injured party can sue for damages for breach of warranty but has no right to repudiate the contract.

The problem with this traditional approach to classifying terms is that it requires a decision to be made about the status of the term at the time when the contract was made. In effect this emphasizes the importance of the intention of the parties, as it should be clear from what was said and written, and from all the surrounding circumstances, whether they intended a particular term to be a condition or warranty. This approach creates business certainty, as it would be possible to say at the outset what remedies would be available for particular breaches. But it takes little account of the realities of business life, where the term breached may be only a warranty, but the consequences could turn out to be very serious. This would leave the victim of the breach in a situation where he or she would be forced to go on with performance of the contract and be limited to a claim for damages.

Consider as an example the problem which arose in *Hong Kong Fir Shipping Co. Ltd v. Kawasaki Kisen Kaisha Ltd*. One of the terms in a contract to charter a ship for two years provided that the ship should be in every way fitted for ordinary cargo service. When the ship was delivered at the beginning of the two-year charter period it was unseaworthy and when it made its first voyage, repairs were needed which took four months to complete. Did the breach amount to a breach of condition? If it did, the charterer could repudiate liability under the contract and thus rid himself of the burden of this unsatisfactory vessel over the remainder of the two-year period *and* sue for damages. If it was only a breach of warranty his claim would be limited to damages and he would be left with the vessel for the remainder of the charter period.

In the Hong Kong Fir case, the Court of Appeal chose a novel solution, preferring instead to say that some terms are 'intermediate' or 'innominate'. Such terms defy initial classification as conditions or warranties and the parties must wait until the scale of the resulting breach is known, at which point the term can be suitably classified. Obviously this produces a much less certain situation between the parties during the currency of the contract.

This development in judicial thinking did not in fact assist the charterer in the Hong Kong Fir case. The term in issue was held to be only a warranty, so the charterer had no right to repudiate the contract. As he had wrongfully repudiated, he was obliged to pay damages to the shipowner.

The same problem of classifying a term was before the Court of Appeal in another charter party case, *The Mihalis Angelos*, where the term related to the date when the ship would be ready to load. In the event, it was nearly a month after that date before the ship was ready. The term was held to be a condition. The court made the point that it was influenced by earlier cases holding similar clauses to be conditions and it was vital to remember that 'one of the important elements of the law is predictability'. One important point emerging from this case is that it may already be established, either by statute or an earlier precedent, that a particular type of clause falls into the category of a condition. Then there is no scope for attempting to classify it as an innominate term. But where a term could be either a condition or a warranty, and it is not clear which the parties intended, and where the range of possible breaches and the scale of the resulting harm is very wide, then the court may classify it as innominate. This classification is theoretically made at the time when the contract is made, but the remedy available for breach of the term will only be established once the scale of that breach is known.

The appropriateness of this 'wait and see' approach is well demonstrated by the facts in the Hong Kong Fir case. There, the clause in question required that the ship was to be 'in every way fitted for cargo service'. These words were so wide in potential interpretation that there could be breaches ranging from failure to supply the vessel with a proper number of anchors, or to put on board medical supplies, to totally defective engines at the other end of the scale.

It is clear that the courts remain anxious about creating business uncertainty. The House of Lords confirmed that the traditional division into conditions and warranties is still acceptable in many cases (*Bunge Corporation v. Tradex Export SA*). They have, however, left the way open to use the more flexible Hong Kong Fir approach in a limited number of cases where this may be justified. Clearly the problem can be largely overcome by the parties being sufficiently specific at the outset about the result of each type of breach. This is a classic example of how rules of contract law can have a planning function.

## Exclusion and limitation clauses

Although many terms in contracts will be concerned with creating obligations, it has become increasingly common for the parties to seek to limit or exclude liability which would otherwise arise under the contract. Exclusion and limitation clauses are frequently encountered in contracts involving car parks, dry cleaners, film processing, package holiday deals, furniture removal and insurance. They may take forms such as 'No liability accepted for any damage caused by flooding' or 'In the event of a breach, our liability is restricted to 10 times the contract price'.

Exclusion clauses fall into two main categories: those negotiated between parties of equal bargaining strength, where such clauses can be seen to facilitate business activity and to allow the parties to allocate risks between themselves and organize which of them should insure; and those imposed on a weaker party by a stronger party who may enjoy a monopolistic position, where the weaker party has no option but to accept. Clauses falling into the latter category are, not unnaturally, unpopular with the courts. In the past the judges strove to protect victims of harsh exclusion clauses, but some of the techniques developed by them to afford protection produced unfortunate results in business cases, sometimes upsetting the very carefully negotiated arrangements between the parties. The courts have never had a power at common law simply to strike out exclusion or limitation clauses on the grounds that they are unfair or unreasonable because contract law proceeds on the assumption that the parties are free to fix the terms of their bargain for themselves.

Judicial dislike of exclusion clauses has always meant that courts will only allow reliance on such a clause if it has effectively become part of the contract. The incorporation of the clause into the contract can occur:

1  By one party signing a contractual document containing the clause, of which he or she would then be deemed to have notice whether it had been read or not.
2  By one party giving reasonable notice of the clause to the other. Reasonable notice may be given in a contractual document but not every piece of paper passing between the parties can be so classified. Generally, contractual documents are those pieces of paper on which a reasonable person would expect to find conditions printed. This would usually include documents such as airline tickets or standard company order forms. It would not include things such as mere receipts for the payment of money. In any case, such a receipt would often be handed over too late, as it is vital that reasonable notice be given before, or at the time when, the contract is made. Once the parties have finished negotiating and finalized their deal it is too late for one of them to try to change it or add to it unilaterally. Where no contractual document passes between the parties the exclusion clause can be incorporated by the reasonable display of a notice at an appropriate place, so that its contents could have been read before the contract was made.
3  By virtue of a 'course of dealings' between the parties. Where two people have dealt with each other regularly, always using the same terms, the court may be prepared to say that adequate notice has been given of any exclusion clause. This principle is only likely to be applied between business parties who have regular commercial dealings.

In cases where the court accepts that the exclusion clause is properly incorporated and forms part of the contract, the judge will next consider what the words of the clause mean, and whether the wording covers the actual facts which have arisen. In the construction or interpretation of the clause the standard policy of the courts is to construe any ambiguity strictly against the interest of the party seeking to rely on the clause. In one case involving motor insurance, a clause in the contract provided that the insurers need not pay out on a claim under the contract if the vehicle was carrying 'an excessive load'. In fact, it was a five-seater car carrying six persons, but the judge held that the word 'load' did not include people.

Prior to 1977, when the Unfair Contract Terms Act was passed, judges had to use these construction and interpretation techniques boldly in order to protect victims of harsh clauses. In consequence, there was a very distorted result in many of the cases. Such striving on the part of judges is much less necessary now since the 1977 Act was passed, and they have been urged by the House of Lords to give the relevant words in an exclusion clause their plain and natural meaning. Nevertheless, the two common law rules relating to incorporation and construction are still relevant, though the provisions of the 1977 Act now govern the majority of problem cases in this area.

## Unfair Contract Terms Act 1977

The scope of this Act is wider than its name would suggest, as it deals with notices limiting liability in the law of tort as well as exclusion clauses in the law of contract. One advantage of introducing statutory control of exclusion clauses is that a flexible approach can be taken, making a distinction between the type of clause negotiated between businessmen on an equal footing, and other clauses 'imposed' by businessmen on consumers. The Act does not apply to all contracts – insurance contracts, for example, are outside its scope. Although the Act is fairly general in its coverage there are other specific Acts of Parliament which also prohibit particular types of exclusion clause.

The main sections of the Act only apply in 'business liability' situations. This phrase is defined as meaning liability for breach of duties arising from things done in the course of business or from the occupation of premises used for business purposes of the occupier (s.1(3)). So, except in sale of goods transactions, which are considered separately (see Chapter 7), the Act does not extend to what may be regarded as private transactions. Once the Act is shown to apply, it has two different techniques for dealing with offending exclusion clauses. Some are rendered totally ineffective however they are drafted. Others can survive and take effect if they satisfy the requirement of reasonableness set out in the Act.

Sections 2 and 3 are the most important general sections of the Act. Section 2 deals with liability for negligence, which is defined by s.1 of the Act as:

1   breach of a contractual obligation to take reasonable care or exercise reasonable skill in the performance of the contract; or
2   breach of common law duty to take reasonable care or exercise reasonable skill; or
3   breach of the common duty of care imposed by the Occupiers' Liability Act 1957.

If negligence results in death or personal injury, then it is not possible to exclude or limit liability either by an exclusion clause in a contract or by a notice displayed generally. So, where a contract contains a term that 'no liability is accepted for death or injury howsoever caused', then if death or injury is caused by negligence, the relevant clause will be rendered totally ineffective.

Section 2 also provides that where loss or damage (other than death or personal injury) is caused by negligence, an exclusion clause *can* take effect *if* it satisfies the requirement of reasonableness. This requirement will be analysed in conjunction with s.3.

Section 3 covers two different situations: first, where one party deals as a consumer and, second, where one party deals on the other's written standard terms of business. Both of these situations require further explanation. Dealing as a consumer means that one party is not making the contract in the course of a business but the other party is. A classic example would be taking your car to be serviced. You deal as a consumer, while the garage deals in the course of a business. Or it could be having your house rewired. You deal as a consumer and the electrician deals in the course of a business. What constitutes 'written standard terms of business' is not entirely clear as the phrase is not defined by the Act.

In either of these two situations s.3 provides that the other party, having committed a breach, cannot exclude or restrict their liability for that breach unless the exclusion clause satisfies the requirement of reasonableness. An example will help to illustrate the working of the section. Imagine that you take items of clothing to be cleaned at a dry-cleaning shop. On the counter is displayed a clear notice to customers that in the event of clothing being lost or damaged, liability is excluded or limited. When you go to collect your clothes, they cannot be found and the shop tries to rely on the exemption clause. It will only be able to do so if it can prove to the court that the clause satisfies the requirement of reasonableness. This means that, in relation to contract terms, it must be a fair and reasonable term to include, given what the parties knew or ought to have known at the time when they made the contract. The burden of showing that it was fair and reasonable is on the person now seeking to rely on the clause. The answer will obviously depend very much on the circumstances of each case, but a judge approaching the problem is given some limited help by the Act itself. It is clear from s.11 that there are cases where the judge must consider whether it was possible for the parties to have protected themselves by insurance. For instance, there are situations where a business may be prepared to quote two prices for a job. A high price will reflect the fact that it is prepared to accept all liability. A low price may be conditional upon it excluding some liability. Naturally, when I consider these prices I may well choose the low price if I can easily and cheaply insure against the risks myself or if I already have insurance which covers those risks.

The Act itself provides guidelines which will sometimes be relevant when determining reasonableness. The court may take into account:

1    The relative bargaining strength of the parties. One factor which might be important here is whether the goods or services could have been obtained easily elsewhere.
2    Did the customer receive some inducement to agree to the term? The example quoted above where a customer chooses the low price is an instance of an inducement. The court will also keep in mind whether a similar contract made with someone else would have been likely to contain an identical term.
3    Did the customer know, or ought he or she reasonably have known, of the existence and extent of the term? Bearing in mind the importance of showing that the term has been properly incorporated into the contract this guideline seems to suggest that there is one standard for deciding on incorporation but a different one for deciding reasonableness. So a clause which is in small print or faint ink might be properly incorporated by, say, signature but may not be reasonable under the 1977 Act.

4    Where the term excludes liability only if a party fails to comply with some condition, was it reasonable to assume when the contract was made that it would be practicable to fulfil the condition? An example might arise where a clause states: 'No liability for any breakages in transit which are not reported to us within five minutes of delivery.' Even when the contract was made it would be clear that such a condition was not practicable.

5    Where the contract relates to a sale of goods which have been specially manufactured, processed or adapted to the buyer's requirements, that may give the seller wider scope to limit the liability.

There are a number of decided cases clarifying the requirement of reasonableness. The decisions are of limited use as precedents as it is important to remember that these cases turn on their own individual facts. However, they do illustrate the factors to take into account when determining reasonableness. *George Mitchell v. Finney Lock Seeds Ltd* makes a useful illustration of the rules. The farmer wanted to plant Dutch winter cabbage in a field of some 60 acres. He was sure that he could make a handsome profit on the vegetable as it would be ready to harvest when home-grown vegetables were in short supply. He obtained the seed from Finney's at a cost of £192. The terms of the contract provided that liability was limited. In the event of the seed being defective, the sellers would refund only the price of the seed. They also excluded 'all liability for any loss or damage arising from the use of any seeds supplied by us and for any consequential loss or damage arising out of such use, or for any other loss or damage whatsoever.'

Whatever the seeds supplied might have been, they were not Dutch winter cabbage. The crop failed and the farmer sued for £63,000. The company was prepared to refund £192. The case turned on whether Finney's could rely on the clause. They could only do so if the clause was a fair and reasonable one. In the House of Lords the judges were conscious that this was the first interpretation of the 'fair and reasonable' provision. They made clear that arriving at a decision as to what is fair and reasonable is not like the exercise of discretion, but the court must put a number of factors on the scales and see on which side the balance comes down. 'There will sometimes be room for a legitimate difference of judicial opinion … and the appellate court should treat the original decision (given by the trial judge) with the utmost respect.' When looking at the facts of the present case, the judges took into account that the farmer had been a regular customer of the seed firm for many years and had had plenty of opportunities to read the exclusion clause. It was not a difficult clause to understand. Similar clauses would be found in every seed firm's contracts, and had never caused farmers as a body to protest through the National Farmers Union. So far as the question of insurance was concerned, the seed firm could have insured against the risk of crop failure without needing to significantly raise the price of seed. The court was also influenced by the fact that the seed firm usually sought to negotiate settlements in cases like this, rather than rely on the strict terms of the exclusion

The law of contract 125

clause. This latter fact was regarded by the court as particularly significant as they thought it pointed to the seed firm themselves recognizing that their own clause was unreasonable. Taking all of these factors into account, the House of Lords held that the clause was not fair and reasonable.

The 1977 Act has been of great importance and value in that the flood of complex, and often irreconcilable, decisions on exclusion clauses has virtually stopped. It is clear that it will be easier to justify a clause which merely limits liability, as opposed to one which seeks to exclude liability completely. Other relevant aspects of the Act are considered separately in relation to misrepresentation, sale of goods and occupier's liability.

## Unfair Terms in Consumer Contracts Regulations 1999

The controls over unfair terms were extended by the Unfair Terms in Consumer Contracts Regulations 1999. These regulations were made to implement the European Directive on Unfair Terms in Consumer Contracts. As a result of these Regulations exclusion clauses may be controlled by two statutory schemes, the Unfair Contract Terms Act 1977 and the Unfair Terms in Consumer Contracts Regulations 1999, but the Regulations also apply to other terms, for example terms requiring a consumer to fulfil their obligations, but a seller or supplier is not under a similar obligation.

Some contracts are not covered by the Regulations, for example employment contracts, contracts concerning succession rights, contracts concerning family law rights and contracts concerning incorporation or organization of companies or partnerships. Assuming that a contract does not fall into an excluded category of contract then the Regulations will apply, to any unfair term in a contract concluded between a seller or supplier and a consumer where the said term has not been individually negotiated.

A seller or supplier means a natural or legal person who sells or supplies goods and services and who in making a contract is acting for purposes relating to his business. A consumer means a natural person who in making a contract is acting for purposes which are outside his or her business. Companies cannot fall within the definition of consumer as they are not natural persons but legal persons. As can be seen the Regulations, generally stated, cover terms in contracts between businesses and non-business consumers (compare the ambit of the Unfair Contract Terms Act 1977).

The Regulations provide that a term has not been individually negotiated where it has been drafted in advance and a consumer has not been able to influence the substance of the term. This covers situations where a consumer is presented with a standard form document and is told by the seller or supplier that they will only contract on the terms as presented. Just because some terms are negotiated, e.g. price, dates of delivery, this does not prevent the Regulations from applying to the rest of the contract, if the remainder can be characterized as a 'pre-formulated standard contract'. If a seller or supplier claims that a term is individually negotiated, then it is for them to prove that.

The Regulations do not apply to terms, providing they are in plain, intelligible language, defining the main subject matter of the contract or the adequacy of consideration. These exclusions are present to ensure freedom of contract and to prevent judicial scrutiny of the adequacy of consideration. This issue was considered by the Supreme Court in *Office of Fair Trading v. Abbey National plc*, where terms levying bank charges were part of the price paid for the services provided by banks and therefore not subject to control by the Regulations.

Having established that the Regulations apply, the next task is to see whether the term is unfair. An unfair term is one which 'contrary to the requirement of good faith, causes a significant imbalance in the parties' rights and obligations arising under the contract to the detriment of the consumer'.

As can be seen there are three elements to the meaning of an unfair term:

(a)   lack of good faith;

(b)   a significant imbalance in the parties' rights and obligations arising under the contract; and

(c)   such imbalance causes detriment to a consumer.

(a)   The meaning of good faith is explained in the European Directive, which states that regard shall be had in particular to, for example, the strength of the bargaining position of the parties and whether the consumer has been offered an inducement to accept the term. These matters are clearly similar to those stated in the Unfair Contract Terms Act 1977.

In *Director General of Fair Trading v. First National Bank*, Lord Bingham in the House of Lords said that good faith requires fair and open dealing. 'Openness requires that the terms should be expressed fully, clearly and legibly, containing no pitfalls or traps. Appropriate prominence should be given to terms which might operate disadvantageously to the customer. Fair dealing requires that the supplier should not, whether deliberately or unconsciously, take advantage of the customer's necessity, indigence, lack of experience, unfamiliarity with the subject matter of the contract, weak bargaining position ...'. A seller or supplier must take into account the interests of the consumer.

(b)   The Regulations give in Schedule 2 an indicative but non-exhaustive list of unfair terms. The list includes: exclusion clauses; terms allowing the seller or supplier to unilaterally alter the terms of the contract; terms requiring a consumer, who fails to perform his obligations, to pay a disproportionately large sum; and terms binding a consumer to other terms of which he or she had 'no real opportunity of becoming acquainted' before the formation of the contract.

If a contract for membership of a health club provides the club may change hours of business at will or refuse admission for any reason whatsoever, then such terms may be classified as unfair terms.

(c)   This requirement indicates that the Regulations are for the benefit of consumers not sellers or suppliers.

A term that is classified as unfair is not binding on the consumer. The Regulations also seek to make contracts more 'user friendly' by placing an obligation upon the seller or supplier to ensure that the terms of a contract are written in plain and intelligible language. If a failure to do this results in doubt, then such doubt is to be resolved in the favour of the consumer.

Finally, the Office of Fair Trading has powers to prevent the continued use of unfair terms and, since the Regulations came into force, the powers have been used widely.

The Law Commissions considered how to unify and simplify the law contained in the Unfair Contract Terms Act 1977 and the Unfair Terms in Consumer Contracts Regulations 1999, and reported in 2005. So far no amending legislation has been introduced into Parliament.

## Misrepresentation and mistake

The basic idea of misrepresentation is a very simple one to grasp. One party makes a false statement which has induced the other to enter the contract. This area of law is regarded as complex, largely due to the number of overlaps with other areas of law, so that it is seldom possible to view a misrepresentation in isolation. The overlaps occur either when misrepresentation induces a mistake, or when the statement made, depending on its relative importance, may be classed as a representation or a term, or because the law of contract and the law of tort have separately developed ideas about negligent statements.

In complex contractual negotiations, both parties will make many statements to each other about the subject matter. We have already considered the problem of classifying terms of a contract. Terms may result from statements made which now form part of the contractual obligations. Alternatively, the statement may be of less importance and may not become part of the contract. Nevertheless, that statement may weigh with one party in influencing him or her to make the contract. He or she is induced by the representation. Some statements may simply have been designed to create a good bargaining atmosphere and these may give rise to no liability at all. These examples are sometimes referred to as 'tradesmen's puffs'. In one case, where a seller described a car as 'a great little goer', no liability could arise from such a claim.

The court must determine into which category a statement falls. When deciding whether statements made prior to contracting are terms or representations they are influenced by factors such as:

- How early in negotiations was the statement made?
- Was an oral statement subsequently included in a later written contract?
- Did the person making the statement have special knowledge on the subject matter of the contract?

The answers to these questions will not necessarily produce a definite solution. Each case will turn on its own facts.

Where a false statement is classed as a representation it is possible for it to be made innocently, negligently or fraudulently, depending on the maker's state of mind. The classification is important because each type of misrepresentation gives rise to different remedies. It is also important to distinguish terms from representations because of the different remedies available. Non-compliance with a term results in breach of the contract. Misrepresentation is a false statement of fact standing outside the contract and not forming part of the obligations and does not, therefore, involve a breach of contract. The difference in the remedies available used to be so great that the distinction between a term and a representation assumed immense importance. Since the changes in the law introduced by the Misrepresentation Act 1967, it is now far less significant.

Figure 6.1 shows how any particular statement can be graded. Take an example like 'The car has only done 20,000 miles' or 'This lorry can carry 30 tons'.

Misrepresentation is defined as a false statement of fact, made by one party to the other, before or at the time of contracting, which is one of the causes that induces the contract. Within the definition, a number of points of difficulty occur:

1    There must be a statement, but the courts have been prepared to imply a statement from behaviour. 'A nod or a wink, or a shake of the head or a smile' from the seller, intended to induce the buyer to believe the existence of a non-existing fact, would be sufficient.

2    The statement must be of fact. This excludes statements of intention and statements of opinion. Statements of opinion give rise to particular problems, as they can be made by someone who has more knowledge and information in the particular bargaining situation. This makes the statement seem especially authoritative. It seems as if the maker of the statement knows facts which justify him or her in holding that

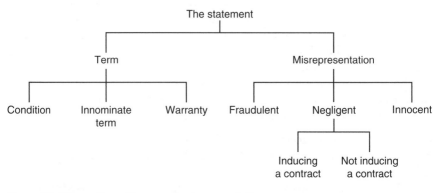

*Figure 6.1  The grading of terms and misrepresentations.*

opinion. In one case a vendor, trying to sell property which was let, described the tenant as 'a most desirable tenant'. The vendor knew that the tenant was in arrears with his rent and earlier rent had only been extracted with considerable difficulty. In these circumstances, knowing the facts as he did, the court held that the vendor could not reasonably hold such an opinion. They regarded his statement as actionable misrepresentation (*Smith v. Land and House Property Corporation*). In that case, the vendor had superior knowledge. He alone knew the problems about the tenant. Compare the situation in *Bissett v. Wilkinson*, where a vendor was selling land in New Zealand. In response to an enquiry from the purchaser, the vendor said that the land could carry 2000 sheep. The vendor had never kept sheep nor had the land ever been used for that purpose, so he had no facts on which to base this opinion. It was only an opinion and not actionable as misrepresentation.

3   Since there must usually be a false statement, it follows that keeping silent cannot usually constitute misrepresentation. If I want to sell my second-hand car, and I say nothing at all about its condition or performance, it will be up to a prospective purchaser to enquire specifically about anything that concerns or worries him or her. This is part of a general principle in contract of 'let the buyer beware' (*caveat emptor*). I would be under no duty to point out faults and defects. Of course, if I am asked specific questions then I should answer honestly, otherwise my statements could amount to misrepresentation. There are some situations, however, where the law imposes a positive duty to speak. In these cases silence can constitute misrepresentation. These situations are:

(a) Where one party tells a half truth but remains silent about some fact which distorts the positive statement made. An example is where a prospective purchaser asked about the farms on an estate, enquiring whether all the farms were let. The vendor replied that they were (a positive statement) but failed to go on to say that all the tenant farmers had given notice to quit (silence distorting the positive representation) (*Dimmock v. Hallett*).

(b) Where the relationship between the parties is fiduciary (a relationship of trust), e.g. solicitor and client, or banker and customer. If one then makes a contract with the other, there is a positive duty to disclose known material facts. This rule is significant when considering undue influence.

(c) Where the contract is one involving utmost good faith. The commonest example of such a contract is insurance. Here the facts are all within the knowledge of one party. On the strength of what he tells the other, that other must decide whether or not to accept the risk and how much to charge as a premium. The law imposes a positive duty in these cases to reveal all material facts. Insurance contracts are discussed in greater detail on page 175.

4      The false statement must actually help to induce the contract. There
       will be many occasions when false statements are made, but they do
       not act as an inducement. This may be because the other party does
       not hear the statement, or does not believe it, or is not in any way
       influenced by it, but makes his decision to enter the contract for quite
       separate reasons.

Where all the relevant features of misrepresentation can be proved, the
injured party will want to know if he or she can escape from the contract,
and also whether he or she has any claim for damages. The answer depends
largely on the type of misrepresentation involved. Was it a statement made
fraudulently, negligently or innocently? Once the maker's frame of mind is
established, the remedies can be considered.

## Fraudulent misrepresentation

This is defined as a false statement made knowingly, without belief
in its truth, or recklessly, careless whether it is true or false. In the very
limited number of cases where this can be proved, the behaviour of the
person making the false statement would amount to the tort of deceit.
Unfortunately, it is very difficult to convince a court that a statement falls
into this category. As one judge commented: 'A charge of fraud is such a
terrible thing to bring against a man that it cannot be maintained unless it is
shown that he has a wicked mind.'

## Negligent misrepresentation

This category of misrepresentation was introduced by s.2(1) of the
Misrepresentation Act 1967, and is one of the factors which has effected a
significant improvement in this area of the law. A statement is negligent if it
was made in circumstances where the maker had no reasonable grounds for
believing it to be true. He or she may have thought that they were making
a true statement but the court will evaluate whether the basis for this belief
was reasonable. This may be a difficult question to resolve, and much may
be at stake. An important commercial case, *Howard Marine and Dredging Co.
Ltd v. Ogden & Sons Ltd* will illustrate the problem.

In order to tender for a large excavation contract with a water authority,
Ogdens needed to hire or buy barges to move spoil out to sea. They
approached Howard Marine and had extensive discussions and negotiations
with Howard's manager. He thought that two particular barges would be
suitable for Ogdens' needs, and he told Ogdens that each barge could carry
1600 tonnes. He was basing his statement on his recollection of information
contained in *Lloyd's Register*. In the company's files about the barges, the
manager could have discovered that the *Lloyd's Register* figure was wrong.
The actual carrying capacity of each barge was nearer 1000 tonnes. Ogdens
were successful in getting the excavation contract, and they immediately

hired the two barges from Howard. As soon as they began using them they realized the vast discrepancy between the manager's statement and the true carrying capacity. Ogdens refused to pay any further hire charges for the barges and wanted to claim damages from Howard. Obviously the court had to classify the statement made. First, they had to consider whether it was of such importance that it could be regarded as a term of the contract. Given the fact that the statement was made quite early in the negotiations, and not subsequently included in the charter party (the written hire contract), the court found that it was not a term. What kind of misrepresentation was it? By a majority verdict only, the Court of Appeal found that this was a negligent misrepresentation and awarded damages under s.2(1) of the Misrepresentation Act 1967. One of the judges thought that it was reasonable for the manager to rely on *Lloyd's Register*, which was normally famous for its detail and accuracy, but the other two judges thought that the manager should have referred to, and checked in, the authoritative file on each barge.

Liability for negligent misrepresentation under s.2(1) of the 1967 Act can only arise if the statement causes the other party to enter into a contract. There is considerable overlap in this area with the law of tort, and negligent statements are a problem area. Just before the 1967 Act was passed the courts had an opportunity to consider what liability, if any, should arise from negligent statements. In the case of *Hedley Byrne v. Heller* the facts involved a company who wanted assurances about the financial stability of a prospective customer. They asked the customer if it was in order to approach the customer's bank for a credit reference. The bank negligently informed the company that the customer was financially sound. Subsequently, the customer went into liquidation, and the company suffered losses in excess of £15,000. As they were unlikely to get their money from the customer, the company was anxious to sue the bank. On these facts, it is clear that the bank's statement had not induced any contract between the bank and the company, so s.2 of the 1967 Act could not cover such a case. That section requires that a person must have entered a contract. Could the company recover the damages against the bank in the tort of negligence? The law on negligence had not developed at that stage to cover such misstatements causing pure financial loss, but the House of Lords in the Hedley Byrne case showed themselves ready, in appropriate circumstances, to develop the law and award damages.

In the mid-1960s, this created a sudden new wealth of possibilities where a person was a victim of a negligent statement. It has become common to refer to negligent statements which result in a contract as negligent misrepresentations, and those which do not result in a contract as negligent misstatements. While it may be to a claimant's advantage to have more possible causes of action, these twin developments in respect of negligent statements make the law more complex than it need be. See page 248 for further treatment of negligent misstatements in the law of tort.

## Innocent misrepresentation

Since the 1967 Act, this is limited to those cases where the maker of the statement honestly believed that his or her statement was true *and* he or she had reasonable grounds for believing it to be true.

## Remedies for misrepresentation

Once it is clear which type of misrepresentation is involved, the next problem is to establish the appropriate remedy. All types of misrepresentation have the same effect on the contract, which is to make the contract voidable at the option of the party misled. This means that the injured party can seek to rescind the contract, to have it set aside as if it had never existed. The parties would then be restored to their original positions as far as the contract is concerned. This is known as the remedy of rescission, and in some circumstances it can be achieved on a self-help basis without any court assistance or intervention. Inevitably such help will often be required, especially where the party at fault will not cooperate. There will be times when a court is unable to order rescission. In principle the remedy is possible for all misrepresentations but will not be available in the following circumstances:

1    Where the court is unable to restore the parties to their original position. This could be the case where goods sold have been consumed, or inextricably mixed with other goods.
2    Where the injured party has affirmed the contract. This means that the injured party is fully aware of the false statement but is prepared to go on with the contract.
3    Where the injured party has delayed too long in seeking rescission. There is no precisely defined time limit, but a long delay once all the facts are known may in itself be an affirmation of the contract.
4    Where third party rights have intervened.
5    Where the court decides to exercise its discretion under s.2(2) of the 1967 Act to award damages in lieu of rescission.

## Damages

When considering the availability of damages for misrepresentation, it must be remembered that the false statement in question may have induced the contract, but it does not form part of the contract obligations. If the statement was regarded as sufficiently important by the parties to form part of the contractual obligations, it will be classified as a term. If it then turns out to be false the appropriate remedies for breach of contract will be available.

As misrepresentations stand outside the contract, it follows that there can be no damages available in the law of contract. Where the statement is fraudulent, damages have always been available in the tort of deceit. Prior to the recognition of negligent misrepresentation, a misrepresentation made without fraud was innocent, for which no damages were available.

Now, if a claimant can prove negligent misrepresentation under s.2(1), the court has power to award damages. It seems clear that these damages will be assessed on a tort basis rather than a contract basis. *Royscot Trust Ltd v. Rogerson* shows a most favourable application of s.2(1) in the assessment of damages. Where the claim is for negligent misstatement within the Hedley Byrne principle, damages for the tort of negligence are available. The difference between tort damages and contract damages is significant, as it could result in quite different calculations. Tort damages are designed to put the claimant into the position he or she would have been in but for the tort; contract damages are designed to put the claimant into the position he or she would have been in if the contract had been properly performed.

It should be noted that in a claim under s.2(1) while a representee must show that a misrepresentation has occurred and damage has been suffered, it is then for the person making the representation, the representor, to prove that they had reasonable grounds for believing the statement to be true and did so believe up to the time that the contract was formed. This makes it easier for the representee as the burden of proof is placed on the representor to establish reasonable grounds (see *Howard Marine and Dredging Co. Ltd v. Ogden & Sons Ltd*, where Howard Marine was unable to show reasonable grounds for the statement as to the capacity of the barges).

## Damages in lieu of rescission
Even if a false statement has been made, the remedy of rescission is sometimes far too extreme to produce a fair result. If I sell you my car, and tell you that it has done 30,000 miles, and you subsequently discover it has done 31,000 miles, it can often be quite satisfactory to award damages in recognition of the false statement. It may seem to give undue protection to the purchaser to allow him or her to back out altogether, although this will often be a question of degree. Under s.2(2) of the 1967 Act, the court has a discretion to award damages in lieu of rescission for negligent and innocent misrepresentation.

## Limiting or excluding liability for misrepresentation
In the same way as it is possible to limit or exclude liability for obligations which would otherwise arise under a contract (see page 120) it is also possible to limit or exclude liability for misrepresentations which induce a contract. The relevant rules are contained in s.3 of the Misrepresentation Act 1967. Section 3 permits exclusion of liability for false statements (negligent or innocent, but not fraudulent), provided that the clause satisfies the requirement of reasonableness (see page 128).

## Mistake
As a broad rule, English law admits few mistakes as affecting the validity of a contract. Often, when one party says: 'I made a mistake,' what this really

means is: 'I failed to watch out to make sure I was making a good bargain.' In such cases, the maxim *caveat emptor* will apply – let the buyer beware.

In some instances where the parties are mistaken, no contract ever comes into existence between them. The extent of their misunderstanding is so great that it prevents them reaching true agreement. In one case the parties believed they had an agreement relating to a cargo of cotton coming to this country on a ship called *The Peerless*, which was sailing from Bombay. By coincidence there were two ships, both called *The Peerless*, both carrying cotton and sailing from Bombay. As it was impossible to say which ship the parties had in mind, it was held that they had never reached a true agreement (*Raffles v. Wichelhaus*).

Sometimes, the state of confusion is caused by one person seeking deliberately to lead the other into making a mistake. This can occur where one person misrepresents his or her identity. Such fraudulent behaviour would be likely to give rise to remedies for misrepresentation. But in cases where it can be shown that the identity of the person was crucial, the mistake about identity may be so important that there is no real agreement. The contract would be void, meaning that no contract has been formed.

# Other factors affecting validity and enforceability

A party seeking to enforce a contract may be unable to do so in cases where it was induced by duress or undue influence, or in cases where there is some illegality. These issues will be considered separately.

## *Duress*

It is hard to imagine that any contracts are made nowadays as a result of threats of physical violence. The old rules about duress would only permit relief if the contract came about because of threatened or actual violence. Originally the result of such behaviour was that the contract was void, i.e. there was no contract at all. Subsequently, the rule was modified. Now, if such facts occur, the injured party could seek to have the contract set aside. The modern effect of duress is thus to render the contract voidable; the contract is valid unless and until it is avoided.

These rules are of such limited significance that they would not be worth mentioning if it were not for the development by the courts in the late twentieth century of principles relating to economic duress. The cases which have developed this theme tend to be complex, but the usual scenario is that one party threatens the other with breach of contract in order to gain some new concessions from the other. For example, one party sees that he has not negotiated a particularly good price for building a ship, so he threatens to abandon the contract unless the price is renegotiated. This will put the other party into an impossible position if he has already contracted to charter the new ship to another party. Unless he agrees to the higher price he will find himself in turn in breach with his charterer. He could resort to litigation against the shipbuilder, but that is costly, uncertain and very time-consuming. The potential loss of business and goodwill may be so great that he reluctantly agrees to pay the higher price. At a later date, would a court allow him to recover the extra sum paid?

The courts have not yet developed a particularly coherent set of principles to solve such a problem. There must be more than mere commercial pressure. To establish economic duress, it is necessary to show that there is a threat which is illegitimate, such as a threat to break a contract, and that the innocent party has entered into the new deal because of the defendant's pressure, which is a significant cause. In making this assessment the courts will have regard to whether or not the innocent party had reasonable alternatives open to him or her. Factors which the court will take into account were set out in a case, *Pao On v. Lau Yiu Long*, 1979:

1    Did the victim protest?
2    Did the victim have any other course of action open to him, e.g. could he have sued?
3    Did the victim receive independent advice?
4    Did the victim take immediate steps to try to avoid the new deal?

This developing area is one which is likely to cause considerable difficulty for the courts, as these situations may arise between business people who are of apparently equal bargaining strength, where the courts have a traditional reluctance to interfere with the contract as agreed between the parties.

## Undue influence

Rules of undue influence were developed in equity because of the narrowness and harshness of the old common law rules about duress. Where there is behaviour falling short of such open violence or threats of harm, equity may grant relief if a contract has been brought about by undue influence. There is no precise definition of this phrase, but it usually occurs where there is a dominant/subservient relationship between two people and the dominant party seeks to take advantage of, or exploit, the other. Undue influence was traditionally classified as actual or presumed.

In *Royal Bank of Scotland v. Etridge (No. 2)*, presumed undue influence was explained by Lord Nicholls. A claimant must prove undue influence: 'proof that the claimant placed trust and confidence in the other party in relation to the management of the claimant's financial affairs, coupled with a transaction which calls for explanation, will normally be sufficient, failing satisfactory evidence to the contrary, to discharge the burden of proof … [T]he court may infer that, in the absence of satisfactory explanation, the transaction could only have been procured by undue influence.' This inference of abuse of influence must then be countered by a defendant.

A relationship of influence is automatically presumed to exist in some relationships, e.g. solicitor and client, or parent and child. In other instances it is not presumed but may be proved to exist. Where a contract is highly favourable to the dominant party it may be avoidable on grounds of undue influence. If so, the contract may be set aside where the transaction calls for explanation and the dominant party is unable to rebut the allegation of undue influence, usually by showing that the innocent party was independently advised.

Some of the cases in this area involve the big banks, where they have obtained signatures on guarantees and mortgages from relatives or debtors, often visiting them in their homes and giving them no chance to receive independent advice. The court must weigh very carefully whether such a transaction was only agreed to because of the bank bringing undue influence to bear. *National Westminster Bank Ltd v. Morgan* is a useful illustration of these rules in practice. Banks must beware of transactions where, for example, a husband seeks a loan for which a bank requires security involving property of the wife. This security may take the form of a charge on the matrimonial home of the husband and wife. In obtaining the signature of the wife to give the bank the necessary security this may be due to undue influence exerted over the wife by the husband. Such undue influence may affect the validity of the transaction between the wife and the bank, if the wife gains no financial advantage from the transaction (*Royal Bank of Scotland plc v. Etridge (No. 2)*). In this situation the bank is said to have constructive notice of the undue influence. To avoid this outcome, banks ought to advise that independent legal advice should be sought before entering into the transaction. A good example of the above is seen in *Barclays Bank plc v. O'Brien*.

## Illegal contracts

The phrase 'illegal contracts' is used here to cover those situations where contracts are either void or illegal at common law or by statute. A wide range of situations is thus encompassed, for which the law offers numerous solutions. Sometimes the illegality involves an immoral contract, such as a contract to commit a crime, or to pervert the course of justice. On some occasions Acts of Parliament designate particular types of contract as unlawful. In these cases the results of making such a contract will frequently be governed by the Act.

There is a general power available to the courts to interfere with contracts where it can be seen that such contracts are contrary to public policy. Such a potentially wide-ranging doctrine must be applied with caution, because the judges must not forget that the parties are largely free to fix their own bargain. Inevitably the definition of 'public policy' must remain somewhat flexible as it will need to shift and change to reflect the changing needs of society. One judge remarked that public policy was 'a very unruly horse and once you get astride it, you can never know where it will carry you'.

Where contracts involve some considerable element of moral wrongdoing, it has been usual to describe them as illegal at common law. Examples would be contracts promoting sexual immorality. In these cases, neither party will be able to enforce the contract. But if one party is the 'less guilty' of the two, he or she may be able to recover money paid or property transferred. This will particularly be the case where one party can be seen to have clearly 'repented'.

Where less moral wrongdoing is involved, for example in contracts in restraint of trade, the approach of the courts is to declare void either the contract as a whole or its offending parts. A contract in restraint of trade is one whereby one party promises or agrees to restrict his or her individual freedom to contract. These clauses are frequently found in employer–employee contracts or in contracts for the sale of a business. They are regarded as important because so many disputes in this area are litigated in the courts.

It is useful first of all to establish why such clauses are regarded as contrary to public policy. Surely, one might argue, if the two parties agree, that should be the end of the matter. That would be to take a pure freedom of contract approach. However, viewed from the point of view of the public interest, it is seen as wrong and contrary to public policy that a person's skill and experience should be rendered effectively useless and sterile by a contract. The courts have, therefore, been active in controlling those contracts in restraint of trade which they regard as being unreasonable. The modern rule is that such an agreement will only be enforced if it is reasonable in the interests of the public and reasonable in the interests of the parties.

When judging what is reasonable, each case will tend to turn on its own individual facts, so precedents are of little value other than as examples. The courts have always made clear that it is not appropriate to incorporate a restraint of trade in every contract of employment. The employer cannot include such a clause merely to prevent competition. He or she may only restrain an employee when there is some legitimate business interest to protect. That interest usually involves trade secrets or business connections, i.e. customers. Take the case of an employee who is engaged as a travelling salesman within a 25-mile radius of London. It may be permissible to include a clause in his contract that when he leaves his present employment he will not solicit any of the employer's customers within that 25-mile radius for a period of one year. It would clearly not be reasonable to restrict the salesman throughout the UK, as he only poses a threat to the employer within a very small area.

When evaluating the reasonableness of a restraint clause, the courts will pay close attention to factors such as the area of the restraint and the time limit involved. The restraint should be no wider and no longer than is strictly necessary to protect the employer's business interests.

It will be clear that there is no precise science about clauses in restraint of trade. An employer who wishes to put such a clause into an employee's contract will have to tread carefully in the drafting to make sure that it goes no further in the restrictions it imposes than will be found acceptable by the courts. If the employer 'gets it wrong' in the drafting, he or she will pay a high price as the clause will be held to be void and will be struck from the contract. The only help the courts will give is to construe a restraint of trade clause in the light of the circumstances existing when the parties made their contract, in order to determine what they must have intended. By interpretation of the clause, it may be possible to limit in scope what otherwise appears to be too far-reaching. An example of this approach can be seen in *Home Counties Dairies Ltd v. Skilton*, where a milk roundsman, once he left his present employers, was to be restrained from selling milk or dairy produce for a period of six months. The phrase 'dairy produce' was so wide that, on a literal interpretation, the man could not even go to work in a supermarket selling yoghurt. The court held that the phrase had to be construed in the light of what the parties must have intended, and should therefore be more confined in its meaning.

The courts tend to concentrate on the issue of reasonableness as between the parties, but it should be borne in mind that rules of public policy are in issue, and public interest is always a factor which could sway a court's decision. This can clearly be seen in the case *Pharmaceutical Society of Great Britain v. Dickson*. The society intended to change the rules for its members, many of whom were owners of small chemists' shops. As a result they would be severely restricted by their professional rules from dealing in non-traditional lines. For example, their shops would have to stop selling goods like paperback books or cassettes. For many small chemists these were high-profit lines which effectively kept the business going. One chemist sought a declaration that such a change in the rules would be an unreasonable restraint of trade. The House of Lords clearly thought so and considered public interest to be a paramount consideration. Such a change in the rules could well have caused many small chemists' shops to close and thus deprived large sections of the public of an important amenity and public service.

Over the years the courts have been vigilant in finding restraint of trade clauses hidden in some unusual situations. In one case, the rules of a pension fund were held to operate as an unreasonable restraint (*Bull v. Pitney Bowes*). The principles have also been applied in cases involving licences for racehorse trainers (*Nagle v. Fielden*) and the transfer system for footballers (*Eastham v. Newcastle United*). These cases clearly demonstrate that the courts will not allow the parties to achieve through a back door what they could not do by a direct method.

Where the courts are prepared to interfere and find a restraint unreasonable it is void and struck from the contract. Where it is only one clause of a much larger contract, the remainder of the contract may be perfectly valid and enforceable. In some cases, especially borderline situations, the inevitable effect of declaring a restraint to be void is that the employer is left with no protection for legitimate business interests. An example of this can be seen in *Commercial Plastics Ltd v. Vincent*, where a research chemist had been working on PVC but only on a limited aspect of its application. The restraint in his contract was drafted to cover too general a field of work on PVC and so the whole restraint was held to be void. The judges felt sure that the employer had genuine interests to protect, and with a properly drafted clause would have secured such protection.

This rigorous approach of the courts stems from the principle that the parties must make their own terms and cannot expect the judges to do this for them. In one limited circumstance the judges may be able to give some assistance. Where the restraint is couched as a series of alternatives, some of which are valid but some of which are too wide and thus unreasonable, the court may be able to sever parts which are excessive. Consider a restraint for '12 months within a 10-mile radius of our shops at Newcastle or York'. If the employee has never worked in the York shop he or she probably poses no threat to the employer there and that aspect of the restraint could therefore be too wide and unreasonable. Using an approach known as 'the blue pencil test' the courts will simply cross out the words 'or York' and leave a valid restraint behind. Of course, this will be no help in a great many cases where the restraint consists of one over-wide restriction, e.g. a restraint for '12 months within 100 miles of our shop in York'. The judges have no power here to cut this down in scope if it should turn out to be too wide to be reasonable.

It is obvious that the rules about restraint are not clear or precise. The main value of putting a restraint clause into a contract is its deterrent effect. It acts as an early warning to an employee that he will not get away with interfering with his employer's interest. But even if no restraint is imposed, or a restraint has been struck out as unreasonable, it should be remembered that employees owe duties to their employers at common law under their contracts of service (see page 201).

# Discharge of the contract

## Discharge by performance

In the normal course of events, both parties to a contract expect that the contract will be properly carried out. Where both parties perform their obligations under the contract exactly and precisely as promised, then the contract is discharged by its complete performance. It follows that if either party does not exactly and precisely perform, there is a breach of contract. Contractual obligations arise by agreement, and it is right that both parties should carry out their obligations in full. A necessary result of this rule is that a person who only partly performs his or her obligations and is to be paid upon completion of these obligations cannot be paid. This is clearly demonstrated in a case involving the installation of central heating. The job was done and the customer had not paid. But the heating failed to give the promised temperatures in the house and the boiler emitted vile fumes. As the installer had not exactly and precisely performed, it was held that he was not entitled to be paid (*Bolton v. Mahadeva*).

That case does demonstrate how harshly the rule of exact and precise performance can operate. Not surprisingly it is subject to a number of exceptions:

1    A party may be paid for partial performance where he or she has substantially performed the contract. So, if one party has performed except for some trivial details, he or she will be entitled to be paid, less the value of the other party's counter-claim in respect of those trivial breaches.

2    A party may be paid where the other party is prepared to accept part performance. In these cases, however, the other party must have a genuine choice whether to accept part performance or not. So where I am due to deliver 10 tons of coal and I arrive with only eight tons, you may agree to accept eight tons. But in the central heating case, the claimant had no such choice. The work was only partly done but it could not be undone. Of course, in those cases where the other party does agree to accept part performance, the obligation to pay will be correspondingly adjusted.

3    A party who has partially performed may be paid as much as the work or performance is worth in cases where the other party will not permit him or her to complete this performance. In this case, of course, the person who will not permit performance is in breach, and the party who has partly performed may either sue for damages in respect of that breach, or claim a reasonable sum for the value of the work already done.

4    Where a contract is divisible, a party who had performed some parts may be paid for them. So, in a contract of employment with salary paid monthly, each complete month could be regarded as a divisible contract. In building contracts, any agreement about progress payments will make it a divisible contract. Note that in relation to 'construction contracts', the Housing Grants, Construction and Regeneration Act 1996 provides that if work is to last for 45 days or more then the party to be paid is entitled to payment by instalments, stage payments or other periodic payments. See also Chapter 8.

Although the rules about precise performance are quite strict, there are, nevertheless, some defences which are regarded as acceptable excuses for non-performance or partial performance. For instance, the parties may have agreed to some lesser performance. The validity of their agreement will be the subject of rules about discharge by agreement.

Or again, the parties may be unable to perform their contract as originally contemplated because of some external factor or event outside their control. The validity of such an excuse is considered under discharge by frustration.

The parties themselves may have taken account of the need to change the rules about performance, e.g. in a building contract more time may be needed if a contractor unearths antiquities, and their contract may make provision for altered performance.

### The time for performance of the contract

It is often said that 'time is of the essence' in a contract, meaning that the time stipulated for performance is a major term, and any breach of that term would entitle the other to repudiate the contract. This will often be true, but each case will turn on its own facts. Have the parties expressly provided for a time for performance? If not, then the contract must be performed in a reasonable time. If a precise time has been fixed, the surrounding circumstances may make it clear that it must be adhered to. If I order a wedding dress to be completed in time for my wedding, time is clearly of the essence. But if I order a new armchair and you undertake to deliver it next week, it may not be so disastrous if you do not make delivery on time. Inevitably, in that situation, the time may come where the patience of the customer starts to wear thin. Then, although time was not originally of the essence in the contract, the customer can make it so by serving notice on the seller. The customer may indicate that, unless he has the chair by the end of the week, he regards the contract as at an end.

### Tender of performance

Sometimes it is only possible to perform the obligations under a contract with the cooperation of the other party. Take the example of a sale of goods where the seller agrees to deliver. He needs the other party to be there to receive the goods. If he offers to perform, i.e. tenders performance, by taking the right amount of goods of the correct description and quality, and the other party will not accept them, then the seller can argue that he has done all he can, and he can then consider himself discharged from his obligations. In *Startup v. McDonald*, a delivery of oil was due to be made before 31 March. The seller arrived with the oil on 31 March, which happened to be a Saturday. It was very late in the evening, and the buyer refused to accept the delivery. This was held to be a valid tender of performance so the seller could recover damages for the non-acceptance of the goods.

Where the obligation under the contract is to pay money, there is only a valid tender if money of the exact amount is offered, made up in such a way as to constitute 'legal tender'. Of course it is quite common for the parties to have agreed expressly to payment in some other way, e.g. by cheque. If no such agreement exists, the other party is entitled to expect money.

### Discharge by agreement

What is created by agreement can be discharged by agreement. Although this statement is largely true, discharging a contract by agreement depends on the operation of the rules of consideration. Effectively, if the parties agree to discharge a contract they are making a new contract. That new contract will be supported by valid consideration if each party is giving up something of value by discharging the first contract. If I agree to sell you goods for £10, and I have not yet delivered and you have not yet paid, then we can discharge that contract by a new agreement. My consideration for the new agreement will be giving up the right to receive £10 and your consideration will be giving up the right to receive the goods.

Obviously, it is not so straightforward to achieve a discharge by agreement if one party has already performed. If I have already delivered the goods, my part of the bargain is complete. It would be possible to discharge our agreement by drawing up a deed, but this would be absurd to contemplate except in commercial cases involving very large sums of money. In these cases, the rules about waiver play an important part. In the example already given, I may agree to forego payment of the £10 due. What would prevent me from changing my mind at a later stage and suing for the money? It could be argued that I am prevented from going back on my earlier promise because of the defence of promissory estoppel (see page 110). As this is an entirely discretionary principle, the other party may feel less than secure in relying on it. I would also be prevented from suing for my £10 if we have made a valid variation of our original contract, supported by new consideration.

## Discharge by frustration

When two parties make a contract their natural expectation is that both of them will perform it. However, in some cases, one party may find it impossible to perform their part of the bargain due to events quite beyond their control. The law of contract basically adopts a harsh principle that if contractual obligations are undertaken, then any non-performance for whatever reason amounts to breach.

However, since the nineteenth century, the courts have developed rules relating to 'frustration' or subsequent impossibility, in an attempt to achieve justice between the parties when things go wrong in an unforeseen way. If one party chooses not to perform because he or she now regards the contract as a bad deal, he or she will be in breach. If the parties themselves have provided an express term to cover some particular event, then the terms of their agreement will take effect. The common law rules developed by the courts provide that if there has been a frustrating event it will have the effect of discharging the future obligations of the parties under the contract. The resulting financial arrangements between the parties will then be dealt with, when appropriate, by the Law Reform (Frustrated Contracts) Act 1943.

The first problem is the difficulty of establishing which events should be regarded as frustrating events. Take as an example the closure of the Suez Canal. If my ship is locked in the canal as a result of the closure, then any contract I have made to hire the ship to you would surely be frustrated.

But if I had a contract to transport your goods, and the closure of the canal simply means that I must use a longer and more costly route, then this is by no means certain to be found to frustrate the contract. In such a case my major complaint would be that I am not going to make a profit on our contract. This factor alone will not influence the court. Contracts inevitably involve risk, and the parties must look out for their own interests. *Davis Contractors Ltd v. Fareham UDC* is a good example of this point. In July 1946 the contractor had agreed to build 78 houses for £92,000, the work to be completed within eight months. Owing to quite unexpected circumstances, and through no fault of either the contractor or the council, there was a serious labour shortage and necessary building materials were extremely difficult to obtain. As a result the job took almost two years to complete, and the contractor incurred extra expense of £17,500. He wanted to argue that the very long delay had frustrated the original contract, and as that would avoid the original contract, he could then claim payment for a reasonable sum. The House of Lords rejected the contractor's argument. One of the judges in the case said: 'The possibility of enough labour and materials not being available was before their eyes and could have been made the subject of special contractual stipulation. Frustration is not to be lightly invoked as the dissolvent of a contract.'

Examples of situations in which the court is prepared to find frustration include:

1   *A forecast event fails to take place.* A classic example was the cancellation of the coronation of Edward VII. Many people had made contracts to hire rooms to view the coronation procession. In one such case, *Krell v. Henry*, the contract for the hire of a room overlooking Pall Mall was held to be frustrated by the cancellation of the coronation. It is evident in a case like this that although the hirer could still use the room on the day in question, it is impossible for the parties to achieve the substantial object of their contract. There is an interesting contrast to be made with the case of *Herne Bay Steamship Co. v. Hutton*, where the contract indirectly arose out of the coronation activities. The Royal Naval Review was to take place at Spithead, and the claimant chartered a boat to cruise around the assembled foreign fleets in the harbour and to see the naval review. Because of the King's illness the review had to be cancelled. But the contract for the charter of the boat was held not to be frustrated, because it was still possible to enjoy a day's sailing, and see all the foreign fleets. In other words, the substantial object of the contract had not been destroyed.

2   *The subject matter of the contract is physically destroyed.* In *Taylor v. Caldwell* the contract provided for the hire of a music hall which burned down before the date for performance of the contract. The fire was accidental and the contract was held to be frustrated. A similar rule for goods which perish after the contract is made but before the

date for delivery is contained in s.7 of the Sale of Goods Act 1979. The contract would be frustrated provided neither party was at fault. The same principle also applies where the person who was to perform the contract dies, or becomes too ill to perform. This rule is only appropriate where the contract calls for the services of one particular person. So, if I hire a famous concert pianist to give a recital, then death or illness of the pianist could frustrate the contract. Where there is an ordinary ongoing contract of employment, short-term illnesses would not usually frustrate the contract. There will inevitably be borderline cases, where a long-term illness rendering the employee unfit to perform his or her former duties may amount to frustration. This sort of example should not give rise to too many difficulties when it is remembered that such an employee's contract could be terminated by giving appropriate notice (see page 216).

3   *Subsequent intervention makes performance of the contract illegal.* This situation can occur when parties have made a perfectly valid contract, but before they perform it some change in the law makes that type of contract unlawful. So, in *Metropolitan Water Board v. Dick Kerr & Co.*, the defendants had undertaken to build a reservoir for the Water Board. They were already at work when they were ordered to stop by the government, which directed them to build a factory instead. The court held that the contract with the Water Board was frustrated. It is true that the defendants could have resumed work on the reservoir very much later, but because of changing costs of materials and labour this would have then been a substantially different task from the one originally undertaken. Obviously the reservoir example can be compared with the council house case (Davis) discussed earlier. In the one case the rules of frustration operated and the defendant was released from what had become a bad bargain. In the other case, Davis Contractors were held to their original contract. How can such a distinction be justified? It seems that in most cases the courts will abide by the traditional view that you must perform as you agreed, but there will be exceptional situations where the increased cost of performance is so great that it can be said to make the contract into something radically different from what was contemplated by the parties. No doubt this allows the courts to find frustration in extreme cases, but it does demonstrate how difficult it is to pin down the scope of the doctrine with any precision.

4   *Frustration affecting contracts concerning land.* It used to be thought that the rules of frustration would not operate in the case of leases or contracts for the sale of land. Two important cases emphasize how significant the rules could be in such instances. In *National Carriers v. Panalpina Ltd* a warehouse was let on a 10-year lease. With five years of the lease still to run, the street giving access to the warehouse was lawfully closed off by the local authority because there was a listed

building in a dangerous condition. It could only be demolished with the consent of the appropriate minister, which took almost two years to obtain. During all that time the tenant was unable to use the premises for warehousing as no lorries could gain access. He stopped paying the rent and claimed frustration of the lease. The House of Lords rejected his claim. One judge said: 'Under the bargain the land has passed from the lessor to the lessee with all its advantages and disadvantages. Why should justice require that a useless site be returned to the lessor rather than remain the property of the lessee?' In *Amalgamated Investment Co. Ltd v. John Walker & Sons Ltd* the facts involved the purchase of a piece of land which the purchaser intended to redevelop. He contracted to pay £1.7 million for the land. On the day after he signed the contract, officials at the Department of the Environment listed a building which was presently on the site as a building of special interest. This made it extremely difficult to go ahead with development plans, and the site without development potential was worth only £250,000. Despite the scale of the financial disaster the purchaser was held to be bound by the contract. Although it was now apparent that he had made a bad bargain, the contract was not frustrated.

From the cases and examples we have looked at, it is possible to state only a very broad principle. Frustration occurs when performance of the contract is impossible or when the parties can no longer achieve the substantial object they had in mind. It is essential that the supervening event should be outside the control of either of the parties. If it comes about through some act or election of one party, or as a result of negligence, then it cannot be frustration. Both of these points are well illustrated in *Lauritzen A. S. v. Wijsmuller B. V.* (the *Super Servant Two*), where the defendants had contracted to transport a drilling rig from the Japanese shipyard where it was being built to Rotterdam. The contract provided that the carriage would be undertaken between 20 June and 20 August, and would be effected using either *Super Servant One* or *Super Servant Two*, both of which were large transportation units. In fact, it was always the intention of the defendants to use *Super Servant Two* to fulfil the contract but, unfortunately, it sank in January. In the meantime, the defendants had made other contracts to use *Super Servant One*, so that vessel was not available to be substituted. The defendants argued that the sinking of the *Super Servant Two* had frustrated the contract. The Court of Appeal held that this was not a supervening event such as to frustrate the contract, because the defendants could have performed without breach by using *Super Servant One*, and it was by their own election that they chose not to do so.

Inevitably, frustration often occurs after the parties have embarked on performance. Their valid contract automatically ends at the moment of the frustrating event, and the parties are therefore discharged from further obligations under it. This can produce a very untidy and unfair situation, as it means that obligations which fell due before the frustrating event are still owed. Take the example of a contract to import and sell £10,000 worth of a particular type of toy, delivery to take place in November, with £2000 payable on the making of the contract, and the remaining £8000 to be paid within one month of delivery. The contract is made in June and £2000 is duly paid. In October the government bans the import of the toy in question. At that point the contract is frustrated. But the buyer has paid £2000 for which he has received nothing. The early approach of the courts to sorting out these problems was to decide that the loss lay where it fell. In an attempt to solve the financial plight of the two parties to a frustrated contract, the Law Reform (Frustrated Contracts) Act 1943 was passed.

The Act makes no attempt to define what frustration is. It simply deals with the consequences. Its two main objectives are to allow for the recovery of money paid and for compensation for partial performance. These will be considered separately.

1    *Recovery of money paid* – the Act provides that all sums paid before the frustrating event shall be recoverable and all sums due before the frustrating event cease to be due. Applying this to the sale of toys example, the £2000 deposit could be recovered. Of course, this may not be particularly fair to the seller, who may have incurred considerable expense in going about the performance of the contract. The Act recognizes this by providing that where expenses have been incurred before the frustrating event, and those expenses were 'in, or for, the purpose of performing the contract', the court may allow a party to recover in respect of those expenses, to the extent that the court thinks is just and equitable. In the toy example, the seller may have made phone calls and sent telexes, and may have had transport expenses and insurance costs. The judge may say that he can keep £500 of the £2000 deposit. Now the buyer is £500 out of pocket. That is an inevitable price which has to be paid when a transaction has gone wrong. The rules of frustration only operate where neither party is at fault, so inevitably the courts are trying to make the best of a bad situation and no solution will ever be perfect. They can only aim to achieve as fair a compromise as possible. An exercise of the courts' discretion in relation to expenses was seen in *Gamerco SA v. ICM/Fair Warning (Agency) Ltd.*

2    *Recovery of compensation for partial performance* – we have already seen that a frustrated contract will have commenced life as a valid contract, under which the parties may have started to perform their obligations. Suppose I have contracted to build you a house for £60,000 payable on completion. If I have completed three-quarters of the work when the government interferes and puts a ban on private building work, can I recover the value of the work I have done? The Act provides that if one party obtains a valuable benefit, by reason of work done in the performance of the

contract by the other, the court may allow the party who has performed to recover a sum which does not exceed the value of the benefit conferred, and which is just in all the circumstances of the case.

The scale of disaster when such frustrating events occur can be seen in *BP Exploration Co. Ltd v. Hunt*. Hunt owned an oil concession in Libya. He did not know if oil would be found there or, if so, how much. He wanted, somehow, to finance the exploration of the oil field but to hold on to a share of any profits if large quantities of oil were found. He contracted with BP to share the field. BP would bear all the exploration costs and, if oil was found, Hunt would repay them from his share of the oil revenues. The exploration work cost a staggering $87 million. A large amount of oil was found and extraction had only just begun when the Libyan government confiscated the land and nationalized the concession. BP had done work in performance. Had that work conferred a valuable benefit on Hunt? Merely exploring for oil gave Hunt no valuable benefit, but its discovery greatly increased the value of his concession before it was nationalized. The judge calculated Hunt's valuable benefit to be the value of the oil already removed and the compensation he would receive from the Libyan government. This formed the upper limit of what he could have awarded to BP. He then had to consider what figure, up to that upper limit, he thought was just in all the circumstances. Although the judge thought that the 'valuable benefit' to Hunt was over $80 million, he awarded $35 million to BP, as it had been an agreed basis of the contract that it would bear the exploration costs. The exploration might have found no oil at all, and in that case it would have had to bear those costs in total with no claim at all against Hunt.

The 1943 Act may not apply if the parties themselves have included a term to cover the results of a frustrating event. It will be a question of construing or interpreting their term, to discover if they intended it to operate in such extreme cases. The rules of the Act have no application to contracts of insurance, so if I insure against sickness, pay a year's premium and then die after the first month, there can be no recovery of part of the premium. In some sale of goods situations the appropriate rules are found in s.7 of the Sale of Goods Act 1979, rather than the 1943 Act.

## Discharge by breach

A breach of contract may occur because one party does not perform his or her obligations at all, or performs them late, or performs them in an unsatisfactory way. One of the major difficulties when looking at discharge by breach of contract is trying to establish exactly which breaches have the effect of discharging obligations under the contract. Where such a major breach can be identified, the injured party may nonetheless want to go ahead with the contract and simply ignore the breach or, more likely, seek compensation for it.

All breaches of contract, however large or small, give the injured party the right to sue for damages. In some cases it may be possible to exercise that right through set-off. Where a painter contracts to decorate the outside of your house for £1000, to be paid on completion of the work, he may commit breaches by not burning off where specified, or by applying only one undercoat instead of two. In such a case, the injured party could solve the problem by deducting something from the £1000 due. In some cases a contract may even provide a schedule of damages for specified breaches, thereby anticipating those which are likely to occur and providing a solution. Such an example is the case where a builder undertakes to complete a job by 30 June, and if he fails to do so, to pay £100 per day (agreed or liquidated damages) in respect of each day's delay thereafter.

A much more difficult question still remains – when does a breach put an end to the contract itself, or permit the innocent party to choose to treat the contract as at an end? And what exactly is meant when we say the contract is 'at an end'? As a general rule, where the party at fault is in breach of a major term (a condition) or in breach of an innominate term where the effects are serious, the innocent party can treat that behaviour as repudiatory, and regard him- or herself as discharged from further performance. Where the breach is of a minor term (a warranty) the injured party is restricted to an action for damages and must continue with performance of the contract.

The various categories of terms have already been examined (see page 118). Let us assume that when the breach occurs it is clear which type of term has been broken. Take the case where you want to ship goods under refrigeration. You enquire whether the ship has refrigeration plant and are assured that it has. On that basis you contract to ship 10 tons of soft fruit under refrigerated conditions. When you arrive to load the fruit, you discover that the ship had no refrigeration equipment. This is likely to be a breach of a major term, a condition. You now have a choice. You can treat it as a repudiatory breach and regard yourself as discharged from further performance, i.e. the obligation to pay the shipping charges. You can also sue for damages in such a case. No doubt you would take this course of action if you were able to find alternative shipping facilities immediately, or if you could sell the fruit in a local market. Where that is not possible, you may decide that you have no option but to ship your fruit in the unrefrigerated ship, thereby choosing to affirm the contract, but of course you can still sue for damages for the breach. As you have affirmed, you have to pay the shipping charges, but if your fruit deteriorates as a result of the breach you can be compensated by the damages.

This example assumes that, even though the breach was major, you were left with a choice of courses of action, either to affirm and go on, or consider the contract at an end. Sometimes that will not be the case, as the very nature of the breach itself may put an end to the contract. Take the case where you engage a central heating contractor to install a system for you for £2000, the work to commence on 1 June. If the contractor informs you in May that he has found more lucrative contracts and does not intend to come to do your job, he is effectively repudiating the contract in advance. You could now wait until June to see if he comes. If he fails to turn up then, this is a case of breach by non-performance. Or you could sue immediately, making clear that you are choosing to treat the contract as at an end. Where these situations occur in practice, the difficulty may be that the injured party needs to respond quickly to minimize the harm and inconvenience, and he may therefore act precipitately without the benefit of proper advice. Yet, once it is clear that the other party does not intend to perform it might be dangerous to wait and see if he will change his mind, because in the meantime the injured party could lose his rights through some supervening frustration.

## Enforcing the contract – remedies for breach

The law provides remedies for breach of contract, but the extent to which purely legal remedies can be truly effective is limited. A remedy is necessary because things have gone wrong. An ideal solution would be for the parties themselves to plan in advance for situations which may arise. They can then impose solutions which are satisfactory to both of them. This contract planning role is very significant in the business and commercial world. The standard form building contract is an example where many contingencies are provided for, without the need to go to court. One advantage of such a system is that it creates a climate within which the parties can stay on reasonably good terms, which is vital if they need to do business with each other frequently. Sometimes the contract planning role can also avoid those situations where court action would probably be useless. There are business situations where the customer is asked to pay a deposit. Not only does this create a situation where the customer is more likely to perform but it also provides a ready small sum of compensation against non-performance. An example would be 'same-day' photographic processing, where deposits are common.

One problem of seeking to enforce a contract through the courts is that the primary and usual remedy for breach is damages. The courts are prepared to order the other party to carry out their obligations, but only in a very limited number of cases. If they will grant a remedy, it is called specific performance. Such an order is not readily available because the court may then need to keep the defendant under constant supervision, and it has neither the desire nor the means to do so. Moreover, in many cases, money will be adequate to compensate the claimant. The types of contract where an order of specific performance may be granted usually relate to sales of land and leases. No two pieces of land are identical, and it is thought that money cannot adequately compensate. That is quite a difficult idea to accept when the 'land' in question may be a semi-detached house on an estate surrounded by dozens of properties which are seemingly identical! The argument about the unique quality of the property can also be used, although less frequently, in relation to sale of goods, for example a unique painting. If the goods are in some way unique, or not readily obtainable, a court may be prepared to grant an order for specific performance. Indeed, they have a discretion to do so under s.52 of the Sale of Goods Act 1979.

As the power to award specific performance is equitable and discretionary, a number of points will be borne in mind by the courts. They will not grant an order if it would operate very harshly on the defendant. Nor will they grant it where a claimant has delayed too long in seeking relief – delay defeats equity. This rule is not precise, and what counts as excessive delay will depend on the facts of each case. The order will not be granted if constant supervision would be required, nor in cases involving personal service, as it is thought that this would be tantamount to slavery. So if I contract with a famous painter to paint my portrait and he backs out of the contract I will not get an order of specific performance. It used to be suggested that an order of specific performance would not be granted in cases involving breach of contract to pay money, as damages would surely be an adequate remedy. This may not be so, and the problem was highlighted in the case of *Beswick v. Beswick*, where a coalman sold his business to his nephew, who was to pay for it by paying the uncle £5 per week for his life and thereafter the money was to be paid to the coalman's widow. After the death of the coalman the payments to the widow were not made. She sued as the personal representative of her late husband and was awarded specific performance of the contract. This was the only way to achieve justice in this case.

There is another form of equitable relief, the injunction, which can be useful in a limited number of contract situations. This is particularly the case where a party to a contract has promised *not* to do a particular thing. A common example is a clause in restraint of trade, where an employee promises not to solicit his or her former employer's customers or a seller of a business promises not to compete. Where a breach of such a clause is proved the court may grant an injunction, but it should be remembered that the relief is entirely discretionary.

## Awarding damages

In theory, an award of damages can be made in respect of every breach of contract, however large or small, but the amount of damages is linked to the loss suffered. The aim of the court when awarding damages is to put the injured party into the position they would have been in if the contract had been properly performed. This compensates him or her for loss of expectation. Somehow, the court has to arrive at an appropriate figure for the damages. This is described as the measure of damages. The figure can only be worked out once it is established for which harm the claimant can recover. He or she can only recover in respect of losses which 'arise naturally, according to the usual course of things, and losses which must have been in the contemplation of the parties when they made the contract. This is the rule of remoteness of damage, which was laid down in the case of *Hadley v. Baxendale*. The claimant's losses must also have been caused by the breach of contract. The two aspects of the remoteness rule can be seen at work in the case of *H. Parsons Ltd v. Uttley Ingham & Co. Ltd*, where the defendants manufactured and installed an animal feed dispensing hopper for Parsons, who was a pig breeder with a class 1 pedigree herd. The installation was carried out after the hopper had been transported by road. During the journey the driver had jammed the hopper ventilator with a rag to stop it rattling and he did not remember to remove it before the hopper was erected. Pig food was poured into the hopper and at first all seemed well. Inevitably, because the ventilator was jammed shut, the pig food eventually turned mouldy. Of course, there was a breach of contract here, and Parsons could undoubtedly claim damages in respect of the mouldy pig food. Unfortunately, the pigs had eaten the food and developed a disease which killed off a large proportion of the herd. In consequence, not only had Parsons lost more than 250 pigs, but he had also sustained the loss of a breeding season. Should he be allowed to recover in respect of the loss of the pigs? The scale of the problem can be clearly seen in this case where the defendants accepted liability with regard to the pig food and were prepared to pay damages of £1802, but Parsons was trying to claim about £30,000. Parsons' claim was successful. The defendants must have contemplated that if they installed a defective hopper which made the food mouldy, the pigs could become ill. It was matterless that the degree of harm (the death of the pigs) was greater, as long as the type of harm (illness) could have been contemplated.

It is clear that the remoteness rules operate very flexibly, but the more the defendant knows about the claimant and his or her requirements, the more he or she is likely to be able to contemplate by way of potential harm. The remoteness rule was recently considered in *Transfield Shipping Inc v. Mercator Shipping Inc* by the House of Lords. It appears that if a loss is wholly out of the ordinary, being disproportionate or unpredictable, and there is a general market understanding as to what can be claimed, then the loss is too remote unless there is an assumption of responsibility for the loss on the part of the defendant. So the rule in *Hadley v. Baxendale* will usually apply but in unusual cases, as described in the preceding sentence, there will need to be an assumption of responsibility for loss to be recoverable.

Once the courts are clear about the harm or loss which falls within the remoteness rule they must then assess an appropriate measure of damages. Sometimes this will be quite straightforward, as in cases for the sale of goods where the seller has failed to supply. The buyer must now look elsewhere for the same goods and may only be able to buy at a higher price. The measure of damages is the difference between the two prices. Or a buyer may refuse to accept delivery of goods properly tendered under a contract. There, the measure may well be the loss of profit on the deal. Or a holidaymaker may cancel a booking for a country cottage late in the day, and the damages will be the loss of profit on the booking.

Sometimes the courts will give damages on the basis of how much it costs to cure a defect in work done instead of difference in value. This is the usual measure in relation to defective or incomplete building work. Cost of cure will not be given if it is excessive when viewed against the benefit accruing to the innocent party. In *Ruxley Electronics and Construction Ltd v. Forsyth*, a contractor had agreed to construct a swimming pool to a depth of 7 feet 6 inches for £17,797, but the pool constructed was only 6 feet 9 inches in depth. The pool was usable and evidence showed that the lack of depth had not decreased the value of the pool. However, to increase the depth of the pool to 7 feet 6 inches would require the demolition of the existing pool and cost £21,500. The House of Lords held that the award of cost of cure damages of £21,500 was out of proportion to the loss suffered by the pool owner and would not be given. Difference in value damages would be appropriate (which in this case would have been nominal as there was no difference in value between the two pools), but the House of Lords restored the original trial judge's award of £2500 for loss of amenity.

Where there has been a breach the injured party cannot just sit back and wallow in their misery. The law imposes a duty to take reasonable steps to mitigate these losses. A claimant cannot recover for any harm which could have been prevented by taking reasonable steps in mitigation. If a buyer refuses to take delivery of goods under a contract, the seller must take reasonable steps to find another buyer. In the holiday cottage example above, he must take reasonable steps to re-let the property. An ironic result of taking reasonable steps to mitigate may be that the claimant actually increases his losses. In the holiday cottage example, he may place advertisements in newspapers and still not manage to re-let the property. He can then include the cost of the advertisements in his claim for damages.

Of course, the claimant is the person who has been wronged so there is a limit to what he or she can be expected to do under the duty to mitigate. As the defendant has committed the breach, he or she is hardly in a strong position to 'call the tune'. This is well illustrated in *Pilkington v. Wood*, where a solicitor gave incorrect advice to a client, in consequence of which the client found that his house had a defective legal title and was virtually unsaleable. When the solicitor was sued for damages he argued that the claimant had failed in his duty to mitigate as he could have solved the problem of the defective title by suing the person who had originally sold him the house. This was hardly asking the claimant to take a reasonable step, as such litigation would have been expensive and the outcome by no means predictable. The claimant was right not to have embarked on such a costly and uncertain procedure.

It used to be a feature of damages for breach of contract that nothing would be awarded for distress, injury to feelings, or disappointment. One justification for this rule may have been that it is very difficult to quantify such losses. The rule has been subject to an exception over recent years in cases where the claimant was contractually entitled to expect pleasure and enjoyment, not pain and distress! However, the sums awarded for distress and disappointment have not tended to be large. *Jarvis v. Swans Tours Ltd* shows the sort of case where an award for disappointment may be appropriate. The claimant had booked a holiday for himself on the strength of a brochure which claimed that he would be joining a house party and special resident host. He would enjoy a welcome party and special yodelling evening, plenty of *gemütlichkeit* (good atmosphere) and, generally, he would be 'in for a great time'. In fact, by the second week of his holiday there were no other guests at all and in every other respect the holiday failed to live up to the brochure claims. Even the yodeller was a local man singing in his working clothes. When awarding him damages for disappointment the judge said: 'Mr Jarvis has only a fortnight's holiday each year. He books it far ahead and looks forward to it all that time. He ought to be compensated for the loss of it.' The court awarded him £125.

It was thought that a contract must have as its *main object* the provision of pleasure or the avoidance of distress for damages for disappointment or distress to be recoverable. In *Farley v. Skinner*, the House of Lords decided that for the award of such damages the contract must have *as an important or major object* of the contract the provision of pleasure or peace of mind. In that case the claimant employed a surveyor to inspect and report on a private dwelling house that the claimant wished to purchase. The defendant surveyor was expressly asked to report on aircraft noise. In the defendant surveyor's report it was negligently stated that the property was unlikely to suffer greatly from such noise. In reliance on the report the property was purchased. On occupying the property the claimant discovered that the property was severely affected by aircraft noise. The House of Lords held that the claimant was entitled to damages for disappointment. It would appear that damages of this type will not usually be available for a negligent survey, as under an ordinary surveying contract there is no promise to provide peace of mind. The important feature of *Farley v. Skinner* was that the surveyor was expressly instructed to investigate aircraft noise, this becoming an important or major object of the contract.

It has already been emphasized that by effective contract planning, the parties can provide in their contract for the sums to be paid in the event of specified breaches, a common example being late performance. Such contractually determined damages are called liquidated damages. The advantages of such a clause are clear. Problems can be resolved quickly and privately, without resort to the courts and, hopefully, in a manner which will preserve good relations between the parties. The parties also know in advance the risk and cost involved in particular breaches. On the other hand, such clauses can cause problems, particularly where the party agreeing to the clause lacks bargaining strength. In that case the result of the clause might be unfavourable to him or her. Either party may miscalculate the real effects of a specified breach and thus be unhappy to be regulated by what turns out to be an unfair clause. Against that, of course, must be set the fact that such clauses are freely negotiated, at least in theory!

The courts themselves are alert to the possibility that such a liquidated damages clause could operate harshly. If there is a dispute about the clause it is open to the court to consider whether it represents a genuine attempt by the parties to estimate their losses, in which case the clause will be enforced even if the estimate is wrong, or a clause inserted by one party to frighten or terrorize the other into performance. In the latter case the clause is classed as a penalty clause and is unenforceable. It is matterless what name the parties give to their clause. The court will determine which category it falls into, using a number of fairly flexible tests:

1   Does the clause provide for the same amount to be paid in respect of different breaches, some of which may be serious while others are trivial? If so, it will be hard to show that the clause was inserted as a genuine pre-estimate of loss.
2   Does the clause provide for damage for breaches involving non-payment of money at a level far higher than the sum unpaid? If you are due to pay £25 per week and I insert a clause saying that for every late payment you must pay me £100, that looks like a penalty.
3   Does the sum provided for by the clause appear to be harsh and unconscionable, given the breaches which are likely to occur?

## Suing on a quantum meruit

*Quantum meruit* means 'as much as it is worth'. Generally, the rules about *quantum meruit* have no particular significance in the law of contract, because the contract itself should fix the amount of any payment due or the method of calculating it. In some cases, however, even the law of contract provides for the payment of a reasonable price. This could happen where minors have been supplied with necessaries. Another example is where goods have been sold and delivered, and accepted, but no price has been fixed (s.8, Sale of Goods Act 1979).

There are times when the parties may think that they have a valid contract. They may have acted in reliance on that belief in performing obligations, only to find that the contract has not taken effect. In such cases if one party confers a benefit on the other, in circumstances where both believe that the benefit will be paid for, then the courts will allow the performer to recover a reasonable sum under the *quantum meruit* rules. This turned out to be a useful solution in *British Steel Corporation v. Cleveland Bridge and Engineering Co. Ltd*, where the parties planned to enter into a contract but it had not been finalized between them. In anticipation of that contract one party began to perform. When a contract failed to materialize the performer was able to be compensated for the work done on a *quantum meruit* basis.

## Limitation periods

The remedy of damages, as the principal relief for breach of contract, must be sought in time, otherwise any claim is statute barred by the Limitation Act 1980. The imposition of time limits is sensible because evidence eventually becomes stale and unreliable and any party at fault must ultimately have some certainty about whether they will be sued or not. In claims for breach of contract the time limit is basically fixed at six years from the date when the cause of action accrued. That usually means six years from the date of the breach of contract. The six-year rule is extended to 12 years where the contract was made by deed.

These rules are fairly rigid except where a claimant fails to discover that he or she has a cause of action because of the defendant's fraud. In these cases time only begins to run when the claimant discovers the fraud, or when he or she could with reasonable diligence have discovered it. 'Fraud' for these purposes could include cases where a defendant has deliberately concealed relevant facts from the claimant. This is what happened in *Applegate v. Moss*, where a builder constructed a house on foundations which did not comply with the contract terms. The house was finished in 1957 but it was 1965 before the defects began to manifest themselves. The house was declared unsafe for habitation. Although eight years had elapsed, and normally a claim would have been statute barred, the time was extended to run from the date when they learnt that the foundations were defective because of the builder's concealment of facts relevant to the claimant's claim – he had covered the foundations up before they could discover how inadequate they were.

These strict time limits do not apply to equitable remedies such as specific performance. But in exercising any discretionary equitable relief, the court works on the maxim that delay defeats equity. It will be a question of fact in each case whether a claimant has unduly delayed.

# 7
# Special contracts

## Agency

The agency relationship lies at the heart of commercial and business life. Agents are used to negotiate the sale of goods or services or to negotiate and conclude the sale of goods or services on behalf of a principal. The relationship of principal and agent arises whenever the principal authorizes an agent to make contracts on his or her behalf. Generally, there will be a contract between the principal and the agent, with the latter being paid for his or her services, but it is possible for the agent to act gratuitously.

Agency is usually created by express or implied agreement, but agreement may be lacking in cases of agency of necessity, or where the agent binds the principal under his or her apparent authority (see page 157). The usual rule is that agents may only bind a principal in contracts which are within the scope of their authority.

Where the agency has been expressly created, whether orally or in writing, the scope of the authority may have been clearly established. More often, however, it is a question of implying whatever authority is customary in the circumstances. So, for example, a board of directors who pass a resolution appointing a managing director of the company impliedly authorize that director to do all the things that fall within the usual scope of the office of managing director.

Where an agent is appointed for a particular purpose, he or she will also have authority to carry out any necessary incidental acts. Where the agency arises as part of a trade, profession or business, the agent has implied authority to perform such acts as are usual in that trade, profession or business. As a general rule, an estate agent has no implied authority to take a deposit in respect of the property to be sold, nor has he or she the right to sign a contract of sale.

### Agency of necessity

Agency of necessity arises where someone acts to protect the interests of another, without any prior authority. In one example, where goods were rescued from a stranded ship, the rescuer then stored them in a warehouse to prevent further deterioration. He was able to recover the costs of

Galbraith's Building and Land Management Law for Students. DOI: 10.1016/B978-0-08-096692-2.00007-5
Copyright © 2010 by Elsevier Ltd

warehousing from the owner of the goods. Similarly, where the owner of a horse failed to collect it promptly from a railway station, the railway company could recover the costs of feeding and stabling the animal.

Of course, in cases such as these, an agency of necessity will only be appropriate if the agent acts reasonably, and in good faith, in circumstances where it was impossible to obtain the owner's instructions. The actions of the agent must be for the benefit of the owner and not for the convenience of the agent. The courts are naturally anxious about imposing liabilities in such cases. Indeed, one judge remarked that 'liabilities are not to be forced upon people behind their backs'.

## Apparent authority

It is clear that agents can always bind principals if they make a contract of a type for which they have actual authority. Additionally, the acts of an agent will bind the principal if they are within the scope of the agent's apparent authority. Simply, this means that an agent may have exceeded his or her real authority but, in the eyes of the third party with whom he or she is negotiating, there was nothing to raise any doubts about the agent's authority to act and to bind the principal. In other words, there was the appearance of authority. This is sometimes regarded as an example of the doctrine of estoppel.

There are a number of ways in which such apparent authority can arise. For example, principals may allow agents to create an impression that they have wider authority than they actually possess. Take the case of a firm of solicitors which has five partners. All partners are agents of each other for all purposes, by virtue of their relationship of partnership. They decide to take in a new young partner, whose name appears alongside the original five on the notepaper. Because the new partner is young and inexperienced, the others tell him that for the first year he cannot buy equipment for the firm to a value greater than £100. Within the first month, the new solicitor places an order on firm's notepaper for a computer for the firm costing £800. Are the partners bound by this contract?

The answer is likely to be yes, because the computer salesman would know that he was dealing with a partner, and could therefore presume that the partner had authority to bind the firm. The young partner was acting within his apparent authority, i.e. as it would appear to a third party. If the other partners wanted to ensure that their young partner did not bind them in such a contract, they would have needed to make known the restrictions on his usual authority.

Apparent authority may also exist where a person allows someone to act as if he or she were an agent. In this instance, unlike the case above, the 'agent' has never been appointed as such and has never had any actual authority. In one example, a debtor came into an accounting office and paid his debt to a person he found in the office there. That person was a friend of the creditor, but was not and never had been an agent, but it would appear to a person coming into the office that he worked there and had authority to receive the money.

Another example of apparent authority occurs where a person has acted as an agent, but has had this authority totally withdrawn, e.g. where the person has been dismissed. Once this happens, the principal should make this clear to third parties, and if he or she has not done so, the former agent may still be able to bind the principal on the basis of apparent authority.

The law is inclined to hold the principal liable for the acts of the 'agent' in these cases of apparent authority for reasons of equity and business confidence.

### Architects and engineers as agents

Although architects and engineers are frequently appointed under standard conditions of engagement which lay down in detail the duties to be carried out, the conditions rarely set out the extent of the agent's authority. The scope of their agency may thus depend upon implied or apparent authority.

### Ratification

In some cases where an agent has no authority to act but has done so, it is possible that the principal will subsequently ratify the transaction, and thus become liable to the third party. Ratification may be express, or it can be implied from the behaviour of the principal. However, there can be no implied ratification without full knowledge of the unauthorized act. If ratification is possible, then it puts all the parties into the same position they would have been in if the agent had been duly authorized to act all along. An undisclosed principal cannot ratify the act of an agent (see below).

### Powers of attorney

Wherever an agent needs to be given authority to execute deeds, e.g. where he or she is given power to sign a transfer of land, then the agent's authority must be conferred by deed. The deed conferring such authority is called a power of attorney.

## Duties of the parties

Where the agency relationship is created by a contract, the duties of the parties will be rather similar to the common law duties between an employer and an employee. The chief duties of the principal are to pay the agent the agreed commission or remuneration, and to indemnify him or her against liabilities reasonably incurred when carrying out his or her duties.

### Paying agreed commission or remuneration

Problems about when payment is due have arisen mainly in relation to estate agents' fees. This is not surprising when the various phrases used in their contracts are examined. Some contracts provide for the fee to be paid if the agent introduces a purchaser. Or the contract might provide for payment if the agent secures an offer. Sometimes the fee is only due on completion of sale. There are many variations on these themes.

The courts had always been inclined to construe these commission clauses very strictly (*contra proferentem*), because it was thought that estate agents enjoyed some advantage over their clients, by being able to impose standard terms on them. Despite the passing of the Estate Agents Act 1979, this aspect of the law, about when commission is payable, was not clarified. Section 18 merely requires that the agent should give the client particulars of the circumstances in which the client will become liable to pay remuneration to the agent. Inevitably, these cases will continue to come before the courts, requiring them to construe in each case when the commission is payable.

### Indemnity

The principal must indemnify the agent against all liabilities and expenses reasonably incurred in the execution of his or her authority. An agent would not, however, be entitled to an indemnity in respect of illegal transactions, or in any situation where liability is caused by his or her own breach of duty.

### Lien

An agent has a lien (a right to hold on to goods belonging to another) over property belonging to the principal until the principal settles all outstanding dues, such as remuneration and expenses, with the agent. Often, this will be a particular lien, where it will relate only to the property in respect of which the dues have been incurred. In some cases the law grants an agent a general lien over all property of the principal. This could be so in the case of solicitors and bankers.

### Duties of the agent

The agent is bound to carry out duties with due care and skill, and the standard of care required will vary. If agents hold themselves out as being skilled in some particular field, then they must carry out the agency duties with the skill of a reasonable practitioner in that field.

The agent cannot normally delegate performance of the duties, unless the contract with the principal expressly or impliedly provides for delegation.

Whilst undertaking these duties, agents must not put themselves into a position where their own interests conflict with the duty they owe to a principal. So, for example, if they are asked to sell goods on behalf of a principal, they should not buy these goods themselves. They would also be in breach if they agreed to act as agent for a third party too.

Strict rules apply if agents have compromised themselves by making a secret profit from the transaction. In some cases, the secret profit may amount to a bribe. Not only must the agent hand over the secret profit to the principal, but also he or she is liable to be dismissed for such behaviour, and would also lose the right to any commission earned on that transaction, and in some cases could be prosecuted under the criminal law.

## Relationship between principal and third party

Once an agent has created a contract between the principal and the third party, his or her role is effectively over. The task has been to act as an intermediary, but in the normal course of events the agent is not a party to the ultimate contract. That contract is made between the principal and the third party.

The rights of the principal to enforce the contract against the third party now depend on whether the existence of the principal was known to the third party when he or she was making the contract. If it was, the principal is referred to as a *disclosed* principal. If not, he or she is an *undisclosed* principal. If the principal was disclosed, he or she could then be either a *named* or an *unnamed* principal.

Generally, a disclosed principal may sue the third party. An undisclosed principal may sue and be sued on a contract made by an agent on his or her behalf. However, in order to be fair to the third party, an undisclosed principal cannot sue the third party if the identity of the principal is material to the contract, and the third party believed he or she was contracting with the agent or if there is an express or implied contractual term that the intervention of an undisclosed principal in not intended or is excluded. If the undisclosed principal sues the third party, the latter can raise any defences that he or she would have been able to plead against the agent. Of course, where an undisclosed principal does decide to sue, his or her identity will then become known.

As for the liability of the principal to the third party, he or she is fully liable on the contract whether the position was disclosed or undisclosed. If the principal gives the agent money to settle with the third party, but the agent fails to do so, the principal is still liable to the third party.

## Relationship between the agent and third party

As a general rule, once an agent has brought about a contract between the principal and the third party, their role is complete, and they should not be liable on the concluded contract, nor should they be able to sue to enforce it. The agent's role is one of communication between the parties. However, in some exceptional situations, the agent can incur liability. This is particularly the case where he or she acts for an undisclosed principal. In these circumstances, the third party will have thought the contract was with the agent, and it would seem very unfair if the third party was deprived of the right to enforce against the agent, when a principal suddenly appears on the scene.

The position of the third party is enhanced in these cases, as he or she may elect whether to sue the agent or the principal. Once the election has been made, however, the result is to discharge the other from liability.

## Termination of the agency relationship

In many cases, the agency relationship exists within the framework of a contract, and where this is so, any method of terminating a contract will terminate the agency relationship. So, for example, the relationship may terminate because the contract is frustrated, or because it has been completely performed, or because it is discharged by a repudiatory breach. It may also be possible to discharge the relationship by either party giving notice. Sometimes, the contract itself may provide for a particular period of notice, but where none is specified, it may be terminated by giving reasonable notice.

In two instances, however, the agent's authority may be irrevocable. The first case is where the agent's authority is coupled with an interest. An example of this could occur where the principal entrusts goods to an agent for sale. The agent advances money to the principal, who authorizes the agent to be repaid out of monies received for the goods. If the principal then tried to withdraw the agent's authority to sell, the agent would have no way to recover the money loaned to the principal. In such a case, this authority is said to be coupled with an interest, and would effectively be irrevocable.

In the second situation, statutory rules under the Mental Capacity Act 2005 provide that once a Lasting Power of Attorney has been drawn up in appropriate form, then even if the principal subsequently becomes incapable, e.g. because of mental illness, the authority of the attorney under the power of attorney is not terminated.

## The Commercial Agents (Council Directive) Regulations 1993

These Regulations came into force on 1 January 1994 and extend the common law rights of commercial agents against their principals. The Regulations only apply to a commercial agent, defined as a self-employed intermediary having continuing authority to sell, or negotiate the sale, of goods on behalf of a principal. The activities of an agent marketing only services, or engaged for a one-off transaction, are therefore not covered by the Regulations.

Where they do apply, the Regulations include the following provisions:

1    The duties owed by agent to principal, and vice versa. These largely reflect the position at common law.
2    The remuneration payable to the agent in the absence of provisions in the agency agreement.
3    The agent's entitlement to commission, both during and after the agency agreement has been terminated, including due dates for payment.
4    Notice periods for terminating the agreement.
5    The agent's right to an indemnity or compensation payment following termination of the agreement.

## The Estate Agents Act 1979

The Act empowers the Office of Fair Trading (OFT) to prohibit an unfit person from undertaking certain or all estate agency work, as defined by the Act, or to warn this person that future breaches of obligations imposed by the Act will result in prohibition. Before exercising its powers under the Act, the OFT must instigate an inquiry into the person's fitness. It may instigate such an inquiry if the person has:

- been convicted of an offence involving fraud, dishonesty or violence, an offence under the Estate Agents Act or other specified offences;
- discriminated in the course of estate agency work;
- failed to comply with obligations imposed by the Act; or
- carried on a specified undesirable practice.

A person may appeal against the OFT's decision to issue a prohibition or warning order. Appeal is made to the First-tier Tribunal.

The Act provides that a person who is bankrupt shall not engage in estate agency work except as the employee of another.

The Act requires estate agents to inform clients of the circumstances in which they will become liable to pay remuneration or make other payments (e.g. advertising costs and expenses). Clients must also be informed what the amount of remuneration or other payments will be or what the method of calculation will be.

The Act also requires estate agents to disclose personal interests that they have (or will have following the proposed transaction) in land, before entering into negotiation for its acquisition or disposal.

The Act lays down requirements in relation to clients' money held by an estate agent. Such money is held on trust for the person entitled to it. It cannot be used to pay off the estate agent's creditors if he or she becomes bankrupt. The Act requires an estate agent to keep properly audited accounts.

Clients' money must be kept in a special client account. Payments out of the account can only be made to persons and for purposes specified by the Act. For example, where an estate agent holds money belonging to the purchaser, he or she is entitled to make authorized payments to the vendor.

### Property Misdescriptions Act 1991

Section 1 of the Act creates an offence of making a false or misleading statement about a prescribed matter in the course of estate agency or property development business, otherwise than in providing conveyancing services. The Act provides that a statement is misleading if what a reasonable person may be expected to infer from it, or from any omission from it, is false. A statement may be made orally, in writing, by conduct, in pictures or by any other method of signifying meaning.

Section 2 provides a defence where a person charged with the offence under s.1 has taken all reasonable steps and exercised all due diligence to avoid committing the offence.

Regulations made under the Act specify the matters where misdescription in the course of estate agency business or property development business will constitute an offence. The schedule lists more than 30 specified matters.

Common examples include misdescriptions of location, view, fixtures, size, proximity to services or facilities, history or age of the property, amount of service charges, easements and covenants.

# Sale of goods

A contract for the sale of goods is defined by the Sale of Goods Act 1979 as one where: 'The seller transfers or agrees to transfer the property in goods to the buyer for a money consideration called the price.' The Act covers a wide range of transactions from selling loaves of bread to buying cars. It must be read in conjunction with all the basic rules of contract, as the Act is not a complete code in itself. Its significance lies in the fact that it sets out:

- the terms to be implied into a sale of goods transaction;
- the extent to which those implied terms can be excluded, subject to the Unfair Contract Terms Act 1977;
- the rules for determining when ownership and risk pass from the seller to the buyer in the absence of express or implied agreement;
- the remedies for the buyer and seller where breaches of contract have occurred.

The Act does not apply to hire purchase transactions, nor to contracts of exchange, contracts for work and materials (e.g. installing central heating), or contracts for the supply of services, where separate statutory provisions apply.

## Implied terms

It is quite common for people to enter into contracts to buy goods without ever speaking to the seller, e.g. when buying from a vending machine, or in a self-service store. If asked what were the terms of their contract, they might reply 'none'. However, such transactions are covered by the Act, and the following terms are automatically included for the buyer's protection.

### 1 Quality of the goods

Section 14 (as amended by the Sale and Supply of Goods Act 1994) provides that goods sold in the course of a business must be of satisfactory quality (i.e. fit to be sold). To be of satisfactory quality the goods must meet the standard that a reasonable person would regard as satisfactory bearing in mind any description applied to them, the price paid and other relevant circumstances. Where the buyer is a consumer, the Sale and Supply of Goods to Consumers Regulations 2002 provide that 'relevant circumstances' include any public statements on the specific characteristics of the goods made about them by the seller, the producer or a representative, particularly in advertising or on labelling. The Act goes on to say that quality includes the state and condition of the goods, whether the goods are fit for all purposes for which goods of the kind in question are commonly supplied, their appearance and finish, freedom from minor defects, safety and durability. This implied term is a condition of the contract, but does not apply if:

(a)    Defects were specifically drawn to the buyer's attention before the contract was made. If a buyer buys a washing machine at a reduced price, having been told that the reduction is because the cabinet is dented, he cannot subsequently try to reject the washing machine because of this damage.

(b)    The buyer examined the goods before the contract was made and the defects ought to have been revealed by that examination.

One of the practical problems of applying the rule about satisfactory quality is the question of how long the goods should remain of satisfactory quality. How long should a pair of shoes last? If they leak at the first wearing, clearly they would seem not to be of satisfactory quality. But what if they leak after six weeks? How often did the buyer wear them in that time? In what weather conditions? Whether goods are satisfactory is a question of fact and, in disputes, it would need to be answered on the individual circumstances.

## 2    Fitness for purpose

Section 14 further provides that where goods are sold in the course of a business, and the buyer makes known the purpose for which he or she requires the goods, then the goods must be reasonably fit for their purpose. What must a buyer do to 'make known the purpose' for which he or she wants the goods? In most cases, where goods have one common purpose, and that is what the buyer wants to use them for, then the answer is nothing in particular, as he or she is regarded as impliedly making known the purpose. It is only where the buyer wants to use goods for some unusual purpose that he or she must make that purpose known in order to benefit from this condition. The protection of this rule about fitness for purpose will not apply if the buyer did not rely on the seller's skill and judgment. This might be shown by the buyer specifically asking for goods by their brand name. In such a case, the inference is that the buyer knows that such goods will satisfy this purpose, i.e. by relying on his or her own skill and judgment.

One obvious advantage of these clear duties imposed on the seller of goods is that the buyer has rights against someone whom can be identified and located. In practice, the seller may try to indicate that he or she is not liable and that the buyer will have to take the matter up with the manufacturer. The buyer could possibly have some legal action against the manufacturer as well by virtue of the tort of negligence, under the Consumer Protection Act 1987, or by virtue of a manufacturer's guarantee under the Sale and Supply of Goods to Consumers Regulations 2002, but he or she should not be fobbed off by the seller in this way. The seller is under a strict liability to sell goods which are of satisfactory quality and reasonably fit for their purpose.

In cases covered by s.14 these conditions are important consumer protections, and they limit the operation of the usual principle of *caveat emptor* (let the buyer beware). That principle is still significant in cases not covered by s.14, e.g. private sales of second-hand goods, because s.14 specifically states that in circumstances outside the rules set out above, there is *no* implied condition or warranty about the quality or fitness for any particular purpose of goods supplied under a contract of sale.

### 3  Sales by description

Section 13 provides that if goods are sold by description there is an implied condition that they must correspond with that description. There is obviously a sale by description in any case in which the buyer has not seen the goods, but they have been described to him or her, e.g. goods selected from a catalogue or by reference to a price list. The Act specifically states that there can still be a sale by description even where the goods are displayed and selected by the buyer. However, although some of the older case law suggested that even a minor discrepancy between the contractual description and the goods (for example, in relation to how they were packaged) gave rise to a right to reject for breach of section 13, more modern case law suggests that what must be determined is whether a particular item in a description constitutes a substantial ingredient of the identity of the goods, and only if it does to treat it as a condition. Note also the effect of the provisions in section 15A of the Sale of Goods Act 1979 (see below).

### 4  Implied terms about title

A person buying goods is naturally anxious to become the proper legal owner of them. Section 12 provides that there is an implied condition that the seller has the right to sell the goods, or will have when the time comes to pass ownership. This condition can never be excluded.

### 5  Sale by sample

The mere fact that a buyer sees a sample during the course of negotiations does not necessarily make the transaction a sale by sample. What is necessary is that the sale should be made by reference to that sample. The parties may well indicate this, e.g. when the buyer orders goods 'equal to sample'. In such a case, s.15 then implies conditions:

(a)  That the goods supplied will correspond with the sample.
(b)  That the goods will be free from any defect making their quality unsatisfactory, which would not be apparent on a reasonable examination of the sample.

Whilst ss.13, 14 and 15 imply conditions into contracts for the sale of goods the remedy of rejection of goods for breach of such a term is curtailed in non-consumer transactions. Where a buyer does not deal as consumer (see page 167), s.15A states that if the breach is so slight that it would be unreasonable for a buyer of goods to reject them, then the breach is not to be treated as a breach of condition but as a breach of warranty.

## 6   Implied terms about payment and delivery

Although the parties may not have negotiated specifically about any of the previous terms, it is likely that they will have thought about payment and delivery. This is reflected by s.10, which applies 'unless a different intention appears'.

Section 10 states that stipulations about time of payment are not 'of the essence', which means that any agreement as to when money is due does not go to the root of the contract. In practice, a buyer who fails to make payment on time can be sued for damages for non-payment, but this behaviour will not allow the whole contract to be repudiated.

As to the time of delivery, s.29 makes it clear that if no time is fixed, delivery must be within a reasonable time if the seller is obliged to send the goods. It is important to note that there is no general rule that a seller must send the goods. Whether it is for the seller to send or the buyer to collect is a question to be determined by the contract. In other words, what was expressly or impliedly agreed between the parties? In many cases, the operation of these rules will be straightforward. Suppose that a buyer purchases a deep-freeze cabinet from a shop which promises, as part of the bargain, 'free delivery within 20 miles'. The buyer lives within the 20-mile area, and agrees a day when the goods will be sent by the seller. In this case, by the terms of the contract, the seller is bound to send the goods. If he or she fails to send them on the appointed day the provision about the delivery date may be a major term, the breach of which would entitle the buyer to repudiate the contract. Alternatively, the failure may be a breach of a minor term, merely entitling the buyer to sue for damages. The solution will depend on the weight attached to the term by the parties themselves. Their contract may make it clear that delivery dates are vital and must be complied with.

Unless otherwise agreed between the parties, delivery of the goods and payment of the price are concurrent conditions (s.28). It is very common practice to exclude the operation of this rule, e.g. in cases where goods are bought on credit and a bill is rendered at the end of the month.

## Exclusion of implied terms

Notwithstanding the extensive 'writing-in' of terms into contracts for the sale of goods, the Sale of Goods Act (SGA) provides by s.55 that any or all of these implied terms may be excluded, subject to the Unfair Contract Terms

Act 1977 (UCTA). When considering the relevance of the UCTA to sale of goods transactions, a distinction must be drawn between those buyers who 'deal as consumer' and those who do not. The UCTA identifies dealing as a consumer occurring when:

1   A person dealing as a consumer *does not* make the contract in the course of business.
2   The seller *does* make the contract in the course of a business.
3   The sale is of goods of a type ordinarily supplied for private use or consumption (UCTA 1977, s.12).

The Sale and Supply of Goods to Consumers Regulations 2002 have amended the definition in s.12 UCTA 1977 to provide that, where the person referred to in point 1 above is an individual, point 3 above is ignored. Therefore, an individual may still be dealing as a consumer even if the goods being bought are not of a type ordinarily supplied for private use or consumption.

So, a man buying a washing machine from an electrical retailer deals as a consumer. A contractor buying supplies from a builders' merchant is an example of a buyer not dealing as a consumer.

## One party dealing as consumer (UCTA 1977, s.6)

The rule in such cases is that there can be no exclusion of any of the terms implied by SGA 1979, ss.12–15 (i.e. no exclusion of the terms relating to title, sales by description, satisfactory quality, reasonable fitness for purpose or sales by sample). Any attempt by a seller to exclude or limit this liability in these areas will be ineffective.

## Neither party dealing as consumer (UCTA 1977, s.6)

Although no exclusion or limitation of SGA 1979, s.12 (title to the goods) is allowed, the seller may limit or exclude this liability under SGA 1979, ss.13–15, if the contract term by which the limitation is imposed satisfies the requirement of reasonableness, i.e. if, given what the parties knew or ought to have known when they made the contract, it was a fair and reasonable term to have included. UCTA 1977 contains guidelines for the application of the reasonableness test, some of which are set out below, which suggest that a court should consider:

- the relative strength of the bargaining position of each party;
- whether the buyer received an inducement to agree to the term;
- whether the buyer knew, or ought reasonably to have known, of the existence of the term;
- whether the goods were manufactured, processed or adapted to the special order of the buyer.

## Passing ownership and risk

A contract for the sale of goods is defined by the Sale of Goods Act 1979 as one where the seller transfers or agrees to transfer the property in goods for a money consideration called the price (s.2). In this definition the word 'property' is used to mean ownership. It is important to realize that ownership can pass at a different time from possession. When making the contract the parties could agree to ownership passing immediately, but no delivery of the goods (i.e. transfer of possession) may take place until a later date. It is important to establish the exact time at which the buyer becomes the owner because:

1    The buyer can only effectively transfer ownership to a third party after becoming the owner.
2    As a general rule, risk passes with the ownership, irrespective of whether the goods have been delivered. This means that if goods ought to be insured, the buyer should do so, despite the fact that he has not yet received them. If the goods are lost or stolen at this stage, the loss will fall on the buyer. But the seller who remains in possession is under a duty to take reasonable care of the goods.

The Sale and Supply of Goods to Consumers Regulations 2002 provides that, where the buyer deals as a consumer, goods remain at the seller's risk until they are delivered to the buyer, irrespective of whether ownership has passed. Where the seller is authorized or required by the contract to send goods to the buyer, the Regulations provide that delivery to a carrier is not deemed to constitute delivery to the buyer. Therefore, a consumer buyer will not be liable for loss or damage during transit in these circumstances.

The parties are free to fix for themselves when ownership will pass but if their contract does not deal with this point, SGA 1979 provides rules to cover the various situations that can arise. The following points should be noted:

1    The time when ownership can pass depends on how the law categorizes the goods in question. If they are unascertained (goods mixed with others and not yet separated or identified as fulfilling the buyer's contract) no property can pass until they become ascertained (goods identified in accordance with the contract after the contract is made). For example, if a builder orders a quantity of timber from a timber merchant who has huge stocks, the goods are unascertained until the merchant selects timber of the correct quantity and specification. Once he makes the selection, the goods become ascertained. Goods may also be specific. These are goods identified and agreed upon by the parties at the time when the contract is made.
2    Ownership of specific or ascertained goods passes at the time agreed by the parties. If they do not make their intention clear, the rules of SGA 1979 s.18 will be applied. These can be summarized as follows:
     (a) In the case of a sale of specific goods which are ready to be delivered, ownership passes when the contract is made.

(b) When specific goods have to be put into a deliverable state (e.g. they must be repaired, cleaned, tested or weighed), ownership passes when this has been done, and the buyer has notice to that effect.

(c) In a sale of unascertained goods ownership passes once appropriate goods are set aside in fulfilment of the particular contract, provided that at this stage the goods have become ascertained.

It will be seen that it is better for the parties themselves to fix when ownership will pass, as the operation of these rules leaves a buyer with little control over the timing.

The SGA 1979 s.19 allows sellers to retain ownership of goods until the buyer has paid for them, even though the seller has parted with possession of the goods. This can be particularly useful when a buyer becomes insolvent, as the seller can insist that he or she still owns the goods in question and can take them back. Otherwise the seller would simply rank as an ordinary creditor in a bankruptcy or liquidation, with little hope of being paid in full. Reserving ownership of goods which are now in the buyer's possession is not without practical problems. One of the first significant cases on this point was *Aluminium Industrie Vaassen BV v. Romalpa Aluminium*. Indeed, it was from this case that 'retention' clauses began to be known as Romalpa clauses. AIV had sold aluminium to Romalpa, on terms that ownership would only vest in Romalpa once the goods had been paid for; that if the foil was used in manufacture, the end product would belong to AIV; and that if this end product was then sold, the proceeds would be due to AIV. This clause was held to be valid and was enforced by the court. Their reasoning was based on the fact that Romalpa were not owners but merely bailees under a bailment of the foil. A bailment is a fiduciary relationship, a relationship of trust which would allow the bailor to recover goods, and also the right to trace the proceeds of sale.

This case appeared to hold out important advantages for retention of ownership clauses, but it must be viewed in the context of later decisions, where the court reached a different conclusion. In *Borden UK Ltd v. Scottish Timber Products Ltd*, the material supplied under the contract was resin to be used in the manufacture of chipboard. An 'ownership retention clause' was inserted. When Scottish Timber went into receivership, Borden wanted to trace their resin (which was now in manufactured chipboard) and trace the proceeds of sale of chipboard which contained their product. They failed on both counts. There could be no bailment here as a bailment implies a right of redelivery. Borden were supplying small quantities at short, regular intervals and must have realized that their product was being incorporated immediately into the manufacturing process. If the clause was intended to create some kind of charge over the chipboard, it needed to be registered, and it was not.

The difference between the cases seems to rest on the fact that no mixture took place in Romalpa, whereas that was inevitable in Borden. Romalpa clauses could still be effective, as long as the seller makes clear that the buyer is a bailee, and no mixing occurs; however, it should be noted that claims to trace sale proceeds in subsequent cases have not been successful. Simpler retention of title clauses, where a seller does not attempt to claim products manufactured from their materials or their proceeds of sale, tend to be more effective.

## Remedies

The remedies available for breach depend on which party is in breach, and what type of breach has been committed:

1   *Seller in breach* – where the seller commits a major breach, e.g. by supplying goods which are defective, and not therefore of satisfactory quality, the buyer ordinarily has the right to repudiate the contract (put an end to it), usually by rejecting goods. The seller also has the right to sue for damages. In practice, the buyer may choose not to reject the goods, and indeed in some cases the Act specifies that they cannot be rejected (see s.15A above, page 166). The loss of the right to reject the goods by virtue of having accepted them (section 35 Sale of Goods Act 1979) is likely to happen when the buyer has purchased specific goods (goods which were identified by both parties at the time when the contract was made), where there was a reasonable chance to inspect the goods, and where they have now been accepted. Acceptance can occur through express intimation or through keeping the goods beyond what is regarded as a reasonable time, without indicating their rejection to the seller. In these cases the right to reject is lost, and the buyer must be content with an action for damages.

    The Sale and Supply of Goods to Consumers Regulations 2002 give consumer buyers a new set of remedies alongside those already available. In essence, these new remedies apply where goods do not conform to the express terms of the contract, or those implied by the SGA 1979. The consumer has rights of repair or replacement or, if repair or replacement is impossible or disproportionately costly, to a reduction in price or the right to rescind the contract. In addition, the Regulations provide that if, within six months of delivery, it becomes apparent that the goods do not conform to the contract, it is presumed that this lack of conformity was present at the time the goods were delivered to the buyer.

2   *Buyer in breach* – the most likely situation is that the buyer has failed to pay for the goods. The seller's remedy will then depend on whether he or she is still in possession of the goods. If so, the right of lien can be exercise over the goods (i.e. holding on to them) with a view to ultimately exercising the right to resell them. If delivery has already been made to the buyer, and ownership of the goods has passed to the buyer, then the

seller may sue for the price of the goods. If ownership has not passed to the buyer, for example because there is a reservation of ownership clause in the contract between seller and buyer, the requirements of which have not been fulfilled, then generally the seller may not sue for the price and will instead claim damages for non-acceptance.

# Supply of goods and services

The Supply of Goods and Services Act 1982 (SGSA) relates to contracts under which ownership in goods will pass, where the contract also provides a service element. The Act also regulates contracts for services only. Obvious examples of contracts falling within the Act are builders who supply materials and do the work, and garages who supply parts and undertake the repair. Less obviously, the supply of goods provision may be important where a 'free gift' comes with a contract: 'Buy a luxury fridge/freezer during January and get this electric coffeemaker free.' The coffeemaker is not a 'sale of goods' because there is no money consideration given for it. But it is a supply of goods.

The Act deals separately with the supply of goods and the supply of services. Where the goods are supplied under a work and materials contract, a number of terms are implied:

1   An implied condition that the seller has a right to transfer the ownership of the goods (s.2).
2   Where goods are transferred by description, there is an implied condition that the goods must correspond with their description (s.3).
3   The goods transferred must be of satisfactory quality, and reasonably fit for their purpose (s.4). These conditions apply only where the goods transferred are supplied in the course of business.

These conditions are couched in terms very similar to those used in the Sale of Goods Act 1979. Very similar terms are incorporated by ss.7–9 in contracts for the hire of goods. These are significant nowadays, as hiring cars, office machinery and heavy plant is very common in business.

As usual, the possibility of excluding these implied terms is limited by the operation of the Unfair Contract Terms Act 1977. Generally in consumer transactions these protective conditions cannot be excluded, but in business transactions an exclusion clause may be acceptable if it satisfies the requirement of reasonableness under the 1977 Act (see Chapter 6).

## *Supply of services*

The 1982 SGSA does not define what is meant by a service, except to make clear that the services provided by professions such as architects and surveyors are included. Obvious examples are activities like car servicing, boiler maintenance, dry cleaning and watch repairing. In any of these cases the customer will feel aggrieved if the job is not done effectively, if it costs too much or if it takes too long. In essence, those are the three areas where the Act now prescribes implied terms. It is possible to exclude the terms implied by these sections, but again, only within the limits provided by the Unfair Contract Terms Act 1977. It should particularly be remembered that where, by negligence, a person causes death or personal injury, liability cannot be excluded or limited by a contract term (s.2).

## Exercising reasonable care and skill

Section 13 states that if a supplier of a service is acting in the course of a business there is an implied term that the service must be carried out with reasonable care and skill. One interesting feature of this section is that it does not prescribe whether the term is a condition (major term) or a warranty (minor term). The standard laid down is apparently imprecise, but it means that a builder must exercise the care and skill of a reasonably competent builder, a surveyor must exercise the care and skill of a reasonably competent surveyor, and so on. The standard is thus set objectively. It is no use asking whether the builder did his best, or even whether he tried reasonably hard given his limitations! He must be measured against a mythical, reasonably competent builder. Section 13 merely codifies the existing common law rules, i.e. similar duties would previously have been implied into such contracts by the courts. As soon as the idea of 'reasonable care and skill' is introduced it is clear that there is the possibility of a claim in the tort of negligence too. Usually it is more straightforward to sue in contract, as the relevant duty is owed by virtue of the contract and need not be specifically proved.

## The time for performance

Section 14 provides that there is an implied term that a supplier acting in the course of business, where no time is fixed by the contract, must perform the service within a reasonable time. Establishing what is a reasonable time is a question of fact, which can sometimes be implied from the nature of the contract. If a firm is engaged to come and spray fruit trees when they are in blossom there is only a short span of time within which the contract can be performed.

## The cost of the service

Section 15 provides that where the contract does not fix the charge or provide a method of fixing it, the customer must pay a reasonable charge. What is reasonable is a question of fact.

# Hire purchase, credit sales and conditional sales

It is often the case that a buyer can only afford to purchase goods if payment can be spread over a period. Historically, hire purchase developed at the end of the nineteenth century as a means of allowing people to buy more expensive durable goods like pianos and sewing machines by spreading payments out over a period, and yet achieving a measure of protection for the seller who remained the owner of the goods until the last payment was made. If a buyer fell into arrears with payments, the seller could 'snatch back' the goods. Where a buyer was found to be over-committed financially, he or she could usually terminate the agreement, but often on terms which were disadvantageous. Recognizing these evils, and the potential inequality of bargaining power between the buyer and the seller in such transactions, Parliament passed a series of Hire Purchase Acts, the rules from which are now consolidated in the Consumer Credit Act 1974.

A number of different ways exist for a customer to buy on credit. Traditionally the law provided several types of transaction, which give flexibility about the terms incorporated and which vary in the protections afforded to the customer. The supplier could choose to use:

1    A hire purchase agreement, where the customer hires goods for a period, with the opportunity to exercise an option to purchase at the end of the period. In this type of agreement, the supplier remains the owner of the goods until the option is exercised. The hirer may terminate the hiring agreement without exercising the option, although this would not usually be wise, as it would prove to be a costly way to hire goods. As the goods remain in the ownership of the supplier, he or she has certain rights to seize back the goods if the hirer defaults on payment. In this type of transaction, terms about satisfactory quality and reasonable fitness for purpose are implied in a form similar to the implied terms of the Sale of Goods Act 1979. HP agreements are controlled either by the Consumer Credit Act 1974 or by rules of common law.

2    A credit sale agreement. These extend credit to the buyer of goods, e.g. where it has been arranged to pay by instalments, but the ownership in the goods passes to the buyer immediately. They are governed mainly by the Sale of Goods Act 1979.

3    A conditional sale agreement, where ownership of the goods remains vested in the seller, but the buyer has no right to terminate the agreement. Because credit is extended to the customer, the Consumer Credit Act 1974 may apply.

Of course, these types of contract are not the only ways in which credit may be given. There has been a dramatic rise in the use of credit cards – either those issued by particular stores for use in their own shops, or those issued by credit card companies for use in any accredited outlet of the company. In either of these cases the sum owed can represent many different contracts made by the debtor. Sometimes, by the terms of his agreement with the credit card company, the debtor must repay at regular monthly intervals and may not exceed a fixed credit limit. In other cases the debtor may owe up to a fixed credit limit and may choose whether to settle in full at the end of the month, in which case no interest is paid, or part of what is owed may be paid and interest run up on the balance.

Inevitably, most organizations which are prepared to extend credit do so only for a return. That return is the interest they charge. The law has long believed that rules are needed to protect people from extortionate bargains, and such rules are all the more vital when seen as part of 'consumer protection'. The rules were dramatically altered by the introduction of the Consumer Credit Act 1974. As its name applies, the Act is of special significance to 'consumers', and the protections of the Act range from licensing people and organizations which offer credit to insisting on sufficiently full information being available, so that a customer can establish the cost of the credit. To give the Act 'real teeth' there are criminal as well as civil sanctions.

For the business community making contracts to purchase supplies or equipment the Act may be of limited significance. Its main provisions only operate where the credit extended is less than £25,000 and the customer is not a corporation. Where the Act does apply there are some very important benefits for consumers, not least that the organization which extends credit, e.g. the finance house, is liable in respect of breaches of the implied terms with regard to satisfactory quality and fitness for the purpose. This could be particularly relevant where the actual supplier of goods is not worth proceeding against.

## Other methods of consumer protection

Many of the protections discussed in this chapter can only be fully effective if a person is prepared to bring a court action to enforce his or her rights. But the public is often in fear of the cost of litigation and, especially with smaller cases, there can be a feeling that the time, trouble and effort involved is disproportionate to the size of the claim. An interesting trend in consumer protection is towards state enforcement of duties. Examples of this can be seen in rules such as those covering food and drugs and weights and measures, which create numerous offences for which prosecutions can be brought. Another is the Trade Descriptions Act 1968, which makes it an offence to apply a false trade description to goods. An example of an offence under this Act is that of the car salesman who falsifies the recorded mileage of a car by interfering with the instrument. If a prosecution is brought against him, the court has power not only to fine the salesman, but also to make him pay compensation to the buyer. From the buyer's point of view the advantage of this is that someone else will take the initiative in bringing court proceedings. The drawback, however, is that compensation may not be nearly so accurately assessed as damages would be by a civil court. Moreover, the buyer may not want money, and might prefer to repudiate the contract and hand back the goods.

The individual's best consumer protection comes from dealing with a reputable trader with high standards who wishes to maintain a reputation.

In an effort to improve trading standards generally the Office of Fair Trading was established in 1973, under the control of the Director General of Fair Trading. The Office encourages groups of traders to prepare and abide by codes of practice, and generally seeks to discourage unfair trading practices.

The Consumer Protection from Unfair Trading Regulations 2008/1277 came into force in May 2009. The Regulations implement an EU Directive concerning unfair business to consumer commercial practices and prohibit unfair commercial practices. A commercial practice is defined in the Regulations as any act, omission, course of conduct, representation or commercial communication (including advertising and marketing) by a trader, which is directly connected with the promotion, sale or supply of a product to or from consumers, whether occurring before, during or after a commercial transaction (if any) in relation to a product. A commercial practice is unfair if it contravenes the requirements of professional diligence and materially distorts or is likely to materially distort the economic behaviour of the average consumer with regard to the product. A commercial practice is also unfair if it is a misleading action or omission or aggressive. In all these cases an effect on a transactional decision of a consumer must be shown; however, there is also a list of specific practices which are always unfair, such as making unsolicited home visits or refusing to leave when requested to do so.

Misleading actions include false information relating to the nature of the product and its main characteristics or marketing of a product which creates confusion with any products, trade marks, trade names or other distinguishing marks of a competitor. Misleading omissions include omitting or hiding material information about the commercial practice. A commercial practice is aggressive if it significantly impairs or is likely to significantly impair the average consumer's freedom of choice or conduct in relation to the product concerned through the use of harassment, coercion or undue influence. The Office of Fair Trading has the duty of enforcing the regulations and contravention of the regulations is a criminal offence.

## Insurance contracts

When a businessperson considers the risks inherent in running a business (e.g. that workmen may injure third parties, the workmen themselves may be injured, business premises may burn, a customer may be injured by a defective product) it takes little imagination to realize that potential claims against the business could be enormous. It is true that some organizations are sufficiently large to bear their own risks. This tends to be the case only where the insurance premiums charged are so high that these large organizations find it more economical to meet claims, if and when they arise, out of their own funds. In all normal cases a firm would be wise to secure insurance cover, and in some circumstances the law makes insurance compulsory, e.g. employer's liability and motor insurance.

Commercial insurance is subdivided into fire, life, accident and marine. It is common to speak of life assurance, while using the word insurance in relation to the other classes. The difference is said to be that with assurance, the risk (i.e. death) is bound to happen sooner or later, whereas insurance covers risks which are foreseeable but which may never occur, e.g. a fire or an accident. In practice, the difference is not important. Two vital principles are fundamental to all insurance contracts.

## 1   Utmost good faith

A person who seeks insurance usually knows all the relevant facts. On the basis of what is revealed the insurers can decide whether to accept the risk (i.e. give insurance cover) and how much to charge by way of premium. Because of this inequality in the bargaining position of the parties the law imposes a duty on a person seeking insurance to disclose all the material facts to the insurer. The duty of utmost good faith (*uberrimae fidei*) only requires disclosure of facts which are material and which are known or ought to have been known by the insured.

(a)   *When are facts material?* The answer seems to be when they would influence a prudent insurer as to whether or not he will accept the risk, or as to the amount of the premium. With motor insurance previous motoring convictions are obviously material. In the case of most types of insurance the fact that another insurer has refused the risk is material. In life insurance previous illnesses may be material.

(b)   *Knowledge of the facts.* The duty on the insured is quite strict. He or she must reveal facts that are known or presumed to be known. This duty cannot be avoid by a deliberate show of ignorance. If the facts within the knowledge of the insured actually reduce the risk, or are facts which are common knowledge so that the insurer would be presumed to know them too, then no disclosure is necessary.

These rules about disclosure recognize the possibility that an insured party may be silent despite being under a duty to speak. They are an exception to the general contract rule that silence does not usually amount to misrepresentation. This is normally the case because of the principle of *caveat emptor* (let the buyer beware). Usually it is up to the parties to negotiate their own bargain and if one of them requires information, he or she must ask the other for it. The law does not generally require the other party to volunteer it. Of course, a person seeking insurance may have made positive statements which turn out to be false. In such a case the normal rules of misrepresentation would operate and the insurance company's remedy would depend on the type of misrepresentation, which can be innocent, negligent or fraudulent.

When there has been non-disclosure of a material fact the remedy of the insurer is to avoid the policy. It is irrelevant that the insurer has not asked a specific question on the proposal form. This rule can sometimes operate very harshly in practice. In one case, *Woolcott v. Sun Alliance of London Insurance*, the plaintiff did not complete the proposal form himself. He was buying a house with the aid of a mortgage from a building society. His mortgage application form indicated that the building society would insure the property. The house was subsequently damaged by fire. When the plaintiff sought to claim on the block policy held by the building society the insurers avoided the policy and refused to pay. The grounds for their refusal were that the plaintiff had been convicted for robbery in 1960 and sentenced to 12 years in prison. They regarded this as a material fact which had not been disclosed to them.

## 2   Insurable interest

If people could insure goods or lives in which they had no legitimate interest the result would be a wager or gambling agreement. It is a necessary ingredient of an insurance contract that the insured must have an insurable interest, i.e. some foreseeable financial loss or liability.

(a)   *Life insurance* – it is accepted that everyone has an insurable interest in their own life and a person may be insured against death or personal accident to an unlimited extent, so long as the premiums can be paid. In certain circumstances a person may have an insurable interest in the life of someone else, e.g. a wife has an unlimited insurable interest in her husband and vice versa; a creditor has an insurable interest in the life of a debtor, up to the amount of the debt.

(b)   *Property insurance* – the owner of property has an unlimited insurable interest up to its full value. In other cases, where a person is not the owner, the law allows insurance to the extent of that person's interest or liability. For example, a building society could insure a building for the amount loaned on mortgage. Where goods are concerned a buyer can insure before delivery is taken, or warehouse workers can insure other people's goods entrusted to them for storage.

(c)   *Liability insurance* – where a person can foresee liability (e.g. arising out of their occupation of premises, or use of a vehicle), he or she has an insurable interest. The law does not allow insurance against criminal liability, as that would be contrary to public policy. So a motorist cannot insure against being fined for careless driving.

## The insurance contract

Subject to the two principles discussed above, the usual rules about the formation of a contract apply. The following points should be noted:

1   *Agreement* – this usually results from the person seeking insurance completing a proposal form (the offer), which is then accepted by the insurers.

2   *Consideration* – insurance contracts are not usually made by deed, so consideration is necessary. This usually consists of paying a premium in return for the promise of a payment out whenever the risk occurs.

3   *Form of the contract* – as a general rule contracts of insurance do not need to be in writing. An oral contract could be valid and enforceable. In practice it is usual to record the terms of the agreement in a policy. Should any dispute arise over the risks which are covered, the words of the policy must be construed (i.e. interpreted) to see whether they cover the facts of the case. It is usually said that the words of a policy should be read so as to give effect to the intention of the parties. This suggests that both play a part in drawing up the policy, whereas the truth is that it will usually be a standard form contract drawn up by the insurers. Mindful of this, the courts will construe any ambiguity against the insurers. It should also be remembered that the Unfair Contract Terms Act 1977 does not apply to insurance contracts. In theory, therefore, insurers are free to limit or exclude their liability.

4   *Terms of the contract* – one feature of insurance contracts is that all terms, major or minor, tend to be called warranties. This is particularly confusing because of other meanings which the word can carry, e.g. in the sense of a guarantee. In any contract situation it is always a difficult question to determine whether a statement which has been made is so important that it becomes part of the contract (i.e. becomes a term), or is of lesser importance so that it is merely a representation. Persons seeking insurance may make a number of statements in answer to questions on the proposal form, e.g. stating that they have no motoring convictions, or that they have never had an operation. Insurers have been in the habit of making such statements form the basis of the contract, by including a note on the proposal form (which is then signed by the insured) to the effect that all the answers will be regarded as warranties. This practice puts the insurers into a most powerful position in relation to the insured. Should there be anything wrongly answered on the form, then, even if the insured has been acting honestly, and even if it is immaterial, an insurer can avoid liability on the whole contract by treating the error as a breach of warranty. This tactic has resulted in one judge commenting: 'I wish I could adequately warn the public against such practices on the part of the insurance offices.' Insurance contracts are not controlled by the Unfair Contract Terms Act 1977, but insurers do adhere to a 'Statement of General Insurance Practice' which resolves some of these difficulties and since January 2005, the Financial Services Authority Regulation, ICOBS, provides amongst other things that it is unreasonable for an insurer to reject a claim on grounds of breach of warranty in the absence of fraud, unless the claim is connected to the warranty, which was material to the risk. The 'Statement' and ICOBS is limited in its scope to private consumers. This seems in line with the thinking behind the Unfair Contract Terms Act 1977, in which protection is given mainly to persons dealing as

consumers. The insurance industry has been sufficiently worried about customer complaints that it has cooperated to appoint an Insurance Ombudsman, who will look into customers' grievances.

5    *Duration of the policy* – often an insurance policy runs for a fixed period, usually a year. There are cases where insurance relates to a specific event (e.g. bad weather insurance for a country show) and the protection lapses once the date of the event has passed. Where the policy is of a continuing type, e.g. life assurance, the insurer must go on accepting the premiums even though the risk inevitably becomes greater year by year. With other types of policy, e.g. motor insurance, the expectation is that it will be renewed, but neither party is obliged to make a fresh contract. If renewal does occur, then the duty to disclose material facts becomes relevant again, because each renewal creates a fresh contract.

## Types of insurance
### 1   Employers' liability insurance
Although sensible employers have always tended to have insurance cover for protection against claims by injured workers, problems have arisen in the past when some employers had no such insurance, and were unable to meet claims for damages from their employees because they had become insolvent. Although an employee injured in an accident at work may have claims under the industrial injuries scheme, the law also provides that damages can be claimed from a negligent employer. That claim is now protected by the requirement of compulsory insurance imposed by the Employer's Liability (Compulsory Insurance) Act 1969. The following main points about the Act should be noted:

(a)    All employers (except local authorities, statutory corporations and nationalized industries) are covered by the rules.
(b)    Employers must maintain insurance against liability for injury or disease sustained by employees in the course of their employment.
(c)    The level of cover required is £5 million in respect of claims from one or more employees arising from any one occurrence.
(d)    The insurance must be made with an authorized insurer. This term is defined by regulations.
(e)    The employer's duty is to maintain an 'approved policy'. The effect of this rule is to make it illegal for the insurer to limit its liability in certain ways.
(f)    The rules are enforced by criminal sanctions. The Act provides for fines of up to £2500 for failure to comply.
(g)    The insurance certificate must be displayed and, if required, produced to an inspector of the Health and Safety Executive.

Limitations on the cover provided under the Act are:

(a)    It only protects employees who work under a contract of service or apprenticeship.

(b)   Only those incidents arising 'out of and in the course of the employment' are covered. The same phrase is used in the industrial injuries scheme, where its interpretation and application has not been without difficulties.

## 2   Public liability insurance

The employers' liability insurance rules only oblige an employer to arrange insurance cover against injury to its employees. Liability may be owed to a third party who is injured by some activity of one of the employees, e.g. if one of the labourers negligently hits a pedestrian with a plank of wood. In such a case the principle of vicarious liability means that the pedestrian is likely to bring an action for negligence against the employer. It is not compulsory for an employer to insure against this type of liability (usually called public liability insurance), although it is very common in practice. Despite not being required by law, such insurance is often specified in building contracts.

It is not only employers who need public liability cover. Anyone may cause death, injury or damage while going about his or her daily life. For example, a pedestrian may cause a motorist to swerve and crash, while a property-owner could be liable if a chimney were to fall on a passer-by. However, public liability cover is often provided as part of another type of policy, e.g. under a householder's policy.

## 3   Insurance against damage to property

This type of insurance usually covers fire damage, but will often extend to damage caused by explosion or lightning. The policy may then be further extended to cover other risks to property caused by, for example, storm, flood or impact. This type of insurance is not compulsory by law, and it is essential that a policy-holder should ensure that all the desired protection is included in the policy. For example, a policy could cover damage caused by fire, yet exclude liability if the fire was caused by explosion.

It is now quite common to extend fire insurance protection to cover consequential loss. Should a fire render a building unusable, a firm's business activities could be severely disrupted. Insurance cover can be provided to cover loss of profits and additional expenses necessarily incurred, e.g. renting alternative property.

## 4   Motor insurance

In the case of motor insurance, certain aspects of insurance cover are compulsory. As a minimum a motorist must have a policy which covers:

- Liability for death or personal injury to third parties, including passengers or damage to property caused by, or arising out of, the use of the vehicle.

- Hospital charges (up to a fixed maximum) in respect of the treatment of an injured party and emergency treatment fees.

It is very unusual for such limited cover to be taken out by a motorist. He or she would normally also want cover against:

- Fire and theft in relation to his or her own vehicle.
- Damage to the vehicle and to the person.

The insurance companies do not find motor insurance altogether profitable. Premiums charged naturally vary with the risk, and most insurers will take into account the type of vehicle, its value, the age and experience of the driver, the locality in which he or she lives, the driver's claims record and any driving convictions. When a third party is injured by an uninsured motorist, or a 'hit and run' driver, the victim may claim from the Motor Insurers' Bureau. Every motor insurer must belong to this organization, which operates under an agreement made with the Minister of Transport.

## Claims under an insurance policy

Once the need to make a claim under a policy arises, the question of construction of the contract becomes important. Is the event which has occurred actually covered by the policy? Naturally the insurers will have based their premium on a fixed amount of cover to be provided, and they will not wish to pay out more than is necessary.

If the loss which has occurred is covered by the policy then, in the case of most types of insurance (except life and personal accident), the principle of indemnity operates. This means that the amount paid out by an insurer should be sufficient to cover the loss sustained, but should not allow the insured person to make a profit. However, even in those cases where the indemnity principle is applicable, the following restrictions on it should be noted:

1   The policy itself may provide for a maximum sum insured. This can provide a complete indemnity if the sum insured is adequate, but a property may have been underinsured, and the policy may contain an average clause which will then operate. The object of such a clause is to 'penalize' the person who has underinsured, since this practice is unfair from the insurer's point of view, preventing a person from receiving a proper premium in return for the risk that is borne. Of course, underinsurance is not necessarily caused by deceitful motives. A person may genuinely fail to realize that the rebuilding of a property will cost much more now because of inflation than when it was purchased some years earlier. Generally, average clauses work on a pro-rata basis. If you underinsure by 50%, your claim will be settled on the basis of 50% of the loss sustained. For example, a house

which could cost £40,000 to rebuild at present-day prices is insured for £20,000. If it is totally destroyed by fire, £20,000 is the maximum payable. If damage amounting to £10,000 is caused (one-quarter of the true insurance value) the insured will be paid one-quarter of the sum insured, i.e. £5000.

2    The policy itself may be subject to an excess. It is quite usual for the contract to specify that the insured will bear an agreed proportion of each claim, often the first £25 or £50. In such cases, the insurance does not provide a full indemnity.

3    An insurer may not be wholly liable if the property has been insured with more than one company. In such a case the contribution principle will operate, whereby other insurers who are covering the same risk share in the cost of paying out. The indemnity principle is thus preserved, as the total settlement figure will only cover the insured's loss and not allow this person to make a profit. As with indemnity, the contribution principle is not appropriate in the case of life insurance. Double insurance consists of covering the same risk by more than one policy. This is not the same as reinsurance. As soon as an insurer has taken on a risk there is always the likelihood that it may have to pay out. In cases where a single accident may be very costly, e.g. the loss of a telecommunications satellite, the insurer may wish to spread the risk. In such a case, it can reinsure part or all of the risk with another insurer. If the risk then occurs, the insured claims against the original insurer in full. The original insurer then claims under its own reinsurance policy.

4    When the risk occurs, an insured person may have a right of action for damages against the person responsible for the injury or for damaging property, as well as the possibility of claiming under the offending person's insurance policy. If a motorist has a vehicle negligently damaged by another motorist, the subrogation principle will apply. If the victim decides to claim from his or her insurers they may in turn claim against the negligent motorist. The practical effect of this would be that one insurance company would be claiming from another. Given the thousands of claims which they deal with, this process would be expensive and time-consuming. It is, therefore, common for insurers to make agreements amongst themselves not to claim from one another under the subrogation rules. In the long run, given the amount of business involved and the administrative savings made, this will usually be a perfectly equitable arrangement. This type of agreement between insurance companies results in the 'knock for knock' arrangements found in motor insurance. The practice does not affect the rights of the insured.

## Assignment

A policy of insurance is a chose in action, i.e. property which can only be claimed or protected by taking action in the courts. As a type of property it is valuable in itself and the insured person may wish to use it as security to raise money, or may wish to transfer the benefit of the insurance to someone else. The assignment may be governed by the rules of the Policies of Assurance Act 1867 in the case of life policies. It must be in writing, and the insurers must be given notice. Note that it is the benefit of the policy which is assigned. Assignments of other types of insurance are effected under the rules of s.136 of the Law of Property Act 1925. This requires the assignment to be absolute, made in writing and notified in writing to the insurer.

# 8
# Special contracts – building contracts

## Control of work under a building contract

Many of the examples used so far to illustrate the working of the law of contract have been simple transactions involving small sums of money where the contract is made and performed within a short period of time. Contrast that with a building contract where the work may be massive in scale, taking months or years to complete and costing millions of pounds. The very scale of the work means that many subcontracts will be involved, as well as the main contract. Despite the complexity of a building contract it is judged by the ordinary general principles of the law of contract. It would be uneconomic to draft a complex new set of terms for every building contract and although this type of contract has not been codified by a statute, standard forms of contract have developed in the construction industry. These have been subject to extensive interpretation by the courts, and amendments to the standard form are issued regularly. Inevitably, the terms of any standard form contract may be seen to lean against the interests of one of the parties, unless representatives of both sides played a part in drawing up the original version.

A building contract is one whereby the contractor or builder undertakes work for an employer or building owner. Building works may be controlled by JCT contracts, issued by a Joint Contracts Tribunal made up of bodies such as building trades employers, the architects' association and local authorities. JCT contracts come in a number of forms, with or without quantities or with approximate quantities. These standard form contracts are long and detailed and have been drawn up by bodies representing wide-ranging experience in the building industry. The parties using such a contract may choose to adapt or vary the specific clauses of the contract and they are entirely free to do so, but with the warning that in such a complex document, many of the provisions may be interdependent and both parties should be clear what are the repercussions of the changes they have made.

Various standard forms of contract exist, reflecting different ways of organizing construction projects. For example, under JCT Standard

Galbraith's Building and Land Management Law for Students. DOI: 10.1016/B978-0-08-096692-2.00008-7

Building Contract 2005 the contractor agrees to carry out the construction of the project, whereas under a Design And Build contract the contractor undertakes obligations to design and construct the works. Different standard form contracts thus allocate risks in different ways.

Standard form contracts may have to be redrafted in the light of changes made to the law. For example, JCT Standard Form of Building Contract 1998 (now JCT Standard Building Contract 2005) incorporated changes made by the Arbitration Act 1996 and the Housing Grants, Construction and Regeneration Act 1996.

It has been seen that as a general principle parties are free to enter into a contract of their choosing, agreeing to the terms that they desire. However, the law increasingly seeks to regulate contracts and construction contracts are no exception. We have seen in Chapter 3 that in the Housing Grants, Construction and Regeneration Act 1996 'construction contracts' must contain adjudication provisions. Equally, such contracts must comply with the provisions of the Act in relation to payment.

First, if work under a construction contract is to last for 45 days or more then the party to be paid is entitled to payment by instalments, stage payments or other periodic payments. The parties are free to agree the amounts and intervals of the payments.

Second, every construction contract is to provide a mechanism for determining what payments become due and when, and provide a final date for payment of any due sum. Again the parties are free to agree these matters. An employer is to give notice, no later than five days after the sum was due, specifying the amount paid or to be paid and the method of calculation.

Third, payment may not be withheld after the contractual final date for payment without a party having first given notice of an intention to withhold payment. A notice must state the amount being withheld and the reason for this, and be given not later than the start of the period of notice prescribed. Parties are free to agree the length of notice.

(Note that should parties fail to agree on the above matters or neglect to deal with them then the Scheme for Construction Contracts will cover such matters – above.)

Fourth, should payment in full not be made by a due date and a party fail to give effective notice of an intention to withhold payment then the other party may, after giving seven days' notice, suspend their performance of the work to be done under the contract. Such suspension does not constitute a breach of contract and the period of suspension is to be disregarded in assessing the due date for completion of work.

Fifth, any contractual provision making payment conditional on a payer being paid by a third party is ineffective, unless the third party becomes insolvent. This provision will have a significant impact where a main contractor is only obliged to pay a subcontractor on being paid by an employer. Clearly if standard form contracts contain 'pay when paid' clauses these will be ineffective and the parties are free to agree terms for payment. Again, if parties fail to reach such agreement then the Scheme for Construction Contracts will apply.

The employer who needs to undertake building work may be totally inexperienced in this field. He may feel that he should have professional advice in making and carrying through the contract. It is a traditional feature of large-scale building work that the employer engages advisers such as architects or contract administrators, who will supervise and control the work in progress. Some standard forms are drawn up in recognition of this state of affairs and many tasks within the contract are specifically allocated to the architect. It is important to remember that the building contract is made between the contractor and the employer. The architect is *not* a party to that contract, but has a quite separate contract with the employer and will often have been employed under the Conditions of Engagement published by the RIBA. These are standard terms providing for the amount to be paid to the architect, and setting out the duties and responsibilities in detail. Where several professional advisers such as structural engineers and quantity surveyors are engaged by the employer, it is usual for them to be under the overall control of the architect. The employer may employ all of these professional advisers under contracts of service but it is more likely that they will be in independent practice, and the employer then engages them under a contract for services.

The professional advisers will undertake significant amounts of work before the shape of the final project is determined. They must establish what the employer wants, the means by which it can practicably be achieved, and whether necessary legal hurdles can be surmounted. Then, the drawings and specifications are prepared and tenders sought. After the contract is placed the architect may be required to supervise the work and certify the payments. Inevitably, the supervision and certification will bring the architect into regular contact with the contractor, but the architect is only the agent of the employer, restricted in authority by the terms of engagement with the employer.

Where large-scale building work is undertaken it is common for much of the day-to-day administration to be carried out by a clerk of the works, who is employed by the employer, and not the contractor. The role is significant as this person is regularly on site. The clerk of works is under the direction of the architect and reports anything which is out of line with the terms of the contract.

An architect owes duties to the employer arising expressly and impliedly from the contract, and also by virtue of duties established by the law of tort. Architects are in the same position as other professional or skilled people in that they will be liable if they do not do their work with reasonable care and skill. A failure to show such care could involve design faults, inadequate site examination, delays in furnishing drawings, disregard of building regulations, an injudicious choice of builder, or lack of adequate supervision of the work. Some of an architect's duties are described as quasi-judicial because certain terms of the standard form contract require settling of matters between the employer and the contractor. This aspect will be considered later in relation to certification.

It should be noted that much construction is now done under design and build standard contracts, for example a JCT 'Design And Build' Contract, under which the contractor will be responsible for checking the quality of its own work.

## Tendering for building contracts

Once it is clear what the scope of the building work is to be, the employer or the architect acting on his or her behalf will invite contractors to tender. The costs of preparing the tender (the offer) are significant, and this cost normally has to be borne by the contractor. The tender reflects how keen the contractor is to get the job. Even if he puts in the lowest tender, he is not automatically assured of success. His tender is only an offer and the employer may accept or reject it. Indeed, the employer may reject all the tenders. Price may not be the only factor by which the offers are judged.

Once a tender is accepted, a valid contract can come into existence immediately, but it sometimes happens that the employer makes acceptance subject to conditions such as the execution of a formal contract or the provision of a surety bond. This can be relevant if the contractor starts work at once, and subsequently it is held that a contract never existed. In each case it will be a question of interpreting the precise words used. It must be remembered that the negotiations for building work may be prolonged and it may be essential, for reasons of time constraints, to get the work under way. This sometimes occurs on the basis of a letter of intent, in which one party indicates to the other that it is likely that a contract will be placed, and inviting the other to commence preliminary work. Generally, letters of intent do not create obligations. These letters of intent are also found between contractors and subcontractors, where the subcontractor has tendered for specialized work (e.g. heating and ventilation) and the contractor has used that tender as a basis for his own tender. He may then indicate in a letter of intent to the selected subcontractor that he will place a contract with the subcontractor if he is successful with his own tender.

Where preliminary work has been requested and done on the basis of a letter of intent, such work may have to be paid for on a *quantum meruit* basis (as much as it is worth) if no contract is eventually placed. Where a subsequent contract is made the effect of it may operate retrospectively to cover work done under the letter of intent. *Trollope and Colls Ltd v. Atomic Power Constructions Ltd* is a case where this occurred. The letter of intent was sent in June 1959 and the contract was not finally made until April 1960, by which time the contractors had undertaken work worth a considerable amount. They contended that they should be paid for this early work on a *quantum meruit* basis, rather than on the contract price basis. The court did not agree. It was held that when the contract was finally concluded in April 1960 it contained an implied term that the contract would govern what had already taken place.

## JCT Standard Building Contracts

Although the JCT Standard Building Contract 2005 has nine sections, eight schedules and runs to over 100 pages in length, it is not in itself the entire contract. Incorporated into the contract are 'the contract documents', which include the contract drawings and bills. The contract commences with the 'Articles of Agreement', which contain the essential agreement between the parties. The parties will also sign 'Articles of Agreement', which is the actual contract between them, and they may additionally by deed. The advantage of making the contract in this way is to extend the limitation period for breach of contract claims from six years in ordinary simple contracts to 12 years for contracts made by deed. The Articles are dated, they name the parties, define and locate the building work, and are signed and witnessed. They impose an obligation on the contractor to complete the work under the Conditions of Contract for the sum agreed (the consideration); specify the architect/contract administrator and the quantity surveyor; and include an adjudication clause and an arbitration agreement (should the parties agree to submit disputes to arbitration). Also to be found in the 'Articles of Agreement' are the 'Contract Particulars' which, as the name suggests, contain further details such as the date for possession of the site, the date for completion of the works, the rate of liquidated damages, the length of the rectification period and whether the parties agree to submit disputes to arbitration. See Chapter 3 for comment on the change in approach to arbitration under JCT 2005.

The Conditions of Contract, referred to in the Articles and therefore incorporated into the contract, contain the essence of the agreement. Clause 2 imposes on the contractor the obligation to carry out and complete the works in accordance with the contract documents. 'Completing' for this purpose means practical completion (defined in section 1 of the Conditions), which is the point at which the rectification period (formerly the defects liability period) commences. This is a significant moment because the contractor then ceases to owe certain obligations, e.g. insurance obligations. As the contractor has to operate in accordance with the contract documents he does not himself undertake any design responsibility. He is not, therefore, undertaking that the building will be suitable for its intended purpose.

The Conditions of Contract under JCT Standard Building Contract 1998 were not laid out in a particularly helpful sequence and the JCT contracts were revised in terms of layout, structure, consistency and clarity of language. The following appraisal of the conditions deals with the most important clauses, but it is not intended to deal exhaustively with all of the conditions. Helpfully, the Conditions commence with section 1, a definitions and interpretation section.

## Possession of the site

Clause 2.4 requires the contractor to be given possession of the site on the agreed date of possession. He must then regularly and diligently proceed with the work, bearing in mind that the architect/contract administrator can issue instructions to postpone work, which could in turn provide the grounds for the contractor to claim an extension of time, and for loss and/or expense.

## On-site provisions

There must be access for the architect/contract administrator and his or her representatives at reasonable times (clause 3). The employer may appoint a clerk of works to act as an inspector on his or her behalf. The clerk of works will be under the supervision of the architect/contract administrator but must be facilitated in the performance of his or her duties by the contractor. Directions given to the contractor by a clerk of works are of no effect unless confirmed by an instruction from the architect/contract administrator within two days (clause 3.4). The contractor is obliged by clause 3.2 to constantly keep a competent person in charge on the site. Such a 'site agent' is entitled to receive architect/contract administrator's instructions on behalf of the contractor.

The architect/contract administrator must provide the contractor with all the requisite levels and measurements so that the work can be carried out (clause 2.10). Materials, goods and workmanship must be as described in the Contract Bills and if requested by the architect/contract administrator, the contractor must produce vouchers to prove that materials and goods correspond with the specification. As a further check the architect/contract administrator may issue an instruction for work to be opened up for inspection or to test goods and materials. The cost of this exercise can be added to the Contract Sum, except that if it proves a contractor to be at fault in some way, he must bear the cost.

Clause 2.24 deals with materials and goods on the site. Such unfixed goods become the property of the employer (assuming that the contractor owns the unfixed goods or is a buyer in possession and the requirements of s.25 of the Sale of Goods Act 1979 are satisfied) only when their value has been included in an interim certificate which has been paid by the employer. Before that time their ownership is determined by the usual rules of sale of goods (see page 168). Once ownership passes to the employer the contractor is under a duty to take reasonable care of the goods in order to protect them against damage or theft.

Should fossils or antiquities be discovered on site then clause 3.22 makes provision for action to be taken by the contractor if he makes such a discovery.

## Progress of the work

It has already been noted that clause 2 requires the contractor to carry out and complete the work in accordance with the contract documents. As the work progresses, unexpected technical problems may occur, or it may not be entirely clear what work is intended to be done under the contract. In these cases the ability to vary the contract is important. Under the JCT Standard Building Contract, authority to order variations is vested in the architect/contract administrator who acts by issuing architect/contract administrator's instructions. Needless to say, instructions can be issued for a range of reasons far wider than just ordering variations but it is convenient to consider these items together.

Section 5 deals with variations. These give valuable flexibility to the contract. It is basically sound to include such a clause as it permits the employer to require a variation as of right, instead of having to rely on the goodwill of the contractor or having to create a quite separate contract furnished with its own consideration to render the variation valid and enforceable. The variations clause clearly demonstrates that the employer is vesting the architect/contract administrator with authority to order variations by which the employer will be bound. When the term 'variation' is used, by clause 5.1, it means the alteration or modification of the design, quality or quantity of the works as shown in the Contract Drawings and described in the Contract Bills. It also covers the alteration, addition to, or omission of, any obligations imposed by the employer with regard to limitations on working hours or space, or the execution or completion of the work in any particular order. Note that a variations clause does not give an employer the right to make any changes to the work to be undertaken. An excessive change would not come within the definition of a variation. Section 5 also contains the rules which govern the adjustment of the Contract Sums following the variation of the work.

## Architect/contract administrator's instructions

Architect/contract administrator's instructions generally, and the power to authorize variations, are dealt with by clause 3.10. The contractor is obliged to comply forthwith with instructions issued by the architect/contract administrator, but they must be on a matter where the architect/contract administrator is empowered by the contract conditions to issue an instruction. These instructions can cover a wide range of situations such as variations, antiquities, defects and making good defects, and removal of items not in accordance with the contract. If a contractor does not comply with an instruction within seven days the employer may engage someone else to do that work, pay them and deduct that sum from money due to the contractor. Where a contractor is not clear under what authority an instruction is given, he can ask the architect/contract administrator to specify this authority and, if dissatisfied, the contractor can seek to invoke the dispute resolution procedures under the contract.

Under clause 3 an architect/contract administrator's instruction is only valid if it is in writing, but the rules do provide for verbal instructions to be confirmed in writing within seven days by the contractor to the architect/ contract administrator. If the architect/contract administrator does not dissent from that written confirmation within a further seven days, this becomes a valid architect/contract administrator's instruction. Alternatively, the architect/contract administrator may confirm in writing within seven days or, in cases where a contractor has complied with an oral instruction, the architect/contract administrator may confirm it in writing prior to the issue of the Final Certificate.

## Adjustments of time

Originally this clause in the contract referred to extensions of time reflecting that usually it was a contractor that needed additional time to complete the works consequent upon a specified event occurring during the operation of the contract. The use of the word adjustment still encompasses the above possibility but also includes the contract's duration being reduced. Variations which have been ordered by the architect/contract administrator may be one of the reasons why a contractor needs an adjustment of time. The power for the architect/contract administrator to grant such an adjustment is contained in clause 2.28. This could be vital to the contractor because if he does not complete on time, he is likely to become liable under a liquidated damages clause. Adjustments of time may not be exclusively for the contractor's benefit. If the employer has failed in some of his or her obligations, e.g. failure to allow access to the site at the proper time, he or she may be keen to see the time for completion extended as the employer without this would be unable to rely on a liquidated damages clause if he or she had been the cause of the delay.

Adjustments are sought by the contractor making a written application to the architect/contract administrator, which must set out the relevant events which he considers justify such an adjustment. These 'relevant events' are set out in detail in clause 2.29 and include exceptionally adverse weather conditions, *force majeure* (e.g. acts of God), civil commotion, strike or lock-out, compliance with architect/contract administrator's instructions, e.g. on variations, non-receipt by the contractor in due time of the necessary drawings or instructions, exercise by the government of statutory powers which affect supplies of labour, goods or fuel essential to the proper execution of the work, or failure or delay on the part of a statutory undertaker in pursuance of a statutory obligation.

Not only must the contractor set out the appropriate relevant event when seeking an adjustment of time, but also must specify the expected effects and the period of expected delay in completing. If the architect/ contract administrator decides to grant an adjustment it is done in writing, specifying such later completion date as is considered fair and reasonable. In this adjustment, the architect/contract administrator must state which relevant events have been taken into account and the extent to which has also been taken into account variations previously ordered that have cut down on the contractor's task by deleting work originally specified. If no adjustment or an inadequate adjustment is granted, the contractor may invoke the dispute resolution procedures under the contract.

## Insurance

Clause 6.4 provides that the contractor must maintain insurance in respect of liabilities imposed by clauses 6.1 and 6.2 – that is, liability for third party claims in respect of personal injury and damage to property where the contractor is at fault. Clauses 6.7, 6.8 and 6.9 (together with Schedule 3) determine who should take out an insurance against 'all risks' where the contract relates to a new building. The employer should insure against the specified perils when the work involves alterations or extensions to an existing building. Clause 6.9 requires the insurances to be effected by joint names policies.

It is eminently sensible to include provisions about insurance in the contract as building work necessarily involves risk. Harm could be caused to people coming on to the site, passers-by, neighbouring landowners or workmen employed on the site. Some insurance is compulsory by law, e.g. employers' liability insurance. Liability for damage or injury may lie on the contractor, or on the employer, sometimes directly and sometimes vicariously, and on some occasions the contractor and employer may be jointly liable.

## Payment for the work

One particular feature of building contracts is that they commonly provide for payment for the work to be made in stages. This is vital when the contractor's expenses and outgoings will be considerable and the overall time for completion may be lengthy. To provide for payments at intervals a system of certification is used. Clause 4.9 provides for the architect/contract administrator to issue interim certificates, usually at monthly intervals unless stated otherwise in the 'Contract Particulars', and the employer must pay the sum due within 14 days. The architect/contract administrator can seek interim valuation from the quantity surveyor whenever it is considered necessary to determine the amount due on an interim certificate. Following the decision in *Sutcliffe v. Thackrah*, it would be wise to adopt this course. There, the architect issued interim certificates on which the employer paid the contractor. The contractor was subsequently dismissed by the employer and became insolvent. The employer discovered that the architect had certified defective work and work that had not been done at all. The consequence of this certification was that overpayments had been made. The House of Lords held that the architect could be liable for damages for his negligence.

When issuing an interim certificate the architect/contract administrator establishes the amount due by calculating the gross valuation, less any retention and the amount of any previous payments. Gross valuation covers the total value of the work properly done by the contractor, together with the value of materials and goods on site. The retention will usually be a maximum of 5% (though occasionally a higher figure is stipulated), and is held 'on trust' by the employer until its release. Half the retention is released after practical completion and the rest after the issue of the certificate of making good of defects at the end of the defects liability period. Not all items are subject to retention, e.g. direct loss due to discovery of antiquities, or losses falling under clause 4.23.

Clause 4.23 deals with the contractor's claims for loss and expense where the progress of the work is likely to be materially affected by factors such as not receiving properly requested instructions or drawings in time, or any other of the 'matters' listed in clause 4.24. If the contractor is aware of such losses or expenses he must make written application to the architect/contract administrator who may determine the amount of such loss and add it to the contract sum. A claim by a contractor under clause 4.23 is without prejudice to other rights. For example, if the contractor is basing his claim on the architect/contract administrator's failure to provide him with the necessary drawings, that would be a breach of contract entitling the contractor to sue the employer for damages.

If sums are added to the contract sum by virtue of clause 4.23, then that adjustment can be included in the calculation of the next interim certificate.

## Practical completion and final payment

Once the architect/contract administrator considers that the work is at the stage of practical completion, where the building can function properly and any outstanding work is only minor, he or she can then issue a certificate of practical completion. This heralds the commencement of the rectification period. The JCT Standard Building Contract specifies six months in the absence of an express statement of another time period inserted by the parties. The architect/contract administrator will deliver a schedule of defects to the contractor no later than 14 days after the end of the rectification period and the contractor must then put the defects right at his own cost, unless the architect/contract administrator is prepared to instruct otherwise. (Of course, these instructions can only relate to defects which manifest themselves within this period. Any which become apparent later must be the subject of a claim for breach of contract under all the normal rules of contract, including limitation periods.) Once defects have been made good, the architect/contract administrator issues a certificate of making good defects. The obligation to make good defects under clause 17 is confined to defects caused by materials or workmanship not in accordance with the contract, or frost damage occurring before practical completion, this being deemed to be the contractor's fault. These certificates of practical completion and on making good defects are significant because they operate to release the retention money, half at each stage.

## Non-completion and breach

By the terms of clause 2.4 the contractor must complete the works on, or before, the completion date. By clause 2.31 the architect/contract administrator must issue a non-completion certificate if the contractor does not complete in time, although the time may have been extended by the architect/contract administrator's instructions. Once the architect/contract administrator has issued the non-completion certificate, the employer may, after written notification to the contractor, deduct or recover liquidated damages calculated at the rate stated in the 'Contract Particulars' to the standard form (clause 2.32).

The above provisions relate only to delays. It is more likely that other breaches of contract will occur, some quite trivial and others causing significant loss. As an ultimate sanction, Section 8 provides for determination by the employer or contractor respectively. But it must be borne in mind that there are adjudication and arbitration arrangements available under the contract. Many disputes between the parties during the currency of the agreement could be resolved through those procedures. Other provisions may assist in remedying minor breaches. For instance, under clause 3.17 and 3.18 the architect/contract administrator can order the opening up and testing and, if necessary, the removal of any work consequently found to be defective.

Clauses 8.4–8.8 are principally relevant when the contractor cannot go on with performance or where he has committed major breaches. Should a contractor go into liquidation the employer has a discretion to give notice to terminate the contractor's employment. If no notice is given then some automatic consequences follow – for example, the contractor's obligations under Article 1 and the conditions to carry out and complete the works are suspended. In other situations, such as failure to proceed regularly and diligently, or persistent failure to remove defective work, the architect/contract administrator may serve a notice specifying the fault. If the contractor continues the default for 14 days the employer may send a written notice within the next 21 days to terminate the contractor's employment. If he or she takes such a step the employer would then have to engage other contractors to complete the work. Any extra cost incurred in their engagement can be recovered from the original contractor. Clause 8.2 specifically provides that notice by the employer must not be given unreasonably or vexatiously.

Clauses 8.9 and 8.10 deal with the contractor terminating his employment with the employer. The contractor effects the determination by serving notice on the employer or the architect/contract administrator. He may do so if the employer fails to honour a certificate, or if the work is suspended for various reasons, and for longer than a certain period, stated in the 'Contract Particulars'. The contractor is then entitled to be paid for work done and materials supplied, and for any loss caused to him (clause 8.12).

Clauses 27 and 28 merely have the effect of determining the employment of the contractor. They do not terminate the contract as a whole, and many of the other terms governing the rights of the parties will still be relevant. The ordinary rules of breach of contract will govern all other situations where either party alleges that the other has not complied with their obligations under the contract.

## Subcontracts

In a large building contract the main contractor is likely to organize that some aspects of the work will be performed by subcontractors. He undertakes the main contract with the employer and then in turn enters into separate contracts with all the individual subcontractors. For the employer the advantage of this system is that he or she gets work done by making only one contract, and obligations in respect of the whole job are owed to him or her by the main contractor, who remains responsible for the work which he lets out on subcontracts.

Generally, the main contractor may only subcontract to the extent permitted by his contract with the employer. Subcontractors will be paid by the main contractor and have no rights against the employer because they do not enjoy privity of contract with him or her. The employer can enhance this position by insisting on choosing the subcontractors, in which case they will be known as nominated subcontractors. Although still found, nomination is increasingly rare. Employers seek to avoid the disadvantages of nomination, e,g, renomination provisions and extensions of time, by merely naming, specifying or requiring the use of one (or a small selection) of what the employer considers to be appropriate specialists. Note that JCT Standard Building Contract 2005 has no provisions for the nomination of subcontractors or suppliers.

Of course, while there may be advantages for the employer in having no direct contract with subcontractors, this has inevitable drawbacks, principally the difficulty of incorporating the terms of the main contract into the subcontract. It may be possible to do this, in so far as it is appropriate, by obligating the subcontractor not to perform in such a way that he causes the main contractor to be in breach. If a subcontractor does default on these obligations the main contractor will, usually, be liable. The choice of subcontractors is thus seen to be of vital importance as a contractor will be anxious in case a subcontractor causes delay or becomes unable to complete their contract. If anything goes wrong the contractor may be involved in two legal actions: being sued by the employer under the main contract, and then himself suing the subcontractor under the separate subcontract.

The Joint Contracts Tribunal used to produce standard forms of collateral agreement, to be entered into by nominated subcontractors and employers, thus creating a direct contract between the two where no privity otherwise existed. This was incredibly cumbersome, as there were three contracts – employer and main contractor; main contractor and subcontractor; and subcontractor and employer – with all the attendant problems of knowing who to sue in respect of breaches. In relation to the JCT Standard Building Contract 2005, these contracts have been discontinued.

# 9
# Employment law

## Employment and self-employment

With a workforce in the UK of many millions, the rules governing
employment are of considerable importance. The period of the 1960s to the
1970s saw a massive increase in new employment rights created by Acts
of Parliament. Previously, the law of employment had developed largely
through case law, and it reflected the attitude that employers could 'hire
and fire' at will, with no recognition of the inequality of bargaining between
employers and employees. To some extent that inequality was remedied
by the growth of the power of trade unions. However, between 1979 and
1997 the trade unions were themselves subject to major changes which
curtailed their activities on behalf of their members and such new rights
as employees received were usually the result of EU developments. The
return of a Labour government in 1997 saw a more relaxed approach to the
role of the trade unions in the workplace and the introduction of various
new individual employment rights, not least as part of the government's
'family friendly' and 'work–life balance' policies. The first decade of this
century has been largely notable for developments led by EU initiatives
in the area of discrimination on the grounds of age, sexual orientation and
religion or belief.

Until recently most of the important protections of employment law
were enjoyed by persons employed under a contract of service. This type
of contract can be contrasted with a contract for services which, simply
stated, is the distinction between the status of an employee and that of a
self-employed person (an independent contractor). Increasingly, however,
employment legislation provides rights to all workers irrespective of their
employment status (e.g. rights to a national minimum wage or to paid
holidays). Even so it may still be necessary to distinguish the employee
from the self-employed person because:

1    Statutory rights (e.g. the right not to be unfairly dismissed and the right
     to be compensated for redundancy) only apply to persons employed
     under a contract of service.
2    The tax arrangements for the two groups of workers are different.
     Employees pay tax by means of the PAYE system, whereas the

Galbraith's Building and Land Management Law for Students. DOI: 10.1016/B978-0-08-096692-2.00009-9
Copyright © 2010 by Elsevier Ltd

self-employed pay tax by half-yearly instalments in arrears under a different tax schedule.

3    The duties implied by law between the parties to the contract are different, tending to be more onerous between employer and employee in a contract of service.

It will be seen that many of the reasons for distinguishing between contracts of service and those for services have financial implications for the employer as well as the employee. It may seem cheaper for the employer to encourage his or her workforce to become self-employed, and by that means avoid the increasingly strict statutory control over contracts of service. However, it is not enough for the parties simply to change the name of their relationship. It is a question of fact whether the contract is one of service or for services. In those cases where the courts think a change has been made simply to take advantage of favourable rules they may find no alteration in the true nature of the relationship.

When the courts have to determine which relationship exists between the parties there has been considerable difficulty in establishing a simple test by which the decision can be made. Traditionally the courts used a simple test of control. Could the employer tell the employee what to do *and* how to do it? If so, there was a sufficient measure of control to give rise to the closer contract of service. The drawback with such an unsophisticated test is that it is quite inadequate to cope with more complex employment situations. The test is only concerned with the ultimate possibility of exercising control. The fact that an employer chooses to let an experienced employee get on with the job is irrelevant. Even so, the test proved to be over simple and by the late 1940s the courts were searching for another test. It is only fair to say that even today the control test still gives a reliable answer in simple employment relationships.

In more complex situations the judges thought an answer might lie in asking whether a person's work was done as an integral part of the employer's business. This 'integration' test was not entirely satisfactory, and at times highly artificial. More recently the judges have come to the conclusion that there is no simple question which will provide a conclusive answer. Rather, it is a matter of looking at all the surrounding circumstances, and the modern approach is said to be a mixed or multiple test. Using this approach, the courts will balance the factors for a contract of service against those for a contract for services, considering as they do so a range of factors, including:

- What are the employer's powers of selection and dismissal?
- What measure of control is exercised by the employer?
- What agreement is there about the method and amount of remuneration?
- What arrangements exist for the payment of tax and National Insurance contributions?
- Who supplies tools and equipment?
- Does the employee bear any economic risk in the enterprise?

- Does the employee hire his or her own helpers?
- Does the employee have any responsibility for investment and management?
- Can the employee profit from sound management in the performance of a task?
- How do the parties themselves view their relationship?
- How are the employees usually engaged in the trade or industry?

The mixed or multiple test may in reality be a kind of 'reasonable person' approach. The difficulty in practice is that a situation may arise where the questions listed above give equivocal answers. Consider the example of *Ready Mixed Concrete Ltd v. MPNI*. On the one side, the lorry driver in question had to wear a uniform, obey the orders of a foreman and obey company rules. All of these features of his contract pointed towards it being a contract of service. By contrast, he was buying the lorry under HP arrangements with the company, and had to bear the cost of servicing it. He also paid his own tax and National Insurance contributions. On these facts he could be seen to be taking some risk. When all the circumstances were taken into account the court found that he was self-employed.

It is matterless what the parties call their arrangement although, when the factors are very finely balanced, some assistance may be given by looking at what the parties have chosen to call it. This must be treated cautiously though, as two contrasting cases will show. In *Massey v. Crown Life Insurance Co.*, one of the branch managers approached Crown Life to suggest a change in his status from employment to self-employment. He took advice about the consequences of the change and his new status was confirmed between the parties, and in a letter sent to the tax office. Subsequently the manager had occasion to regret his decision when certain employment rights were no longer available to him. He sought to prove that, in reality, he was still engaged under a contract of service. The Court of Appeal did not agree. Given how carefully the question of change of status had been approached, the statement by the parties that he was self-employed was of great weight and significance. Compare Massey's case with *Ferguson v. Dawson and Partners Ltd*. Ferguson was injured on a building site and wished to sue his employers for breach of statutory duty. To succeed in his claim he needed to show that he was engaged under a contract of service. The evidence showed that Ferguson had been brought along to the site by some of his friends and had been taken on in an extremely informal manner by the foreman, who had informed him that he would be working 'on the lump', i.e. on a self-employed basis. No doubt Ferguson was happy with that situation while everything was going well as he had scope to keep ahead of the taxman and avoid paying appropriate National Insurance contributions! Despite the agreement of the parties to the 'lump' arrangement, the Court of Appeal held that the reality of this situation was a contract of service. It had none of the carefully worked out and evidenced formality of Massey's arrangements, so the view of the parties could hardly be decisive.

Although there are numerous Acts of Parliament conferring rights on employees, none of them gives an exhaustive definition of a contract of service. One usual formulation is to define an employee as a person who works under a contract of employment. That in turn is defined as a contract of service whether express or implied. And as that phrase is not further defined, the rules developed by the judges remain significant.

To a certain extent it could be said the courts follow in the wake of the drafting practices of employment lawyers as they seek to find new ways of ensuring for their client companies that workers are excluded from employment protection legislation. Thus, once it had been established by the House of Lords in *Carmichael v. National Power plc* that a contract of employment required as an irreducible minimum a 'mutuality of obligations' (i.e. that the employee must be ready and willing to work and that the employer had an obligation to provide work), it was no surprise to find employers inserting clauses into contracts that provided that there was no obligation on the worker to turn up or on the employer to offer work (*Stevedoring and Haulage Services Ltd v. Fuller*). In the same way clauses providing that personal service is not necessary but that work can be delegated to a substitute have been held to exclude that relationship from one of employer–employee (*Express & Echo Publications Ltd v. Tanton*). Another way in which employers sought to avoid responsibility for members of their workforce was by hiring via an agency. This raises the issue of what is the relationship between agency workers and the 'client' firms to which they are sent. The traditional view seemed to be that, whatever the relationship between the agency and the worker, the latter was not an employee of the client. There are conflicting decisions in this area but at least two recent cases have held out the view that, at least where there is a long-standing relationship and where a worker is recruited because of specific skills, the worker is an employee of the client, thus defeating the anticipated outcome (*Motorola Ltd v. Davidson* and *Franks v. Reuters Ltd*). Indeed, in *Brooke Street Bureau (UK) Ltd v. Dacas* and *Cable Wireless plc v. Muscat*, the Court of Appeal confirmed that in such cases it would be surprising if the 'end-user' did not have such powers of control and direction over a worker supplied by an agency as to make that worker an employee. However, in *James v. Greenwich LBC*, a subsequent Court of Appeal decision has cast doubt on whether a client of an agency becomes the direct employer simply because there is a long-standing relationship, particularly if the agency relationship still exists and if payment to the worker is made via the agency.

Recent cases have been notable for attempts by 'employers' to insert clauses into their contracts with their 'workers' to the effect that the work did not need to be done by the immediate contractor but might in the alternative be performed by a substitute. On the whole such arrangements have been viewed by the tribunals and courts as being 'shams' seeking to hide the true nature of the relationship and thereby avoid statutory entitlements (see *Redrow Homes (Yorkshire) Ltd v. Wright*; *Redrow Homes (Yorkshire) Ltd v. Buckborough*). In *Protectacoat Firthglow v. Szilagyi* and *Autoclenz Ltd v. Belcher*, an approach which focuses on the reality of the relationship and not on the legal form was confirmed. In the former, on commencement with Protectacoat the claimant was required to enter a partnership agreement with another worker and to supply services to Protectacoat involving the application of exterior textured coating to buildings. Despite terms in the working agreement to the contrary, the claimant and his 'partner' were provided with a company van and the fuel to drive it and with the necessary tools. On arrival at the sites to which the claimant was sent he was instructed to state that he was employed by Protectacoat. Each morning he had to attend the company's yard and return there on completion of each job. It is perhaps not surprising that the Court of Appeal should agree with the employment tribunal that the partnership agreement was a sham.

In light of the difficulties faced by so-called *atypical* workers, the casual or home worker, the worker employed under a zero-hour contract or the agency worker, the Employment Relations Act 1999 introduced power whereby the existing statutory rights given to employees could by statutory instrument be extended to other groups of workers. Should this be done it would achieve for these workers the sort of protection afforded to casual building workers by the court in *Byrne Brothers (Formwork) Ltd v. Baird*. Here the workers, clearly Schedule D labour-only subcontractors, were still able to claim paid holidays when laid off over Christmas and the New Year because the relevant regulations, the Working Time Regulations 1998, applied to all workers. The court took the view that the claimants satisfied the statutory definition of the term 'worker' because they were dependent on the particular firm for work and were thus not in business on their own account. Moreover, they succeeded notwithstanding a clause that specified that a substitute worker could be provided. Under the Agency Workers Directive which must be implemented by December 2011, temporary agency workers who have been 'employed' for at least 12 weeks are entitled to receive basic working and employment conditions on a par with those which would apply to a worker who had been recruited directly.

## Creation of a contract of employment

Although contracts of employment are governed by the ordinary rules of contract, it is important to consider the role and impact of collective bargaining. This is the name given to negotiations between trade unions and employers, usually concerning the establishment of better terms and conditions of employment for union members. Where a union (or group of unions) has established bargaining rights with an employer (or group or federation of employers), their joint negotiations may result in a collective agreement, which in practice fixes the terms of the employment of relevant employees. Such agreements, together with the increased statutory control of employment, now severely limit the true 'freedom of contract' of the actual parties, the employer and the employee. An employee may ask: 'How can an agreement (the collective agreement), to which I was not a party, affect my individual relationship with the employer?' The answer is that the collective agreement itself cannot alter the terms. But such agreement may well become 'incorporated' into the contract, thereby altering it, and have direct force and application to the employer and employee as a result. Two simple methods by which incorporation could occur are:

1 Express incorporation where, by the terms of the original contract, the parties agree to be bound by the appropriate collective agreement for the time being in force.
2 Incorporation by usage, where the employer and employees have acted on the strength of the terms of the collective agreement, e.g. the employer has paid the increased hourly rate referred to in the collective agreement and the employee has accepted the increase.

A significant portion (since 1980 a diminishing portion) of the working population still belongs to various trade unions, but the impact of collective bargaining in the early 1990s was less extensive, with a move towards individual bargaining. Where collective bargaining is still prevalent, it can have an effect on the terms and conditions of everyone in a firm or industry, whether members of the union or not.

Inevitably, a collective agreement will not be an exhaustive statement of all the terms of the contract. Indeed, it has already been indicated that collective negotiation has no place in some contracts. In the higher echelons of employment, where staff have desirable skills and talents to offer, the terms of the contract may be fixed by much more individual negotiation. In either case, many of the terms will be implied into the contract by statute or by common law. There is even scope for terms to be implied by custom. These will be considered in detail later.

## Form of the contract

Apart from exceptional situations, such as that of apprenticeship, there are no rules requiring a contract of employment to be made in writing. However, since 1963, many employees have enjoyed the right to receive a written statement containing the main terms of their contract. The Employment Rights Act 1996 (ERA 1996) requires an employer to issue this statement within two months of the start of employment and it must contain:

1 Names of the parties and date of commencement of employment.
2 The job title and the place of work.
3 Details about pay, including the amount or method of calculating the amount and the intervals at which it is to be paid. (A separate right for employees to receive an itemized pay statement, setting out the purpose and amount of deductions, is provided by ERA 1996.)
4 Details about hours of work and overtime arrangements.

5   Details about holidays and holiday pay.
6   Details of rules covering sickness and sick pay.
7   Details about pensions.
8   Details about disciplinary rules and grievance procedure.
9   Details about the period of notice to be given by employer or employee to terminate the contract.

The ERA 1996 specifically states that if there are no agreed particulars under any heading, that fact must be stated. This rather emphasizes a feature of employment which is not always appreciated. Holidays are usually a question for agreement between the parties, and if no holidays have been agreed, then at common law none are due. However, the implementation of the Working Time Directive 1993 has led to the introduction of paid annual holidays and controls over maximum working hours (see below).

With limited exceptions the information required in the written statement must be contained in a single document. Since the Employment Act 2002 this may in fact be a written contract of employment (or a letter of engagement) and it should be noted that there is an important distinction between a written statement and a written contract, the former providing evidence, though not conclusive evidence, of some of the terms of the contract, the latter binding the parties subject to exceptional circumstances (*Gascol Conversions Ltd v. Mercer* and *Systems Floors (UK) Ltd v. Daniel*). But not all employees are entitled to receive a statement and in particular the following are not eligible:

1   Employees whose employment lasts for less than one month.
2   Employees who ordinarily work abroad.
3   Mariners.

The ERA 1996 requires the information given to the employee under the statement to be kept up to date and any changes to be notified within one month. This does not give an employer a statutory right to vary the contract but merely obliges him or her to notify employees of any lawful changes that may be made to their terms of employment. It would be extremely rare for an employer to have the right to unilaterally vary the contract; normally alterations to terms must be consensual (i.e. agreed by both parties).

If an employer fails to give a written statement, in whole or in part, as required by these provisions or to notify of agreed changes, an aggrieved employee may make a reference of the issue to an employment tribunal. If a dispute arises as to the particulars which have been given, either party may refer the issue to a tribunal. However, it seems that tribunals have limited powers, whatever the dispute, in that they are not empowered to interpret a statement but only to rule on what should have been included on the evidence before it. Moreover, where there is no clear evidence of a relevant term, it has been suggested that tribunals must not 'invent' terms based on what would be reasonable (*Eagland v. British Telecommunications plc*).

If the means of enforcement seem ineffective, some greater protection may be afforded to employees by the introduction in the Employment Act 2002 of a right to receive between two and four weeks' pay where in specified tribunal proceedings an employee establishes a good claim and it is discovered, irrespective of the nature of the claim, that an employer is in breach of the duty to provide a statement. It should be noted that the Act does not introduce a separate free-standing right to claim compensation for failure to provide a written statement.

Unfortunately employment tribunals have no enforcement powers and if an employer continues to default by refusing to comply with the statement issued by a tribunal, an employee will have to enforce his or her terms by a separate action in the county court or raise the matter in a subsequent relevant substantive claim before a tribunal and request the tribunal to exercise its new powers under the Employment Act 2002.

## Duties of the parties

Once a contract of employment has come into existence, the very nature of that relationship imposes certain duties on both parties, whether they expressly negotiated on these points or not. These duties are summarized below.

### Employer's duties

1  *A duty to pay the agreed wages.* Under the National Minimum Wage Act 1998 and its attendant regulations every worker in the UK is entitled to receive at least the minimum fixed rate of pay (£5.93 per hour from October 2010). There are lower rates for those who are under 22 but older than 18, and those under 18 but older than 16. There are detailed and complex regulations which deal with the implementation of this basic right. In the same way the Working Time Regulations 1998 (implementing the 1993 Working Time Directive) contain detailed provisions regulating the maximum working week (generally set at 48 hours), rest breaks, night working and paid annual holidays. The latter measure adopts the same broad coverage as the 1998 Act in that it applies to workers generally (i.e. not being confined to employees but also including those employed under any other contract whereby they undertake to do or perform personally any work or services for another party to the contract not being a client or customer of any profession or business undertaking carried on by that individual). The 1998 Regulations, which have already been subject to amendment, allow for a number of derogations which significantly weaken the protection given by them. Thus, for example, the provisions on the maximum working week do not apply to those whose working hours are not predetermined or may be determined by themselves. The Regulations give managing executives or other

persons with autonomous decision-making powers as illustrations but the vagueness of the provisions may allow for wider groups to fall within this derogation. Enforcement of the rights introduced by this measure is limited, only the provisions dealing with rest breaks and paid holidays being directly enforceable by way of tribunal complaint. The control of maximum hours and night work is prima facie an issue for the Health and Safety Executive (see Chapter 11), although according to the court in *Barber v. RJB Mining (UK) Ltd* an action for breach of contract may arise where an employee has been required to work more than the 48-hour average without his or her agreement. Indeed, the court suggested that such an employee may even work out that he or she has exceeded these working hours for a particular reference period and lawfully stop work until the next reference period commences. Even so, given the various loopholes, there is a suspicion that the 1998 Regulations have made little, if any, difference to the culture of long working hours in the UK.

The amount of wages paid to a worker may also need to reflect the policy of the Equal Pay Act 1970 and the related EU legislation (Article 141, the Equal Pay Directive and the Equal Treatment Directive). The Act implies an 'equality clause' into all contracts of employment (broadly defined to include not only employees but those employed to personally execute any work or labour), whereby men and women who are engaged on 'like work', or 'work rated as equivalent' or 'work of equal value' should receive equal treatment with regard to all terms and conditions of employment, including pay. Comparisons are limited to the treatment of members of the opposite sex employed in the same employment (i.e. employed by the same or an associated employer at the same establishment or at another establishment where common terms and conditions are observed). Even where a claimant has cleared these hurdles it is open to the employer to raise a defence that any different treatment is genuinely based upon a material difference or factor such as age, length of service or qualifications. Claims for equal treatment are brought before employment tribunals which may rule that a woman's contract may be modified or varied so as to be no less favourable in the future than that of her male comparator and may also award arrears of pay which normally may extend back to six years before the claim was brought. (See below Equality Act 2010.)

English law has long concerned itself with *how* wages are paid, rather than how much is paid. The old Truck Acts 1831–1896, regulating deductions from wages and methods of payment to persons employed in manual labour, have now been repealed but in their place the ERA 1996 covers all workers. The Act is largely concerned with deductions from wages for reasons like bad workmanship. Such deductions are permitted so long as they are made under the terms of the contract, and those terms must have been notified to the employee in writing.

Of course, an employer may always deduct those amounts which are permitted or authorized by statute, e.g. tax and social security contributions. There are also court orders which permit deductions from earnings, called attachment of earnings orders. The actual method of paying the wages, e.g. cash, cheque or directly into a bank account, is a matter for agreement between the parties. Once the method is fixed one party may not change it unilaterally without the consent of the other.

2  *A duty to indemnify the employee for expenses reasonably incurred during the employment.* This point will often be covered by express terms about expenses. In other cases, if an employee is out of pocket as a result of his or her work, much may depend on how the expenses arose. Where the employee is doing the work in a lawful way and acting under an employer's authority, he or she should be reimbursed. If, however, the employee chooses an unlawful method of performance, e.g. employed to drive but driving in excess of the speed limit, then the employee must bear the cost. A more specialized aspect of the duty to indemnify is the rule that if an employee commits a tort in the course of the employment, the employer is usually vicariously liable. The full extent of this rule is considered in Chapter 10.

3  *A duty to take reasonable care for the safety of the employee.* This duty is linked with the whole question of health and safety and is considered in detail in Chapter 11.

4  *A duty to show mutual respect to the employee.* This is an example of a duty which has emerged more recently in employment law, largely from suggestions made in cases being heard before employment tribunals, although it is a duty which has now been recognized by the House of Lords. It can be said to mean that employers must treat their employees with appropriate respect and consideration. It is an example of how far the law of employment has moved in recent years from the old philosophy of the employer being able to 'hire and fire' at will. Claims of breach of the implied duty of mutual trust and respect are typically brought by employees alleging constructive unfair dismissal (see below). Such claims may range from a failure to respond to a complaint about lack of adequate safety equipment, or a failure properly to investigate an allegation of harassment, to the imposition of a disciplinary penalty disproportionate to an offence.

It is usually argued that in most cases an employer is under no legal duty to provide work. An old case contains the comment: 'If I pay my cook she cannot complain if I choose to eat all my meals out.' In a number of important ways this rule is of practical significance:

1  It allows an employer to pay wages in lieu of notice. The employer does not usually have to provide work.
2  When there is no work an employer may lay off an employee (N.B. this may only be done without pay if there is a term of the contract to

that effect), although where this occurs over a sufficiently long period the employee may be eligible to claim a redundancy payment.

3    If no work and no pay are available this may amount to the employee being suspended. The usual rule is that an employee may only be suspended without pay if his or her contract expressly or impliedly provides for this. Suspension with pay normally causes no problems because the employer is under no duty to provide work.

4    If an employee has no work and no pay the employer may have a statutory duty to make guarantee payments of fixed amounts for a specific number of days (ERA 1996).

An employer is under no duty to provide a reference or testimonial, even though an employee may suffer in consequence of this decision. This can be particularly harsh when an employee secures a new job which is dependent on satisfactory references. The employer may feel justified in refusing to give a reference because anything he or she writes could open up possibilities of legal liability, not least because under the Data Protection Act 1998 any reference may be accessed by its subject from the recipient. For example, the employee may sue if the reference is seen to be defamatory. Employers should have little to fear on this point, as an action for defamation can be defeated by the defences of truth or qualified privilege, i.e. the employer can show that without malice or bad motive he was simply communicating information to a person who had a genuine interest in receiving it. Nowadays, an employer is more likely to fear that statements in a reference could expose him or her to liability for negligent misstatement. This issue is considered in more detail in Chapter 10. Moreover, an employer may be estopped from denying the truth of statements made in a reference. This can be relevant in cases in the employment tribunal for unfair dismissal. If the employer has written a glowing reference in the hope that a mediocre employee will move on, he or she may subsequently wish to dismiss the employee. In any later proceedings it will be difficult for the employer to deny the truth of that glowing testimonial!

## Employee's duties

1    *A duty to render personal service.*
2    *A duty to render faithful service.* This means that if the employee in any way damages the employer's interests he or she is effectively in breach of duty. Where such a breach occurs there may be justification for the employer to dismiss the employee. Sometimes a breach may cause an employer serious ongoing harm, and he or she may wish to seek an injunction to restrain the particular activity of the employee. In other cases it may also be appropriate for the employer to seek damages from the employee. Some areas of particular importance emerge from this general duty of faithful service:

(a)  Employees should not put themselves in a position where their duty to their employer conflicts with their own self-interest. In *Boston Deep Sea Fishing Co. v. Ansell*, an employee placed orders for supplies on behalf of his employers. He chose to give the orders to firms which rewarded him for putting business their way. Those firms may well have had the competitive edge anyway, but the 'rewards' may have clouded Ansell's judgment. He was dismissed for this breach of duty.

(b)  Employees should not make a secret profit from their employment. This will often be closely linked with the duty in (a). In Ansell's case he had failed to disclose to his employers what he was making 'on the side'. That was quite a crude example of this problem, which crops up regularly in business and commercial life. Employees who are in a position to place large orders for supplies may be courted by hopeful companies. They should take care in accepting any gifts or favours, for they lay themselves open to being in breach of duty, even if their judgement was not affected. Many large organizations tackle the problem by imposing express rules in the contracts of employment, forbidding acceptance of gifts.

(c)  Employees must not compete against their employers. An example of such competition might be the employee working as a boiler service engineer who suggests to customers that they approach him personally in his off-duty hours, and he will then undertake their work on his own behalf. In a case like this the employee could be restrained by an injunction and his behaviour would almost certainly warrant dismissal. The reason here is obvious. His spare-time activities are seriously detrimental to his employer. There is nothing to prevent an employee engaging in a second job as spare-time work so long as it does not interfere with the duties he or she owes to his employers. This rule, that an employee must not compete against the employer, is also relevant when the employment terminates. In appropriate cases, the duty of faithful service can continue to be owed *after* the employment relationship has ended. In such cases it is common to include a 'restraint of trade' clause in the contract, which will be enforceable against the employee so long as it is reasonable (see Chapter 6).

## Patents, designs and copyrights

The issue of competing against the employer can also arise where an employee invents something in the course of his or her employment. It could be very unfair if he or she then leaves the job, and seeks to patent the invention and profit from it. The rules covering such a situation are now contained in the Patents Act 1977 (as amended by the Patents Act 2004). Before the Act was passed, an employer who thought it likely that an employee might invent something would put an express term into the contract, requiring the employee to assign (transfer) the rights in the invention to the employer. This often produced a very unfair result, with the employee personally getting no reward even if the employer made handsome profits out of the invention. The 1977 Act aims to deal with questions of ownership of the invention and the right of the employee to be rewarded. The invention belongs to the employer if it was made by an employee in the course of his or her normal duties, or in the course of duties specifically assigned to him or her (s.39). Any other inventions belong to the employee. If an employer then exploits the invention of an employee, the employee can receive compensation from the court if he or she can show that a patent has been granted which is of 'outstanding benefit' to the employer (s.40). The Act states that the compensation must secure for the employee a 'fair share' of the benefit derived by the employer, bearing in mind the nature of the employee's duties, the effort, skill and advice of others, and the contribution of the employer in terms of facilities and other assistance. Any attempt to reduce the rights of the employee by agreeing otherwise in the contract is absolutely unenforceable. In *Kelly and Chiu v. GE Healthcare Ltd*, the court assessed the fair share to be awarded to the inventor employees at £1.5 million, 3% of the £50 million benefit that had gone to the employer. As *Kelly* demonstrates, one problem with the Act is that an employee may have to wait many years before being able to establish an outstanding benefit. A second difficulty may arise where employees are reluctant to bring compensation claims whilst still working for the employer.

Similar problems can also arise in relation to copyright. The Copyright Act 1956 was repealed and replaced by the Copyright Designs and Patents Act 1988, which developed out of a government White Paper on intellectual property and innovation. Not surprisingly, the period since the 1956 Act had seen much change and progress, especially in the computer field, and new rules were needed to take account of these changes. Where a literary, dramatic, musical or artistic work is made by an employee in the course of employment, his or her employer will be the first owner of any copyright in the work, subject to any agreement to the contrary. The work in question could be sales and promotional literature, or a users' manual, or an installers' guide. It is important to bear in mind that such apparently mundane publications are governed by these rules in the same way as novels and pop songs. So complex is this area that the general public is only likely to be aware of the controversy about whether the Act should have included a levy on blank recording tapes.

Design rights are dealt with slightly differently by the 1988 Act. Designs are protected on principles similar to copyrights but lasting only for 15 years as against the 50-year period for copyrights. The designer is the first owner of any design right in a design, but if the design has been commissioned for a fee, the person who commissioned the design is entitled to any design right in it. Generally, the rules are the same as those for copyright – when

the design was created in the course of the employment the employer will
be the first owner of any design right in the design.

3    *A duty to obey the lawful and reasonable orders of the employer.* For an order to
     be lawful it must not contravene the general law nor must it be contrary
     to what was agreed by the parties. Employees could therefore refuse to
     obey an order which required them to contravene safety regulations,
     or one requiring them to work in another part of the country when
     their contract plainly does not anticipate that they should be mobile.
     It may be more difficult to determine whether an order is reasonable.
     Each case will turn on its own individual facts. Clearly an order is not
     reasonable if it exposes the employee to grave risk. Where the employee
     refuses to obey a lawful and reasonable order he or she is in breach of
     contract, but that may not necessarily be a 'fair reason' for dismissal.
     In *Wilson v. IDR Construction Ltd*, the employee refused, in breach of
     his contract, to move to another construction site. Given his past work
     record, the genuine reason the employee gave for not wishing to move,
     and the employer's failure to allow the employee to explain, it was held
     to be unfair to dismiss him for that reason. Cases like this highlight the
     significant difference between a common law breach of contract and
     the statutory protections for unfair dismissal. 'Technical' breaches have
     to be viewed in the overall context of the contract.

4    *A duty to carry out tasks with reasonable care and to take reasonable care of
     the employer's property.* Again, any behaviour short of reasonable care
     is technically a breach of the contract, but whether it would justify
     dismissing the employee depends on all the circumstances. In any well-
     structured employment situation the disciplinary procedure drawn up by
     the employers should provide for sanctions such as written warnings, so
     that every act of misbehaviour does not necessarily warrant dismissal.

## Statutory rights for employees

The duties previously considered arise by virtue of the employer–employee
relationship. The past 30 years have seen an explosion of new rights and
duties created by statute, but it should be remembered that Acts of Parliament
had played an important role in some spheres of employment for much
longer, e.g. health, safety and welfare, and methods of paying wages.

Statutory rights are mainly granted to employees, thereby redressing to
some extent the inequality of bargaining power between a single employee
and a powerful employer. Some important examples include:

*   The right to a written statement of terms.
*   The right to receive an itemized pay statement.
*   The right, in appropriate circumstances, to receive a guarantee
    payment or a medical suspension payment.
*   The right to a minimum period of notice.

- The right not to be unfairly dismissed, or made redundant, without compensation.
- The right to request written reasons for dismissal.
- The right to maternity leave or maternity pay.
- The right not to be discriminated against.
- The right to paid annual leave.
- The right to paid paternity leave.
- The right to request flexible working.
- The right to be accompanied at disciplinary and grievance hearings.
- Rights in connection with trade union membership.

Most of these statutory protections are confined to persons who work under a contract of service, and each right has a number of qualifying conditions attached to it.

Generally, each right will lay down a period of continuous service which must have been served before the right accrues. These periods vary from one month in the case of guarantee payments to two years for redundancy payments. The concept of 'continuous service' is, therefore, vital to establishing eligibility. In any normal working life there will be times when an employee is not at work, e.g. through holidays or leave of absence or illness. If continuity of service was broken on each occasion, none of these statutory rights would ever be earned. The law provides that where a week is covered by the contract of employment, it will not break the continuity of service.

Not only must the rules establish when there is continuous employment, they must also determine whether the weeks in question count or not. Take the example of someone employed for just over two years who now needs to consider eligibility for a redundancy payment. During that two years the employee in question has six weeks of paid holiday, two separate weeks of sickness, one week's absence with the permission of the employer when the employee's mother died, and five weeks spent on strike. It could be critical to the two years' rule to establish which of these periods count and which do not. In fact, all of them except for the period on strike would be added in. By law, strikes do not break continuity but the time spent on strike (or absent from work through a lock-out) cannot be counted in.

Rules exist to govern the continuity of employment where a business is sold. Where the business is sold as a going concern, employment with the old and the new owner can count as one continuous period. Inevitably the rules about continuity are complex but employees get the benefit of a presumption that their service is continuous; it is for the employer to prove a break in continuity. The rules are contained in the Employment Rights Act 1996 and the Transfer of Undertakings (Protection of Employment) Regulations 2006 (TUPE).

The latter regulations, which were originally enacted in 1981 to implement the Acquired Rights Directive, have in fact played a significant role in the protection of employees faced with the transfer of the undertaking in which they are employed. One area where this protection has been particularly welcomed relates to the 'privatization' of services such as school meals, hospital cleaning, building services or refuse collection. Government policy in the 1980s and 1990s led to such services being put out to tender. Where this has occurred and, indeed, where at the end of the initial contract a further tender has either been won by another contractor or the service has reverted to public provision (a 'service provision change'), TUPE has operated to protect those employed in the service from being forced to accept inferior terms and conditions of employment or from dismissal.

## Protection against discrimination

It is now 40 years since the first of the modern discrimination legislation aimed at protecting individuals from unlawful discrimination was passed by Parliament, and in the intervening period the extent of the protection provided by statute and the attendant case law interpreting these measures has grown exponentially. Thus this legislation included:

- Equal Pay Act 1970
- Sex Discrimination Acts 1975 and 1986
- Race Relations Act 1976
- Disability Discrimination Acts 1995 and 2005
- Part-time Workers (Prevention of Less Favourable Treatment) Regulations 2000
- Fixed Term Employees (Prevention of Less Favourable Treatment) Regulations 2002
- Disability Discrimination Act 1995 (Amendment) Regulations 2003
- Employment Equality (Sexual Orientation) Regulations 2003
- Employment Equality (Religion or Belief) Regulations 2003
- Employment Equality (Age) Regulations 2006
- Equality Act 2006.

In addition one should not forget the impact, albeit indirect, of the European Convention of Human Rights and of the Human Rights Act 1998. As well as the individual rights protected by discrimination legislation, public authorities also had legal duties relating to the main prohibited areas of gender, race and disability. Under these separate duties public authorities had to promote equality of opportunity, good relations and positive attitudes, and eliminate harassment and unlawful discrimination. Codes of Practice were drawn up which gave practical guidance on how to meet the legal requirements of the equality duties. As with other Codes, the contents could be taken into account by tribunals and courts in determining whether a breach of duty had occurred. As a consequence of the Equality Act 2006, responsibility for promoting equality, human rights and an effective legislative framework now falls to the newly created Equality and Human Rights Commission (EHRC). This body also provides advice and guidance and seeks to raise awareness of a person's rights. Its remit extends in fact beyond employment to discrimination in relation to the provision of goods, facilities, services, housing and education, although here it is employment which is of concern. The EHRC has taken over the work and responsibilities previously performed by three separate bodies, the Equal Opportunities Commission, the Commission for Racial Equality and the Disability Rights Commission. However, it has a wider role in that it is charged with protecting, enforcing and promoting equality across seven areas: age, disability, gender, race, religion or belief, sexual orientation and gender reassignment. It also has a role in protecting human rights and promoting good relations in society.

One measure that the EHRC has played a significant part in influencing is the Equality Act 2010, which is due to come into force in October 2010 and which seeks to harmonize all the existing legislation noted above in one Act with a single approach to all the present heads of discrimination, which will in future be referred to as *'protected characteristics'*, specifically: age; disability; gender reassignment; marriage and civil partnership; pregnancy and maternity; race; religion or belief; sex; sexual orientation.

In some cases it may be obvious that a person has a protected characteristic (for example, that a person is a man or a woman) but in others such as disability the Act provides a detailed definition of what amounts to disability. In short a person (P) has a disability if P has a physical or mental impairment which has a substantial and long-term adverse effect on P's ability to carry out normal day-to-day activities. One schedule of the Act, various statutory regulations and a set of guidance notes supplement this definition. The Act has also been seized upon as an opportunity to introduce both new concepts and tidy up existing provisions. It has supposedly been written in plain English such that the man or woman in the street will be better able to understand their rights. Only time will tell whether this last objective has been achieved.

1    As with earlier legislation the 2010 Act provides a definition of discrimination ('prohibited conduct') which may take different forms and then identifies the situations where acts of discrimination because of the protected characteristics are made unlawful. Thus prohibited conduct means:
- Direct discrimination
- Indirect discrimination
- Harassment
- Victimization
- Combined discrimination – dual characteristics
- Discrimination arising from disability and failure to make reasonable adjustments.

These terms are further defined by the Act. Thus the definition of *direct discrimination* states: 'A person (A) discriminates against another (B) if, because of a protected characteristic, A treats B less favourably than A treats or would treat others.' This definition is wide enough to include less favourable treatment of a person (B) because B is associated with a third person who has a protected characteristic, for example a mother who is the carer of a disabled relative. It also covers a situation where B is perceived to have a protected characteristic (for example, a religious belief or a particular sexual orientation) but in fact does not follow that perceived belief or does not have that perceived orientation.

*Indirect discrimination* is, on the face of it, less obvious than direct discrimination. The latter might arise where an employer simply refuses to employ or to promote a person (B) because the employer believes that B, a woman, is not strong enough to do the work or would not wish to do the work because it is particularly dirty. A case of indirect discrimination might involve a more subtle, less obviously discriminatory approach by an employer. The Act defines indirect discrimination as the application by A to B of a provision, criterion or practice which is discriminatory in relation to a relevant protected characteristic of B's. A provision, criterion or practice is discriminatory if:

(a)  A applies, or would apply, it to persons with whom B does not share the characteristic;

(b)  it puts, or would put, persons with whom B shares the characteristic at a particular disadvantage when compared with persons with whom B does not share it;

(c)  it puts, or would put, B at that disadvantage; and

(d)  A cannot show it to be a proportionate means of achieving a legitimate aim.

An example of indirect discrimination might arise where an employer has a policy that employees will be selected for promotion on the basis of their prior willingness to agree to do overtime. B, a woman (a protected characteristic), is not able to do overtime because of childcare commitments. If it can be shown (which is probable) that women are more likely than men to be responsible for childcare, then B and other women are put at a particular disadvantage compared to men. It would then be up to the employer to show that the policy of selecting for promotion on this basis was a proportionate means of achieving a legitimate aim.

*Harassment* is a further example of prohibited conduct and is one that attracted much attention under the earlier legislation. Three types of harassment are defined by the Act. A harasses B if A enages in unwanted conduct related to a protected characteristic with the purpose or effect of violating B's dignity, or creating an intimidating, hostile, degrading, humiliating or offensive environment for B. In determining whether the unwanted conduct has the prohibited effect a tribunal must consider all the circumstances including the perception of B as well as whether B's reaction was a reasonable one. The second type of harassment involves similar considerations to the first, but here the harassment involves any form of unwanted verbal, non-verbal or physical conduct of a sexual nature. Lastly it is harassment under the Act for A to treat B less favourably because B has either rejected or submitted to harassment consisting of the other two types. This is true whether the conduct is that of A or some other person. A is aware that material of a sexual nature is displayed in the work's restroom but does not require it to be removed when B requests that this be done. If this results in the restroom becoming an offensive environment for B, then harassment has occurred. Similarly if B, a junior employee, rejects the advances of his female manager and does not, as a result, receive a pay rise when other employees do at the end of the year, there has been harassment. Indeed, if B had complained about his treatment and did not receive a pay rise or perhaps was not promoted when he might reasonably have expected to be, a further act of prohibited conduct would have taken place, namely *victimization*. Under the Act A victimizes B if A subjects B to a detriment because B does a protected act or A believes that B has done, or may do, a protected act. Protected acts include bringing proceedings under the Equality Act or alleging that A or another person (B's manager in the earlier example) has contravened the Act.

*Combined discrimination – dual characteristics* adds to the prohibited conduct covered by the Act by providing a potential remedy for a person who suffers discrimination on more than one ground. Thus, if B is treated less favourably by A because she is a woman and a Muslim, B will be able to bring a combined claim rather than have to proceed to establish separate claims. However, marriage and civil partnership, and pregnancy and maternity, which are protected characteristics in the Act, are not potential grounds of combined discrimination.

*Disability discrimination* – in addition to the prohibited conduct outlined above there are two other claims that might arise in relation to a disabled person. Discrimination takes place where A treats B unfavourably because of something arising in consequence of B's disability and A cannot show that the treatment is a proportionate means of achieving a legitimate aim. However, A will not be liable if A did not know, and could not reasonably have been expected to know, that B had the disability. A further duty arises for an employer to make reasonable adjustments in relation to disabled workers where (i) a provision, criterion or practice, (ii) a physical feature or (iii) the lack of an auxiliary aid puts them at a substantial disadvantage.

2    Once an act of prohibited conduct (one of the types of discrimination as defined above) is established, the next stage is to identify which acts of discrimination are rendered unlawful by the Equality Act. So far as employment is concerned, an employer must not discriminate against an applicant because of a protected characteristic in selection arrangements, as to the terms of a job offer or by refusing to offer a job; and once in employment an employer must not discriminate in access to promotion, transfer or training or in receipt of other benefits, facilities or services. An employee dismissed or subjected to any other detriment because of a protected characteristic could also complain under the Act. In the same way an employer must not victimize or harass a job applicant

or an employee. An employer may also be liable for harassment of its employees by a third party if it fails to take such steps as would have been reasonably practicable to prevent that harassment. Thus, A is the main contractor on a building site where C works as a subcontractor. B, an employee of A, is gay. When C discovers this, B is subjected to abuse and cat-calling. If A fails to take steps to deal with C's behaviour, it is likely that A would be liable in any complaint brought by B. However, A would only be liable if B has been subjected to such behaviour, whether by C or anyone else, on at least three occasions.

3    Exceptionally, where an unlawful act of discrimination has occurred, the employer (A) may be able to excuse itself by establishing that, given the nature or context of the work, having a particular protected characteristic is an occupational requirement. A would also need to show that specifying the requirement was a proportionate means of achieving a legitimate aim and that the claimant (B) does not meet the requirement. Thus A, a fashion house, may legitimately specify that B, a male, does not meet the requirement to be female to model its dresses.

4    Where a person wishes to pursue a claim for sex discrimination it is possible to complain to the EHRC, who may carry out an investigation into an employer's policies, and who may give advice and assistance to individual claimants. In exceptional cases, where a major point of principle is at stake, the EHRC may finance the bringing of a case. This could be especially useful when it is borne in mind that individual claimants are not eligible for legal aid for employment tribunal cases.

The more usual enforcement procedure is for an aggrieved person to commence proceedings in the employment tribunal, usually within three months of the date of the act of discrimination. There is no upper limit on the amount of compensation that may be awarded. At one time a serious problem facing the claimant was that he or she had to bear the burden of proving discrimination. However, this has not been the case for most employment claims since 2001 and the EqualityAct extends that position to all such claims to provide that, once a claimant has proved facts indicating that an act of unlawful discrimination has occurred, it is for the employer to prove that it did not commit that act. B, a qualified surveyor of African descent, applies for a post with employer A. B is not short-listed but subsequently discovers that the only successful applicants are white and that they have less experience and fewer qualifications than he does. If B can establish these facts, A must at least respond otherwise a tribunal will find against it.

Where discrimination is related to the amount to be paid for the work, the issue would be resolved under the Equal Pay Act 1970. That Act provided for equal treatment regarding terms and conditions whenever men and women were doing like work, work rated as equivalent or work of equal value. An employer could pay at different rates only if it could show that it was doing so because of a 'genuine material difference' which was not just the difference in sex. The employer might have pointed to factors like length of service, previous experience or additional qualifications. Equal pay disputes were also resolved in the employment tribunal. The Equality Act re-enacts the 1970 Act with very little change. Comparisons under the new Act are with a colleague of the opposite sex doing like work, work rated as equivalent or work of equal value. An employer can defend differences in pay if it can establish a 'material factor' which is not the difference in sex. If the material factor is shown to be indirectly discriminatory on the grounds of sex, an employer can still justify this but only on the basis that this factor is a proportionate means of meeting a legitimate aim. Such a defence might cover a bonus payment paid to night-shift workers who turn out to be predominately male because female employees with childcare responsibilities do not wish to do the antisocial hours. One measure designed to improve the success rate for claimants is the outlawing of secrecy clauses in employment contracts. Thus, being involved in discussions about a colleague's pay becomes a 'protected act' for victimization purposes. Pay transparency is also to be improved by requiring employers with 250 or more employees to publish an annual pay audit, although this obligation is unlikely to arise before 2013 for private sector organizations. Public sector bodies will have a lower employment threshold and an earlier commencement date.

## Termination of contract of employment

The parties to a contract of employment do not usually fix a date for its termination, unless they have negotiated a fixed-term contract. If the duration of the contract is not fixed, the law provides for either party to end the contract by giving notice to the other. Many cases are now covered by rules providing for fixed minimum periods of notice, although it is still open to the parties to agree longer periods if they wish. The ERA 1996 provides that if an employer wishes to terminate, he or she must give a minimum of one week's notice for each year of continuous employment, up to a maximum of 12 weeks' notice. If an employee wishes to terminate he or she must give a minimum of one week's notice, although the contract may specify a longer period. If an employee fails to give the proper period of notice he or she is in breach of contract. Theoretically he or she could be sued for damages by the employer. This may seem a little unlikely, but it should be remembered that the employer may be holding moneys due to an employee, e.g. accrued holiday pay, bonuses or expenses, or a week's pay lying on. The employer could certainly delay making those payments to the employee. What is more, the employer could also refuse to give a reference.

There are no particular legal rules about how notice must be given to be effective. This depends on the terms of the contract. It is common to require notice to be in writing, and often the contract will specify that the period of notice must be calculated by reference to a pay day. So a weekly-paid employee who normally gets paid on a Thursday must give notice to expire on a Thursday.

Nothing in these rules prevents either party from terminating with no notice at all, if the behaviour of the other party justifies such action. When an employee receives no notice, this is referred to as summary or instant dismissal. The question will often arise whether such behaviour on the part of the employer is appropriate. An employee who is the victim of such action may be able to pursue a claim for unfair dismissal.

## Unfair dismissal

A statutory right to claim for unfair dismissal was introduced in 1971. Before then, an employee who had been dismissed was limited to a claim against his or her employer in a civil action in the county court for wrongful dismissal. In many cases, such an action was worthless because the county court could only award damages, and these were often limited to an amount equal to the ex-employee's wages for the period of notice he or she should have received. If proper notice had been given, there was little point in an employee bringing an action for wrongful dismissal. However, this common law remedy is still available and may sometimes be useful. This is particularly likely to be the case where the employee is highly paid, as there is no maximum limit on damages for wrongful dismissal, or where the employee is outside the three-month time limit for pursuing a claim of unfair dismissal, or where the employee does not yet have a sufficient period of continuous employment to be within the statutory unfair dismissal rules.

Prior to 1971, even where the proper notice had been given, an employee might still feel a sense of grievance about the dismissal, especially if there had been no good reason for it. The common law attitude at that time still reflected attitudes that the employer could 'hire and fire' at will. A long-serving and loyal employee could be dismissed for no reason at all, and would have no claim, so long as the employer had given the appropriate period of notice. Unfair dismissal protection was created to change this situation. An employee now has a statutory right not to be unfairly dismissed. An aggrieved employee can seek a remedy in the employment tribunal. The remedy emphasizes an idea which has developed in employment law, that an employee has a 'property right' in the job. If this is lost when the employee is not at fault (e.g. unfair dismissal or redundancy), then he or she should be protected or compensated.

The current rules on unfair dismissal are to be found in the Employment Rights Act 1996. Employees who may claim their protection are those working under contracts of service, except:

1   Persons employed outside Great Britain (the test for which is not entirely clear, as can be seen from *Lawson v. Serco*).
2   Persons employed under an illegal contract.
3   Persons who have not yet served one year continuously with their employers.

## Definition of dismissal

To be able to proceed with a claim for unfair dismissal, an employee must show that he or she has been dismissed. There can sometimes be doubt whether a dismissal has occurred. Take the case where an employee simply walks out, or an employer asks the employee for their resignation and this is reluctantly given. Section 95 defines dismissal as:

1   Termination of the contract by the employer with or without notice, or
2   Where employment is under a limited-term contract and that term expires without being renewed under the same contract, or
3   Termination by the employee in circumstances justified by the employer's conduct. This situation is called constructive dismissal. The employer's behaviour must amount to a breach which is sufficiently grave so that it entitles the employee to be treated as discharged from further performance of the contract. Not every breach by an employer can be regarded as so significant. Much will depend on the factual circumstances, but a simple example is the case where the employer proposes to reduce an employee's wages without negotiation. It will always be a question of fact whether a constructive dismissal has occurred.

These cases pose a real dilemma for employees. They are faced with some kind of breach by the employer (e.g. the unilateral imposition of a new shift system) and if they leave, there is the complication of needing to prove constructive dismissal. If employees delay taking any action the employer could argue that they had impliedly consented to the change by staying on and working under the new arrangements. To a certain extent the development of the implied duty of mutual trust and respect has assisted employees faced with unjustified actions by their employers. Tribunals have accepted that conduct which is in breach of that duty justifies an employee leaving.

## Fair reasons for dismissal

The Act has given employees who are within its scope the right not to be unfairly dismissed, and lists the reasons why a dismissal may be fair (s.98):

1   *Reasons connected with the employee's capability or qualifications.* Capability is broadly defined to include skill, aptitude and health, amongst other things. Qualifications includes degrees and diplomas as well as technical or professional qualifications relevant to the particular job.
2   *Reasons connected with the conduct of the employee.* Misconduct may range from very trivial acts to acts which can be classed as gross. Inevitably, these will vary in different employment situations. Swearing may be regarded as fairly trivial on the shopfloor of a factory or construction site but may be regarded as extremely serious on a hotel reception desk or in the offices of a professional person. Generally, the more senior an employee, in terms of age and position in the firm, the

more the employer can expect from him or her. The misbehaviour may take place at work, but in appropriate cases, an employer may dismiss because of conduct occurring outside work, where the conduct could have an adverse effect on the employer's business.

3    *Retirement of the employee.*

4    *Redundancy, where the employee was fairly selected for redundancy.* The ideal situation for employers faced with inevitable redundancies is to deal with the problem by reference to an established procedure, agreed at some previous time with the appropriate unions. Otherwise, employers may be able to show they acted fairly by applying the principle of 'last in, first out'. This approach may still leave employers having to choose some from a group who all commenced work together. In that case, they could take account of experience and ability, and the hardship likely to be caused by the redundancy. Employees who think they were unfairly selected can claim unfair dismissal.

5    *Where the continued employment of the employee would cause the employer to be in breach of a statutory duty.*

6    *Some other substantial reason justifying dismissal.* It is difficult to spell out in detail in the Act all the circumstances where dismissal can be fair, so some residual provision is inevitable. Under this heading, a number of cases have involved employees unreasonably holding out against variations to their contracts of employment. The employer cannot unilaterally change the terms of employment. Indeed, to attempt to do so would be a breach of contract. But where the commercial needs of a business dictate that it must be run in some different way, the employer may need to approach employees to seek their approval for altered manning levels, or new shift patterns. In such cases, if the employer has set about negotiating the changes in a reasonable way, and one employee holds out against the change, the employer may have little alternative but to dismiss that employee.

Establishing a reason capable of being fair is not sufficient in itself. The employer must also prove that it is fair in all the circumstances of the case to use that particular reason for justifying dismissal. For example, swearing at the employer is a reason capable of being fair, but whether it is fair to dismiss for swearing in any particular case would depend on the words used, the circumstances surrounding the offence, any provocation, previous treatment of employees for swearing, any works rules, any warnings received by the offending employee, and any other relevant factors. The key question facing the employment tribunal is to determine whether the employer behaved as a reasonable employer would have done. In many cases this will involve the tribunal asking three questions:

- Did the employer carry out a reasonable investigation?
- Did the employer follow a reasonable procedure?
- Was the sanction of dismissal a reasonable one in all the circumstances?

As far as reasonable investigations are concerned it is a well-settled principle from the case of *British Home Store Ltd v. Burchell* that an employer must establish the fact of the belief that the employee had been guilty of an act of misconduct (e.g. swearing at a manager or customer), must show reasonable grounds to sustain that belief and there must have been a reasonable investigation to justify that belief.

With respect to the question of reasonable procedure one factor which may be relevant is the extent to which the employer followed its own disciplinary practices and procedures. These should form part of the information given to employees in their written statement. A sensible employer will have disciplinary procedures suitable for the working situation which are in general conformity with the Code of Practice on Disciplinary and Grievance Procedures (2009) issued by ACAS (Advisory, Conciliation and Arbitration Service). The House of Lords has confirmed that, in unfair dismissal cases, an employer cannot ordinarily excuse a failure to follow a fair procedure by claiming that the outcome, the employee's dismissal, would have been the same even if it had done so – *Polkey v. A. E. Dayton Services*. The impact of *Polkey* was later overturned by statutory changes but has in effect been reinstated now those changes have themselves been repealed (see below).

As regards reasonableness of sanctions it has been laid down, in what has become known as the 'range of reasonable responses' test, that there is a band of reasonableness within which one employer might reasonably dismiss an employee whilst another would quite reasonably keep him or her on. It depends entirely on the circumstances of the case whether dismissal is one of the penalties which a reasonable employer would impose. If no reasonable employer would have dismissed, then the dismissal is unfair. But if a reasonable employer might reasonably have dismissed, then dismissal is fair. This approach is a long-standing one and was confirmed by the Court of Appeal in *Post Office v. Foley* in 2000. In essence it is an attempt to ensure that tribunals do not substitute their own view of what is reasonable for that of the employer. It is seen by some as allowing too much managerial discretion to employers at the expense of employees.

## Disciplinary rules

The existence of a right to claim, coupled with the possible cost of the claim to the employer if the employee succeeds, has forced employers to give more detailed consideration to their internal disciplinary and grievance procedures. These procedures must be communicated to employees who are entitled to receive a written statement.

When drafting disciplinary rules, employers can seek guidance from ACAS. The rules should aim to encourage better standards in employees, and should not be concerned exclusively with imposing sanctions. Where sanctions are necessary it is better to provide a range appropriate to the different forms of misbehaviour. Not all misconduct necessarily warrants instant dismissal. It may be more appropriate for an employee to receive oral or written warnings, or some other form of 'punishment', such as demotion.

The ACAS Code of Practice on Disciplinary and Grievance Procedures stresses that disciplinary rules and procedures are necessary for promoting fairness and transparency in the treatment of individuals. They also help the organization to operate effectively. Employees must know what standard of conduct is expected from them. The responsibility for formulating a policy for maintaining discipline lies with the management, who should aim to secure involvement at all levels when formulating new rules. Rules should not be so general as to be meaningless, but rather they should specify clearly and precisely what is necessary for the efficient and safe performance of work, and the maintenance of satisfactory working relations. Employees should know the likely consequences of breaking the rules, particularly where such a breach is likely to lead to summary dismissal. Disciplinary procedures should be viewed as a means of encouraging improved conduct in individuals. Great emphasis is laid on fairness in operating the procedures. This means telling an individual what the complaint is, and allowing him or her the opportunity to state their case before a decision is reached; allowing the employee the right to be accompanied by a trade union official or a fellow employee; ensuring that no decision is reached until the case has been fairly investigated; and ensuring that the employee is given an explanation for any penalty imposed. An employer should keep proper records of all actions taken under the disciplinary rules. Good employers will normally provide for breaches of the disciplinary rules to be struck from an employee's record after completing a further specified satisfactory period of service. Employees should be allowed to appeal. Under the Employment Act 2008 a tribunal may increase an award made to an employee by up to 25% if the tribunal considers that the employer has unreasonably failed to comply with the ACAS Code of Practice. An award may be reduced by a similar amount if an employee unreasonably fails to comply with the Code.

## Statutory dismissal, disciplinary and grievance procedures
In response to concerns about the increasingly expensive and time-consuming nature of employment disputes, the Employment Act 2002 and accompanying regulations introduced statutory minimum dismissal and disciplinary procedures from October 2004. The new procedures became implied terms of every contract of employment. However, dissatisfaction with these statutory procedures led to their repeal as from April 2009 by the Employment Act 2008. The law in this area has thus returned essentially to where it was before October 2004 as set out above.

# Remedies for unfair dismissal

Claims by an employee under the unfair dismissal rules are made to an employment tribunal. An action should usually be commenced within three months of the effective date of termination of the employment. Claims are initially referred to a Conciliation Officer of ACAS, whose duty it is to mediate between the parties to see if a settlement can be reached. At this stage the employer may agree to take the employee back, or may agree to pay a sum by way of compensation which is acceptable to the employee. A large proportion of claims are resolved at this stage. Since 2001 it has in fact been possible for the parties to agree to forego their rights to a tribunal hearing and to accept compulsory arbitration via an ACAS appointed arbitrator.

Before a full tribunal hearing takes place, it may be appropriate to hold a pre-trial assessment to consider whether the case has any real substance. The case is not decided at this stage but if it is clear that one party has no real chance of success, the tribunal may warn that party that costs could be awarded against them. Not surprisingly, a number of applications to the tribunal are withdrawn at this stage. If no withdrawal occurs, the case proceeds to a full hearing. An employee found to have been unfairly dismissed may be awarded:

1    *Reinstatement.* The employee's job is restored and he or she is treated in all respects as though no dismissal ever occurred.
2    *Re-engagement.* This gives the employee a job comparable to his or her former job.
3    *Compensation.* All awards consist of two elements, the *basic* award and the *compensatory* award. The calculation of the *basic* award depends on a combination of factors, namely age, length of continuous service and the amount of a week's pay. An employee's working life is divided into three periods: years worked between 18 and 21, years worked between 22 and 40, and years worked over the age of 41. A maximum of 20 years of employment is taken into account and a financial cut-off point on what constitutes a week's pay is fixed. This is increased from time to time to take account of rising wage levels. Once the relevant facts are known, the basic award is calculated by awarding half a week's pay for each year of service between 18 and 21, one week's pay for each year of service between 22 and 40, and one and a half weeks' pay for each year of service over the age of 41. The basic award may be reduced if the employee has caused or contributed to his or her dismissal, or if the employee has been guilty of bad conduct before dismissal which is only revealed after the dismissal takes effect. The *compensatory* award is an amount which the tribunal considers just and equitable bearing in mind a claimant's losses and expenses, including loss of future earnings. The calculation is often speculative, since it is impossible to know how long a claimant may remain unemployed. The limit on the maximum amount of a compensatory award is reviewed annually. This part of an employee's compensation may be reduced where his or her conduct has contributed

to the dismissal, or where he or she has not taken reasonable steps to minimize losses, e.g. a failure to seek new employment. The maximum figure for the compensatory award may be increased exceptionally where additional awards can be made. An additional award is relevant if an employer refuses to comply with an order to reinstate or re-engage the employee. The amount could be between 26 and 52 weeks' pay, over and above the basic and compensatory awards.

Note that both reinstatement and re-engagement are discretionary remedies. The tribunal must consider the wishes of the parties and the practicability of compliance before making such an order. Where either order is made, if an employer refuses to comply with it they will be obliged to pay an additional amount of compensation.

## Tribunal procedure

Procedure at the tribunal hearing may be relatively formal, perhaps as a result of the 'adversarial' nature of these proceedings. The tribunal consists of a legally qualified chairman (an 'employment judge'), who sits with two lay members who are intended to be representative of employers and employees respectively. All three play an equal part in the decision-making process. Evidence is given on oath. Although the tribunal may take a more informal approach than a court, it is usually better to bring witnesses who can speak directly about the facts than to rely on hearsay evidence. Applicants should bring along relevant documentary evidence, such as their written statement of terms, or their written reasons for dismissal, or documents relating to their financial position which may be relevant when calculating the compensatory award. It may be difficult to persuade people to come to give evidence, or to get hold of documents considered vital to the case. In these circumstances an applicant can ask the tribunal to grant a witness summons order, or a discovery of documents order.

The parties will each be invited to present their case and to call any necessary witnesses, who may be cross-examined and who may also be questioned by members of the tribunal. Where an applicant is unrepresented, the members can do much to assist by the type of questions they put. The length of each case will depend on factors such as the number of witnesses and the complexity of the evidence, but it will not normally last more than two days. The tribunal retires to consider its decision, which is given to the parties, with full reasons.

Although large sums of money are at stake, and the law to be applied is complex and constantly changing, no legal aid is available to pay for representation in the tribunal. A claimant who is dissatisfied with the outcome of the hearing may have a right to appeal. Appeals on a point of law lie to the Employment Appeal Tribunal, then in turn to the Court of Appeal and the Supreme Court.

## Redundancy payments

A dismissal is due to redundancy if:

1    the employer has ceased to carry on business for the purposes of which the employee was employed; or
2    the employer has ceased to carry on business in the place where the employee was employed; or
3    the requirements of the business for employees to carry out work of a particular kind have ceased or diminished.

Where a redundancy situation exists, there is no entitlement to a payment if the employee is dismissed for misconduct, or if he or she unreasonably refuses suitable alternative employment which has been offered in accordance with the ERA 1996. Employees who have been given notice because of redundancy are entitled to reasonable time off with pay during working hours to look for a new job or to make arrangements for retraining. What is a reasonable amount of time off will vary according to the differing circumstances of employers and employees. During such absences employees should be paid at their normal rate.

Payments are calculated by reference to rules similar to those used to calculate a basic award for unfair dismissal. Many employees are disappointed at the size of the payment they receive. This may be because they have read about especially generous redundancy schemes operating for particular industries, or because they have failed to understand the basis on which the calculation is made. Vital to the calculation of both redundancy payment and unfair dismissal compensation is the concept of 'a week's pay'. Where an employee has normal working hours and the pay does not vary from week to week, a week's pay is simply the basic weekly wage. Overtime earnings are not included unless overtime is obligatory under the contract of employment. If the employee has variable earnings because of a shift pattern, or piece rates, or productivity bonuses, these are usually averaged over a 12-week period.

An employer is under a duty to consult with trade unions or other representatives if sizeable numbers are to be made redundant over a short period. This duty is quite separate from the obligation to notify the Department for Business, Innovation and Skills. If employers fail to consult, a special protective award may be made against them, which will safeguard the remuneration of employees for the period for which it is made.

Hearing redundancy and unfair dismissal claims forms a significant part of the work of employment tribunals. However, the trend has been to increase this workload in employment-related spheres. Since 1974, the tribunals have heard appeals against Prohibition and Improvement Notices issued under the Health and Safety at Work etc. Act 1974. The equal pay and sex discrimination rules can also be enforced in the employment tribunal. Extensions to the work of the tribunals were made by the Employment Act 1980 in relation to unreasonable exclusion or expulsion from a trade union. The jurisdiction of tribunals has been extended to breach of contract claims arising or outstanding on the termination of an employee's employment with any appeal to the Employment Appeal Tribunal. Indeed, each new statutory right adds to the jurisdiction and workload of the tribunals. However, arbitration via ACAS is a possibility in cases of unfair dismissal and requests for flexible working.

This brief account of employment law is sufficient to show it to be a complex and constantly changing area. Many firms do not have a legal or human resources department which can guide them through the maze of rules. However, the Department for Business, Innovation and Skills publishes a large range of explanatory booklets, and the officers of ACAS are always ready to give help and advice on questions which are within their particular sphere.

## Employees and trade unions

Employees governed by a contract of service may enjoy a further contractual relationship which is of major significance in employment, i.e. membership of a trade union. Several million people belong to trade unions in this country, ranging from the very large general unions like UNITE (a merger of the Transport and General Workers Union and Amicus) to small specialist unions with quite small numbers of members.

The position of the trade unions under the law is unusual, sometimes accounted for because of the history of their development. Trade unions could only begin to exist and operate as they do today once they were freed from the problems of laws like the Combination Acts and the civil and criminal rules about conspiracy. These shackles were not removed until 1871, when the growth of the movement into what it is today could get under way. After conceding the right of trade unions to exist and organize workers, the law originally did very little to interfere with, or control, their internal activities. The result of this early policy of non-interference was that trade unions kept control of their internal affairs, to a considerable extent, until the passing of the Trade Union Act 1984. Under provisions introduced by that Act (now to be found in the Trade Union and Labour Relations (Consolidation) Act 1992), unions are required to hold secret ballots for trade union election, secret ballots before industrial action, and secret ballots about political funding, if any. Of the statutory rules relating to trade unions, most have been framed to regulate their external activities, e.g. political activities and industrial activities. The main reason why a union engages in industrial action is to support the claims it is making for its members in collective negotiations with employers. The process of voluntary collective bargaining is recognized in this country as one of the principal means of fixing terms of employment. If a union is to be in a position to bargain effectively with an employer, it must gain recognition for that purpose. Between 1975 and 1980 the rules of the Employment Protection Act 1975 permitted trade unions to use the services of the Advisory, Conciliation and Arbitration Service (ACAS) to gain recognition from employers who were reluctant to grant bargaining rights. These procedures proved ineffective, and were repealed by the Employment Act 1980. For a period, a union wishing to be recognized for bargaining purposes had to establish that right for itself, by taking industrial action if necessary. Moreover, under the provisions of the Trade Union and Labour Relations (Consolidation) Act (1992) (TULRCA) as amended by the Trade Union Reform and Employment Rights Act 1993 (TURERA), ACAS was no longer under a duty to encourage the extension and reform of collective bargaining.

However, the Employment Relations Act 1999 reintroduced a statutory recognition procedure by adding a new schedule to TULRCA 1992. The procedure, which is involved and complex, places an emphasis on securing voluntary arrangements in the first instance, with resort to the Central Arbitration Committee (CAC) only if that fails. The role of the CAC is to determine what is an appropriate 'bargaining unit' and whether there is majority support for collective bargaining by the union from those working in that unit. If necessary the level of support may be determined by a ballot. If this reveals majority support and those voting in favour constitute at least 40% of the workers in the bargaining unit, a declaration is issued by the CAC that the union is entitled to recognition.

After a review of the operation of the Employment Relations Act 1999, the government passed the Employment Relations Act 2004, which introduced measures to modify the operation of the statutory recognition procedure. The Act also enabled regulations to be made regarding the right of employees, or their representatives, to be informed or consulted by their employer on a range of issues relating to their employer's undertaking. These regulations (the Information and Consultation of Employees Regulations 2004), which apply to undertakings employing 50 or more employees, implement the Information and Consultation Directive 2002.

## Consolidation of legislation

Throughout the remainder of this chapter, it should be remembered that the Trade Union and Labour Relations (Consolidation) Act 1992 consolidates the Employment Acts of 1980, 1982, 1988 and 1989, the Trade Union and Labour Relations Act 1974 and the Trade Union Act 1984. However, since 1992 there have been various measures that have 'undermined' this consolidation of collective employment law, notably the Employment Relations Act 1999 and the Employment Relations Act 2004.

## Definition of a trade union

The modern definition of a trade union was found in the Trade Union and Labour Relations Act 1974, now consolidated into the Trade Union and Labour Relations (Consolidation) Act 1992: 'An organization consisting wholly or mainly of workers ... whose principal purposes include the regulation of relations between workers and employers.' It will be realized that trade unions are defined both by reference to their membership and their purposes. No formal act of creation is necessary. Although an organization complying with this definition qualifies for the name 'trade union', the best legal rights and protections are afforded only to 'independent trade unions' (ITUs). The status of being independent is achieved by a trade union applying to an official called the Certification Officer (CO) for a Certificate of Independence.

As a first stage towards achieving certification, a union applies to the CO to be listed. Listing is conclusive proof of status of a trade union. The next is to apply for a Certificate of Independence. The list of advantages conferred by independence is impressive, particularly where the union is also recognized by the employer for bargaining purposes. It includes:

1    An independent trade union has the statutory right to receive certain information for bargaining purposes.
2    ITUs have the right to be consulted about proposed redundancies.
3    ITUs have the right to be informed and consulted about transfers of undertakings.
4    ITUs can appoint safety representatives under the Health and Safety at Work etc. Act 1974.
5    Members of ITUs may have time off work to take part in union activities.
6    Members of ITUs have the right not to be dismissed because of taking part in union activities at an appropriate time.

When the CO considers a union's application for a Certificate of Independence, he or she must take into account the criteria laid down by the 1992 Act. These provide that the union must not be under the domination or control of the employer, nor liable to interference from the employer. In other words, the advantages of independence should only be conferred on genuine unions, which can exercise a proper, vigorous and independent role on behalf of their members. Unions least likely to gain certificates are 'house unions' or staff associations which have often emerged from a firm's social club.

The CO is also permitted to hear the views of other unions about granting the certificate. Another union may have a vested interest in the outcome, as its own position may be jeopardized by the emergence of a new independent union. This could particularly affect the recruitment of new members.

As the Certificate of Independence is the key to claiming all the statutory rights and protections, a union which is refused this status may wish to appeal against the CO's decision. A right of appeal against the refusal of a certificate lies to the Employment Appeal Tribunal. If the application is still unsuccessful the union may apply again, after it has put right those matters specified as reasons for refusal of the certificate.

## The legal status of a trade union

Strictly speaking, a trade union is an unincorporated association but so many special rules have been created in relation to unions that their position more closely resembles that of a corporation. The trade union is an association of individuals, bound together by their rule book, which is effectively the terms of a contract between all the members. The greatest difficulties caused by lack of corporate status would usually be connected with holding property or making contracts. Special rules in TULRCA cover these points. A trade union may make contracts in its own name, and its property will be vested in trustees.

## The rule book

Generally a union is free to draft its own rules subject to the following limitations:

1   The rules must not offend the discrimination laws or EU regulations.
2   The rules may not exclude the principles of natural justice, nor may they seek to oust the jurisdiction of the courts.
3   Every member has a right under TULRCA not to be subjected to unjustifiable discipline.
4   The rules of a union which has a political fund must provide for members to be able to opt out of contributing to the political fund.
5   The rules may not exclude the duty of every trade union to secure that its principal executive committee has been directly elected by the membership.

As the rule book constitutes the terms of a contract between the members, the courts have power to enforce the rules if an aggrieved member brings an action. It must, however, be borne in mind that not all the rules may be contained in the rule book; some may exist because of custom or practice.

## Membership of a trade union

The right to become and remain a trade union member has been very significant, especially when a 'closed shop' was in operation at a place of work. Basically that meant that the worker must either be, or become, a trade union member. The closed shop (which the law calls a union membership agreement) has been gradually eroded by legislation, most recently by the Employment Act 1990. All statutory support for closed shops has now been removed, although they can still exist or be created by agreement between employers and employees.

When an applicant applies to a union for membership, the situation is controlled by the union's rule book (which also controls questions of the discipline of members, including possible expulsion from the union). It has already been seen that trade unions have great control over the contents of their rule book. If the rules are clear, unambiguous and properly applied, there should be no need for recourse to the courts for their interpretation by an applicant refused admission.

As the consequences of an exclusion or expulsion from membership of a union can be so grave, the excluded worker may refer his or her case to an employment tribunal as the worker has a general right not to be excluded or expelled from a union (TULRCA 1992 as amended by TURERA 1993). However, in *ASLEF v. UK*, the European Court of Human Rights held that a union which was opposed to racism and fascism was not obliged to admit into membership an individual who was a member of the British National Party. As a result of this decision TULCRA 1992 was amended by the Employment Act 2008 to qualify the right of an individual not to be excluded or expelled from union membership where that individual is or was a member of a political party and such membership is contrary to the union's rules or objectives.

## Freedom to join a trade union

Normally all workers are free to join a union if they choose to do so. (The exceptions to this rule are members of the police and armed forces, who cannot form themselves into unions.) What is more, workers are often free to choose between several appropriate unions. This apparent choice was once restricted where the Bridlington Agreement operated but at the level of the individual members this agreement has been effectively unenforceable since TURERA 1993. However, the Bridlington Agreement, made in 1939 between unions affiliated to the TUC, has continued to regulate at the collective level inter-union competition for members, and the transfer of members between affiliated unions. The main object is to prevent unions from 'poaching' members from each other. Where a union wishes to complain about such activities, the dispute will be referred to a Disputes Committee of the TUC. Again, this is a body with no legal sanctions to impose. If it finds the claim of poaching is well founded, it will usually tell the offending union to pay financial compensation to the complainant union. The Bridlington Agreement may contravene Article 11 of the European Convention on Human Rights, which guarantees freedom of association. If this is so, then the Bridlington Agreement cannot affect the rights of individuals to join the trade union of their choice. Indeed, the introduction of the general right to union membership (TULRCA 1992 as amended) did weaken the Bridlington Agreement. Nowadays, changes in patterns of employment have resulted in many employers wanting to deal with only one union. In such a case it may be a question of competing unions holding a 'beauty contest' to determine which union should be recognized by the employer. Examples of this have occurred in the print and motor industries.

## 'Union only' contracts

It used to be quite common for organizations like local authorities to demand of all their suppliers that they should allow their workers to be union members. Indeed, some authorities would go so far as to refuse to invite tenders from suppliers who did not comply with this rule. Since 1982, any term in a contract for the supply of goods or services is void if it purports to require that work done under the contract shall only be done by unionized or non-unionized labour.

A new tort was also created (see TULRCA 1992). It is a breach of statutory duty if, on grounds of union membership or non-membership, a person:

1    fails to include a particular person's name on an approved list of suppliers; or
2    excludes a person from tendering; or
3    terminates a contract for the supply of goods or services.

Similar liabilities may arise from any actions taken to impose union recognition requirements in contracts for the supply of goods or services. It is also unlawful to refuse to deal with suppliers or customers merely because that person does not recognize or consult with a trade union.

## Expulsion from membership

A union's rule book will normally provide a range of disciplinary measures, to cover various acts of misbehaviour by its members. Misbehaviour can range from failure to pay union dues, to breach of vague rules like 'behaviour prejudicial to the union'. The discipline imposed can range from a fine, or temporary withdrawal of benefits, to ultimate expulsion from the union. The remedies for a worker who wishes to contest an expulsion are:

1    *To bring an action against the union in the ordinary courts* – this can succeed if the union has acted in breach of its rules, or contrary to the principles of natural justice (e.g. by failing to give the expelled worker the opportunity to put his or her side of the case).
2    *To bring an action against the union in an employment tribunal* – under TULRCA 1992 as amended, where the member has been subjected to unjustifiable discipline. The concept of unjustifiable discipline was introduced in 1988, and could include imposing a fine, suspension or expulsion on a member for refusing to take part in industrial action.

## Industrial activities of trade unions

Industrial action represents the means by which employees can collectively bring pressure to bear upon an employer. The theory of this concerted action is that it produces an equality of bargaining power between workers and employers which could never exist between one individual employee and the employer. Common reasons for using industrial action are support of claims for higher wages or demonstrating protest at unsafe working conditions.

Although the strike is the ultimate form of industrial action, a union may achieve results by using a work-to-rule, a go-slow, a boycott of customers or suppliers, or a refusal to work contractual overtime. If a strike is considered to be the most appropriate action, it may not be necessary for the union to call out all its members. Results can often be achieved quickly with selective stoppages, especially if these are by key workers, e.g. those operating computers, or involved in the collection of money for the employer.

All forms of industrial action can give rise to the question: 'Are the workers concerned in breach of their contract of employment?' The answer to this is not easy. For example, employees working to rule could give little cause for complaint if they were meticulously observing the rules laid down in their contract, but if they seek to give those rules an unrealistic meaning, so that they are wilfully obstructing their employer's business, then they may be in breach of contract. If there was a breach, one of the remedies would be for the employer to sue the offending employees for damages for breach of contract. Moreover, on those occasions where employees take selective action and perform only part of their duties, the courts have allowed employers to deduct a 'reasonable sum' representing the duties not carried out. Indeed, where an employer makes it clear that part performance is not acceptable, it may lawfully refuse to pay anything at all even if an employee has performed the great majority of his or her duties. However, any employer conscious of the need to preserve good industrial relations is unlikely to resort to such steps. It is far more likely that an employer who believes its business interests are being wrongly interfered with will want to take action in tort against the union and its officials. A number of specific torts (e.g. interference with contract, and conspiracy, often known as the economic torts) relate almost exclusively to trade union activities. Other torts like trespass and public nuisance can easily be committed while participating in industrial action.

An example of a typical situation will show how these possibilities could occur. A union in discussion with an employer about new wage rates feels that negotiations are making no progress and that the employer could offer more. The employees who are union members are called out on strike after a ballot. The union members account for about 60% of the workforce. While some employees continue to work, the effect of the strike is not fully felt, so the union officials organize a picket line and attempt to persuade those still working not to cross the picket line (i.e. persuade them to break their contracts of employment). What is more, officials realize that the impact of their action can be increased by persuading lorries bringing supplies to turn back without making delivery (i.e. causing a breach of commercial contracts). The number of pickets may well get out of control, spilling on to the road and causing it to be blocked, or into the employer's premises, constituting trespass.

The possibility of the employer suing successfully in tort, either for an injunction to stop the commission of the torts, or for damages, is restricted because of the immunities enjoyed by trade unions, their officials and their members. Since 1906, these immunities have existed if the person committing the tort was acting in contemplation or furtherance of a trade dispute. Because of the protection it affords against tort actions, this is frequently referred to as the golden formula, although it is subject to more recent limitations in its scope. Two elements of the formula require consideration:

1   When is there a trade dispute?
2   What constitutes acting in contemplation or furtherance of it?

## Trade disputes

A definition of a trade dispute is contained in TULRCA 1992. It must be a dispute between workers and their employer, which relates wholly or mainly to one or more of the following matters:

1   The terms and conditions of employment.
2   The physical conditions in which people work.
3   The engagement or non-engagement of workers.
4   The termination of employment or suspension of workers.
5   The allocation of work between workers.
6   The question of discipline.
7   The membership or non-membership of a trade union by any worker.
8   The facilities for officials of trade unions.
9   The machinery for negotiation or consultation relating to any of these matters.
10   The question of recognition of a trade union.

The trade dispute may also relate to matters outside Great Britain if the workers involved are affected by those matters. This is of considerable significance where companies may be multinational, or working on contracts abroad, engaging cheap local labour.

There are many complex cases on the full meaning of the definition, because an employer who sees business severely affected by industrial action would naturally be keen to prove that the golden formula should not apply on the ground that there was no trade dispute. An interesting example where this argument succeeded involved the BBC. The BBC planned to televise the 1978 Cup Final at Wembley, and to 'beam' their transmission to many foreign countries. An anti-apartheid movement approached the union organizing the TV engineers and asked for support in not beaming the programme to South Africa. If the engineers were to agree to this, and refuse to carry out the orders of the BBC, they would be in breach of their contracts of employment. The BBC in turn would inevitably be in breach of a number of its commercial contracts with countries overseas who would not then receive the programme. Was the proposed action of the TV engineers in contemplation or furtherance of a trade dispute? The Court of Appeal decided that no trade dispute existed here, although the engineers could easily have brought themselves within the definition. If they had asked the BBC to insert a new term in their contracts which provided that they would not be required to take part in broadcasts to South Africa, the BBC would almost certainly have refused. Then the engineers could have said that their dispute was related to 'terms and conditions of employment' (*BBC v. Hearn*).

## Contemplation or furtherance of a dispute

Although the judges in the cases have often emphasized that acts can only 'further a trade dispute' if they are not too remote from the dispute, or if they are reasonably capable of furthering the dispute when viewed objectively, these approaches inevitably limit the scope of the golden formula. The matter was considered by the House of Lords twice, in 1979 and 1980, and on both occasions the court confirmed that the proper approach to interpreting the word 'furtherance' is to apply a subjective test. Did the person in question honestly and genuinely believe that these acts would further the dispute? If so, he must have the benefit of the golden formula, however unwise or damaging his actions may be.

## Limitations of the immunity

The whole question of the scope of trade union immunity from actions in tort is politically very contentious, and developments in this field have continued apace. The present position can be summarized as follows:

1   *The position of the trade union itself.* The union can be liable in tort if the act in question is authorized or endorsed by the president or general secretary, or the principal executive committee or any other committee or official whether employed by the union or not. Where the trade union is successfully sued for damages, the damages awarded are subject to upper limits dependent on the number of members. A union with less than 5000 members may have to pay damages up to £10,000, whereas a union with more than 100,000 members may have to face a damages bill for up to £250,000.
2   *The need for a ballot of the members.* Immunities will only be extended to industrial action which has been preceded by a secret ballot. The ballot must have taken place within four weeks of the action being taken and a majority of those voting must support the proposed action. Immunities will be lost if no ballot is held or if it does not conform to the statutory rules.
3   *Secondary action.* When an official or member commits a tort, he or she will be protected if it is one of the torts protected by the golden formula. Which torts are so protected has varied from time to time. With the exception of lawful picketing, secondary action does not attract immunity. The law thus protects employers far removed from the actual dispute, who may find themselves being used as unwilling pawns by a trade union experiencing difficulty in achieving its objectives by more direct means.
4   Further restrictions on the immunity from liability in tort are contained in TULRCA 1992. Thus, there is no immunity for tortious actions

where the reason for the action is that the employer is employing a person who is not a trade union member or who is not a member of a particular union. Anyone who commits an industrial tort while seeking to impose or enforce a closed shop will not have the protection of the golden formula.

5   Where the member or official concerned is picketing, limited immunity is granted as long as:
   (a)  he or she pickets in contemplation or furtherance of a trade dispute;
   (b)  he or she pickets only in the right places – this is now usually limited to the member's own place of work;
   (c)  he or she pickets for the right purposes – these are limited to peacefully obtaining or communicating information or peacefully persuading anyone to work or not to work.

One difficulty about the immunity in relation to picketing is that it is extremely restricted, and bears no relation to the realities of the picket line. Although some changes in the law were made in 1980, they did not put right the worst practical problems, namely that there is no right to stop vehicles and there is no control over the number of pickets. The 1980 changes were principally aimed at putting a stop to secondary picketing, i.e. picketing on premises other than those of the employer in dispute. It was hoped that the other problems could be resolved by the issue of a Code of Practice. The most recent code was published by the Secretary of State for Employment in 1992. The rights to freedom of expression (art. 10) and to peaceful assembly (art. 11) under the European Convention also arguably support a basic right to picket. In reality, however, these rights may have very little impact on the legality of picketing.

By its very nature picketing is a public activity which tends to involve the commission of criminal offences, e.g. obstruction of the highway, causing a breach of the peace. Although the immunity is so worded as to cover criminal behaviour ('it shall be lawful'), the scope of the protection is extremely narrow, as proved by the cases on the point. Thus, there was no protection for a picket charged with obstruction because he had tried to detain a lorry driver for long enough to speak to him (*DPP v. Broome*), nor for the pickets who blocked a road by circling round continuously in it (*Tynan v. Balmer*).

Picketing must also be conducted in compliance with the provisions of the Public Order Act 1986 (as amended). Section 14 of the Act empowers the police to impose conditions on 'public assemblies', i.e. assemblies of more than 20 people in a public place in the open air. The police can issue directions as to the place, the duration and numbers involved. The power of the police exists where it appears that the assembly may result in serious public disorder, serious damage to property, or serious disruption to the life of the community. The police also have power to act where they believe the purpose of the assembly is to intimidate others.

# 10
# The law of tort

## Introduction

The basis of liability in tort is probably one of the most difficult legal ideas for the layman to grasp. The word itself is of French origin and means a wrong. Basically, the law of tort is concerned with situations where the behaviour of one party causes, or threatens to cause, harm to the interests of another party. The rules of the law of tort determine when one party can be compensated for the behaviour of another. The law of tort is limited in its scope and although new torts do evolve from time to time, there is not necessarily a legal remedy for every wrong suffered. A claimant will only succeed in an action if he or she can show that the defendant's behaviour falls into a specified situation covered by the law of tort. Those specified situations are then given identifying names such as the tort of defamation, the tort of nuisance, the tort of negligence and the tort of trespass.

Much of the law of tort has developed through the cases. Some of the rules have now been given statutory force, for example the Law Reform (Contributory Negligence) Act 1945, the Occupiers' Liability Acts of 1957 and 1984, and the Torts (Interference with Goods) Act 1977. But case law continues to play a very significant role in the development of tort principles. Case law allows the rules to operate very flexibly and to be applied to widely differing situations. The cost of that flexibility will sometimes be a lack of certainty. While the rules are continuing to evolve, it may be difficult to say with any precision whether liability will arise.

Compensating claimants under the law of tort is a long-established part of our legal system, with cases dating back to the thirteenth and fourteenth centuries. The earliest forms of tort usually gave protection to the person or to property. These were the various forms of trespass, where the injury which had been inflicted was direct. As society became more sophisticated, the law of tort developed to give protection against less direct forms of harm. Torts such as negligence and the torts designed to protect economic interests then evolved.

### Tort distinguished from crime and contract

There are circumstances where exactly the same act or behaviour may give rise to liability in both the law of tort and criminal law. Take, for example, the

Galbraith's Building and Land Management Law for Students. DOI: 10.1016/B978-0-08-096692-2.00010-5

case of a motorist who causes death by dangerous driving. He or she may be charged with offences in the criminal courts, and at the same time sued for damages on behalf of the deceased victim or by the victim's dependants. Criminal law is concerned with the preservation of order in society, whereas the law of tort is concerned to compensate victims who have suffered harm. Although it is easy to state the differing functions of criminal law and the law of tort, nowadays these differences are blurred in practice. For example, the criminal courts have power in certain circumstances to make compensation orders to victims as well as meting out punishment to the offender.

The same set of circumstances can also give rise to liability in both the law of contract and the law of tort. For example, a person who buys a defective hedgecutter which causes him to cut off his finger may sue the supplier for breach of contract, or the manufacturer in tort under the Consumer Protection Act 1987 or in the tort of negligence. The most significant difference between contract and tort is that contractual liability is based on consent between the parties, a form of arranged liability, whereas tort liability is imposed by law. Although this distinction may be true in essence, again the boundaries between the two areas are not so clear. In the law of contract many of the obligations are in fact imposed by law. For example, in a sale of goods situation conditions are imposed by the Sale of Goods Act 1979 and the Sale and Supply of Goods Act 1994, and the parties are not free to exclude them. Where tort and contractual liability overlap, claimants can take advantage of whichever claim is more favourable to them. That may depend upon the limitation rules governing time limits for bringing actions and the rules for the calculation of damages.

Although there is considerable emphasis on the compensatory function of the law of tort, an award of damages is not the only remedy when a tort has been committed. A claimant may wish to put a stop to some form of behaviour being carried on by the defendant, for example repeated excessive noise. He may ask the court to grant an injunction in his favour. An injunction is an order of the court telling the defendant to do, or refrain from doing, certain things. A defendant who fails to obey an injunction is in contempt of court.

When suing in tort a claimant does not have to specify the particular tort on which he or she is relying. In practice, the conduct being complained about may amount to more than one tort. For example, the same behaviour may amount to both the tort of nuisance and the tort of negligence. If more than one tort has been committed the claimant will not recover double damages; he or she will simply have the protection of alternative reasons why the claim should succeed.

## Tort compared with other systems of compensation

In order to succeed in a tort action the claimant must usually show fault on the part of the defendant. To do so, it is likely that complex, lengthy and costly court proceedings will need to be brought. If a claimant is successful in those proceedings there is no certainty that the defendant will have the means to satisfy the judgment made against him or her, i.e. he or she may be unable to pay the damages. As a general rule in English law, there is no obligation to insure against the possibility of tort claims, although there are exceptions to this rule, e.g statutory provisions which require a motorist to insure in respect of tortious liability to third parties, or an employer in respect of liability to employees. In view of the problems of proving fault, it has sometimes been argued that victims of accidents should receive compensation from some kind of central insurance fund on a 'no-fault' basis, i.e. without having to prove liability against a defendant. A Royal Commission considered this idea and its Report was published in 1978. It recommended that the no-fault principle should be extended to apply to motor vehicle accidents, but it did not recommend abolition of tort actions in other cases of personal injury.

There is some experience of operating a no-fault insurance system in the UK. The industrial injuries scheme, which forms part of our social security system, allows a workman who is injured in an accident at work to receive disablement benefits if he suffered personal injury caused by accident arising out of, and in the course of, his employment, whether or not his employer was at fault. Such an injured workman may also successfully sue his employers for damages in the tort of negligence if he can prove that the employer was at fault. In assessing his damages within the law of tort some account will be taken of social security payments received. The same workman may also have private insurance cover where the payments made to him are unaffected by benefits received under the social security system or through tort damages.

## Trespass

Trespass is commonly thought to be a tort relating to land, but the law also recognizes trespass to the person and trespass to goods. Trespass to the person occurs if force is directly and intentionally inflicted on another person. For this purpose force means any physical contact. It would equally be a trespass to punch someone on the nose or to give them an unwanted kiss. It is a special feature of all forms of trespass that an action can be brought even if no damage has been suffered. This tort is said to be actionable *per se*, of itself, without proof of damage. Of course, a person who has suffered no real harm will usually receive nominal damages. People often consent to actions which would otherwise constitute a trespass to the person, e.g. where one willingly holds out a hand for another to shake it, where a patient willingly opens his mouth to allow a dentist to extract his teeth. Where consent has been expressly or impliedly given it is a complete defence to an action. The defence of consent is usually known by the Latin maxim *volenti non fit injuria*: he who consents cannot complain of the injury.

Where there is a trespass to the person, the same behaviour could also be a crime. For example, a punch on the nose could also result in a prosecution for assault.

## Trespass to goods

Originally, there were a number of forms of tortious action which could be brought to protect goods. The law in this area was reformed by the Torts (Interference with Goods) Act 1977. Wrongful interference with goods can take the form of trespass or conversion. Trespass to goods is an intentional and direct act of interference by a defendant with goods in the claimant's possession. The emphasis here is to protect possession. Frequently, the person in possession will be the owner of the goods in question. The defendant's act must be direct and intentional, so tearing a page from your book, or throwing a stone at your car, or breaking a window are all examples of trespass to goods. In these cases the person inflicting the damage had no intention to deny the other person's right to possess the items. If the defendant does commit some act which denies a claimant's right to possess the goods this may amount to conversion. So if the defendant took the book from the claimant, rather than simply tearing some pages out of it, he has converted the book to his own use. The principal remedy for any wrongful interference with goods is damages, but where a defendant is still in possession or control of the goods, s.3 of the 1977 Act may permit the claimant to obtain an order for delivery of the goods and payment of any consequential damages. Where a claimant seeks only damages these are based on the value of the chattel together with any consequential loss suffered.

## Trespass to land

This tort consists of an unjustifiable interference with the possession of land. The interference may consist of walking over another's land, or throwing things on to the land, or placing a ladder against the surrounding wall or even swinging a crane jib over the land. For the purposes of this tort, land means the surface of the earth itself, anything which is a fixture on it (for example, a building), the air space above it to the extent necessary for the use and enjoyment of the land, and the subsoil below the land. Aeroplanes flying over land would constitute a trespass if the activity were not permitted by Act of Parliament.

Trespass protects possession of land. Where a person is rightfully in possession he or she may bring an action against even the owner of the land. An example could occur if the landlord rented premises to a tenant without reserving the right of entry to those premises. Entry without the tenant's permission would be considered as trespass. The tort is actionable without proof of damage. In those cases where no harm has occurred the claimant may be seeking an injunction, for example to restrain a neighbour who regularly takes a short cut through the plaintiff's garden. Cases such as *Anchor Brewhouse Developments Ltd v. Berkley House Ltd* (1987) show that the courts are prepared to grant injunctions to prevent oversailing by tower cranes, even where no damage is caused.

An apparent trespass may be justified because the person is on the land with express or implied permission or under statutory authority or in exercise of a right of way. An example of an implied permission to enter premises is the open door of the supermarket. Having entered the shop, however, if a shopper creates a disturbance he or she may be asked to leave and would then become a trespasser if refusing to do so. There are numerous examples of persons having the right to enter premises without the occupier's permission. Acts of Parliament confer powers on officials such as factory inspectors, gas and electricity board officials and public health officials, who have the right to enter premises if particular circumstances exist. It is not a defence to trespass to plead that the defendant did not know he or she was trespassing. Even if a person has lost their way, they could be liable.

There is a general misconception that trespass is a crime. This is not usually the case but in a limited number of cases Acts of Parliament can make trespass a criminal offence. Typical examples would be trespass on railway lines, sidings or embankments. The notice 'Trespassers will be prosecuted' is, therefore, usually meaningless except in these limited cases. Changes have occurred, however, in the law of trespass to deal with the position of squatters and itinerant groups of hippies travelling to the Summer Solstice festivals – see the Criminal Justice and Public Order Act 1994.

It is obvious that construction work could involve constant danger of trespass. A contractor would be wise to seek permission for acts which would otherwise cause a breach of these rules. A claimant who has been dispossessed of land by a trespasser may bring a possession claim for recovery of the land and may combine it with an action for damages. If the claimant remains in possession, he or she may claim damages and, if necessary, an injunction to restrain further acts of trespass. The help of the court is not always needed in cases of trespass. The owner of the land can ask the trespasser to leave. If the trespasser does not do so, the owner may then use reasonable force to eject him or her. What is reasonable will vary with the circumstances of each case.

# Nuisance

For the purposes of the law, nuisances fall into three categories: public nuisance, private nuisance and statutory nuisance.

## Public nuisance

Public nuisance is primarily a crime, and consists of a nuisance which 'materially affects the reasonable comfort and convenience of a class of her Majesty's subjects' (*AG v. PYA Quarries*). Thus, where numerous acts of private nuisance do not cause comon injury to a section of the public, they do not amount to public nuisance (*R v. Rimmington*). Examples of public nuisance include obstructing the public highway, allowing smoke from a burning field to blow across the highway, polluting a public water supply or creating some projection over the highway. An unruly march or demonstration which causes obstruction of the highway could also constitute a public nuisance. Normally such behaviour is controlled by criminal prosecution, but it may be possible for a person affected by the public nuisance to sue in tort. A claimant would need to show that he or she suffered special and particular harm as a result of the public nuisance, greater than that suffered by ordinary members of the public. Where this can be proved, he or she may succeed either in recovering damages, or in obtaining an injunction, or both. Sometimes, the behaviour of the defendant will constitute both a public and a private nuisance and, in such a case, the individual affected will be in an even stronger position to sue. In one case, quarrying operations had caused all the houses in the neighbourhood to be affected by dust and there was held to be both public and private nuisance. A right to sue for public nuisance may be important in those cases where a claimant would have no right to sue in private nuisance, for example in those cases where there is no interest in land to protect.

## Private nuisance

The tort of private nuisance is committed when one person unlawfully interferes with another's use or enjoyment of land. The interference must be unreasonable for the tort to arise. The range of activities which can amount to private nuisance is very wide and could include harm caused by smells, fumes, vibrations, noise, dust, encroachment by tree roots, escape of sewage and the keeping of noisy or smelly animals. Essentially, the law of private nuisance attempts to reconcile conflicting interests. An Englishman's home is his castle, and he is free to do within it as he pleases, so long as he does not spoil his neighbour's enjoyment of his own land. In a recent case, *Hunter v. Canary Wharf*, it was established that erecting a building that interfered with a neighbour's television reception was not a nuisance.

In seeking to apply this 'live and let live' approach, and when assessing whether the interference is unreasonableness, the court takes a number of factors into account. The duration of the interference is relevant, not only in establishing the existence of a nuisance but also in determining which remedy is more appropriate – damages or an injunction. Where the occurrence is an isolated event or of a temporary nature, that in itself may be evidence that there is not a sufficiently substantial interference to constitute nuisance. As the court takes all the circumstances into account, duration alone is not conclusive. Consider the case of a firm demolishing a building. If the work will be over in a matter of weeks, it may be reasonable to expect the owners of adjoining properties to put up with the noise and dirt over that time. But if the contractor works 24 hours each day, seven days each week, with the constant noise of machinery operating, vehicles coming and going and the glare of arc lamps at night, these activities may well constitute a nuisance. Certainly in *Andreae v. Selfridge and Co. Ltd*, a claimant hotel owner was able to recover damages from a demolition contractor who had created excessive amounts of noise and dust.

The nature of the locality where the alleged nuisance is occurring will be relevant. One judge summed up this aspect of the tort by saying: 'What would be a nuisance in Belgrave Square would not necessarily be so in Bermondsey.' A person keeping pigs on his land might do so without complaint in the country but might be liable for nuisance if he kept them in a town garden.

Although locality may be an important factor, the courts will be more inclined to find a nuisance if the interference has caused actual harm to property (in which case locality is generally irrelevant) as opposed to mere interference with the occupier's comfort and enjoyment of his or her land. A claimant is far more likely to succeed where actual physical damage can be shown. If he or she is simply complaining about behaviour which causes personal discomfort in the use of the property, then he or she must show that the interference is substantial, and one which ordinary people of ordinary sensitivity would complain about. The essence of nuisance is to achieve a balance. If the defendant has behaved reasonably, the claimant cannot complain if he or she has suffered damage only because his or her own property is extraordinarily sensitive, or the use he or she wishes to make of it is in some way unusual or special. If a defendant's behaviour would cause no harm to ordinary plants and vegetation, then there is no cause of action for a claimant who has suffered harm only to his or her extra-sensitive orchids. But if the defendant's behaviour would have damaged even the daisies on the lawn, then the claimant can recover for the damage to this sensitive property as well as to any ordinary property. It is said that you cannot increase your neighbour's liability by using your own property in some special or extraordinary way.

When the court is assessing the reasonableness of the interference it will take into account that the defendant may be doing something which is of value to the general public, for example making early morning deliveries of milk. A rather greater degree of noise and disturbance may then be acceptable before the behaviour amounts to nuisance. This does not mean that such behaviour can never amount to nuisance. Where it does, the public utility or public interest question may be relevant in determining whether an injunction should be granted. That point was in issue in two instructive cases, *Miller v. Jackson* and *Kennaway v. Thompson*.

In *Miller v. Jackson*, the case turned on the playing of cricket on a village pitch which had been in use since 1905. Adjacent land was developed for private housing and Mr and Mrs Miller purchased a house directly in line with the wicket. They then discovered that they had to live under very trying conditions during the cricket season, when it was impossible to use their garden if play or practice was taking place. Although balls had come into the garden and bounced off the house, no one had yet been struck. On these facts there was indeed a nuisance but, partly because of the public interest, an injunction was refused.

In *Kennaway v. Thompson*, the nuisance arose out of water sporting activities on a lake adjacent to the claimant's house. The noise from powerful motor boats was excessive and the court agreed that the activities did indeed constitute a nuisance. Taking into account the public interest in taking part in such recreations, the court decided that some balance could be achieved between the conflicting interests of the claimant and the club users. An injunction was granted, not in terms of a total restraint on water sports but limiting the size of boats which could be used and the number of times each year when competitions could be staged.

If the defendant can be shown to have acted out of malice, that will usually be conclusive as to the existence of a nuisance. In one case where the defendant lived next door to a music teacher, whenever a music lesson took place he would bang tin trays, whistle and shout. His acts were shown to have been done only for the purpose of annoying and interrupting the claimant, and were held to amount to a nuisance.

## Liability for nuisance

Where a potential nuisance situation exists, problems arise as to who can bring an action, and against whom that action must be brought. Private nuisance is designed to protect a person's use or enjoyment of land. A claimant must, therefore, be able to show some legal interest in the land, whether as owner or tenant. No action in private nuisance is available to a mere licensee (a person who only has permission to be on the land but no legal rights to the land itself). A mere lodger or guest cannot sue, although if either has suffered personal injury he or she may have a right of action in another tort, for example negligence. Actions in public nuisance are not restricted by this 'interest in land' requirement.

However, the Human Rights Act 1998 may in the future have an impact on the rule that the claimant must have a legal interest in the land to sue for private nuisance, in view of the provisions relating to respect for private and family life and home.

## Who can be sued

The occupier of the land from which the nuisance arose is the obvious person on whom to impose liability. But there may be a difficulty if the occupier did not actually cause the nuisance. It seems that any of the following persons may be liable:

1   The actual creator of the nuisance. What is more, he remains liable even if he has disposed of the land and is no longer in a position to

put the matter right. The reasoning behind this is that he committed the wrong in the first place and should remain liable.

2   The occupier of the land, if the nuisance was created by himself or his employees, in which case he is vicariously liable for their actions. Although an employer is not vicariously liable for the acts of independent contractors, an occupier who engages a contractor to do construction work on his land may well be primarily liable for any nuisance that the contractor creates, if the occupier should have realized that the work was likely to cause a nuisance.

3   A trespasser who creates the nuisance is liable but may not be easily found or identified. The occupier will be liable for a nuisance created by a trespasser, or resulting from an act of nature if, once he knows about it, he adopts it or continues it (a continuing nuisance). In effect this means that if the occupier fails to take reasonably prompt and sensible steps to stop the nuisance, he will be liable.

An example which shows how the occupier can become liable for an act of nature is a case involving a tree which was struck by lightning and which then caught fire. The occupier called someone in to chop it down, but took no steps to put the fire out, leaving it to burn itself out. The fire spread to the claimant's property. The defendant was liable in nuisance for the damage caused. A case involving a hotel in Scarborough is another example of a continuing nuisance, where land on a cliff top that was owned by the council fell away, taking the neighbouring hotel along with it. It was alleged that the local council had known about the risk of this happening but had done nothing to prevent it. The court decided that the council was not liable because it could not have known about the risk to neighbouring land without undertaking an expensive survey of the area.

4   The tenant occupier. Where premises are leased by a landlord to a tenant, the tenant as the occupier seems the obvious person to be liable for a nuisance arising from the premises. However, the landlord is also made liable in a number of circumstances, particularly where the nuisance existed when he let the premises or where it is caused by disrepair of the premises for which he is responsible.

Unlike trespass, where a claimant does not need to prove damage, an action in nuisance is only possible where damage can be proved. The damage may be to the property itself (e.g. some nuisance which causes windows to break, or masonry to crack) or may consist of interference with the claimant's enjoyment of his or her property.

Even where all the various elements of the tort of private nuisance can be proved by the claimant, defendants may escape liability if they have a valid defence. For example, they may be able to plead that the conduct complained of is authorized by an Act of Parliament. This defence can often be used by local authorities, or public authorities who have been authorized by statute to conduct certain operations. Whether the defence of statutory authority will succeed depends partly on whether the statute lays down a duty or merely a power. Much will depend upon the precise wording of the particular piece of legislation. However, in 1981 the House of Lords showed themselves prepared to give a wide interpretation to the words of the statute in the case of *Allen v. Gulf Oil Refining Ltd*. In that case, the Act granted the authority to acquire land and build an oil refinery there. It was held that the defendant's powers must be taken to include the power to operate the refinery on the land and such operations would not, therefore, establish any liability in nuisance unless negligence could be proved.

The effect of the statutory authority defence may well be affected by the European Convention on Human Rights. The Convention, together with the Human Rights Act 1998, requires that statutory authority is reviewed in accordance with human rights law, such as the right to privacy under Article 8 of the Convention.

Defendants may also be able to show that they had a prescriptive right to do the act complained of. Rights to commit a nuisance may be acquired under the Prescription Act 1832 if defendants can show that they have committed the nuisance openly and continuously for more than 20 years. This is limited to those cases where the right is capable of existing as an easement, which is a question to be determined in accordance with land law principles.

Claimants may also have consented to the state of affairs (*volenti non fit injuria*), but this defence would not operate where claimants acquired the land with knowledge of an existing nuisance. It is no defence to argue that claimants came to the nuisance – in other words that they must accept the state of affairs prevailing when they arrived. A defendant's behaviour will still be tested by the stated principles and could be found to amount to nuisance.

## Remedies

It is true that the law allows a person affected by a nuisance to take appropriate steps to put a stop to it, but this form of self-help should not be lightly undertaken. The judges have said that this is not a remedy which the law favours and is usually inadvisable. The sort of problem a claimant could face is shown in one case where branches from the defendant's apple trees encroached over the claimant's land. The claimant cut off the branches, stripped off the apples and then sold the fruit. Although he was within his rights to lop the branches, he had no right to keep them, or the fruit. He was successfully sued by his neighbour for having misappropriated the fruit!

It will usually be much safer for a claimant to seek an injunction when wanting to put a stop to a nuisance. The granting of an injunction is entirely at the discretion of the court and may be refused. That may happen where the nuisance occurs only very infrequently or in circumstances where the injunction would cause undue hardship to the defendant. Damages are always available as a remedy to compensate the claimant for harm suffered as a result of nuisance.

This account of the rules of private nuisance will be sufficient to convince many people that they would be foolhardy to embark on private litigation. Bringing an action may well prove costly in terms of time and money. If the action is against a large company or organization, a private individual may well feel at a disadvantage because of the resources and expertise available to the other side. There can be significant problems of proof, especially in cases where it is hard to show a link between 'cause' and 'effect'. If these arguments are valid in small-scale instances of nuisance, they will be all the more so when the nuisance complained of is creating extensive environmental damage. In such a case it may be especially unrealistic to expect one private individual to sue, and the rules of statutory nuisance may then be more than ever relevant.


## Statutory nuisance

Most claimants simply want to put an end to the state of affairs causing the nuisance, but may be put off by the thought of bringing a costly and time-consuming individual court action, which may not improve relations with their neighbours. In many cases, the state of affairs about which they wish to complain may also constitute a statutory nuisance under the Environmental Protection Act 1990, which came into force on 1 January 1991. The Act replaces many of the rules previously found in the Public Health Act 1936 and the Control of Pollution Act 1974. Examples of nuisance covered by the 1990 Act include: premises in such a state as to be prejudicial to health or a nuisance; accumulations or deposits or animals which are prejudicial to health or a nuisance; emissions of smoke, gas, fumes, dust, steam or other effluvia; noise or vibration. The Clean Neighborhoods and Environment Act 2005 introduced two new statutory nuisances: insects emanating from premises and artificial light. In order for a state of affairs to be a statutory nuisance it must either be prejudicial to health or a nuisance. A state of affairs is prejudicial to health if it is injurious to health or likely to cause injury to health. The nuisance element of the phrase can only be satisfied by showing what amounts to a nuisance at common law (see page 239). A statutory nuisance cannot exist where there is merely an interference with the personal comfort of the occupiers, although noise that has an irritating quality has been held to constitute a statutory nuisance.

A local authority is under a positive duty to make inspections within its area to establish whether any statutory nuisance exists. Additionally, the local authority must take such steps as are reasonably practicable to investigate complaints of statutory nuisance made by persons living within the local authority's area. The inspections are made by Environmental Health Officers, who have the right of entry to property for detection purposes.

Where a statutory nuisance exists, or is likely to occur or recur, the local authority must serve an abatement notice on the person creating the nuisance or, in some cases, on the owner of the property from which the nuisance emanates. The notice indicates what must be done and a time limit for action. Non-compliance with such an abatement notice is a criminal offence, for which the offender can be fined on a continuing daily basis until compliance. Alternatively, the local authority may undertake the work necessary and recover its expenses.

The 1990 Act includes a right of appeal against an abatement notice. The appeal must be lodged within 21 days and will be heard by the magistrates' court.

Where there is evidence of a statutory nuisance and the local authority declines to act, or is itself responsible for the nuisance, any person aggrieved may bring proceedings in the magistrates' court. In such instances, the aggrieved person must give at least 21 days' notice in writing to the defendant, stating the intention to bring proceedings, and outlining the matters complained of. When the case comes before the magistrates' court, the court has power to make a nuisance abatement order, as well as imposing a fine of up to £5000 and ordering reimbursement of necessary expenses. Failure to comply with a nuisance abatement order is in itself a criminal offence. The court may order the local authority to perform the necessary work under the order where a defendant has defaulted.

The rules under the Environmental Protection Act do not give rise to any liability in tort. A person suffering harm from a nuisance would still need to pursue a claim in private nuisance if a large claim in damages was in issue. However, in smaller problems, the possibility of an award of criminal compensation under the 1990 Act may suffice. The advantage of the statutory rules is that, where the local authority is prepared to take action, the state of affairs constituting the nuisance can be put right without any need for individual action.

## The rule in *Rylands v. Fletcher*

The type of liability created by this tort is strict liability. This means that if a set of given facts fits the requirements of the rules, then the defendant will be liable, whether he or she took care or not. This tort is always known by the name of the case which gave rise to the particular form of liability. An occupier who engages in hazardous activities on his or her property will be liable where those activities injure or cause harm to someone. The same set of facts may also give rise to an action in negligence. The importance of the *Rylands v. Fletcher* rule was that it pre-dated the general development of negligence principles. Nowadays, it is often suggested that the rule is of decreased significance, and certainly its technical requirements mean that it is difficult to prove.

The rule is usually stated as follows:

> Where a person, for his or her own purposes, brings on to his or her land, and collects and keeps there anything likely to do mischief if it escapes, he or she must keep it at their peril, and if not doing so, is liable for all the damage which is a natural consequence of the escape.

The use of the word 'escape' is naturally suggestive of animals, which could come within the rule. But liability in respect of animals may also exist under the Animals Act 1971 and the torts of negligence and nuisance.

The rule involves a number of points:

1. The dangerous 'thing' must be brought by the defendant on to his or her land. It follows that no liability can arise from an escape of something which is naturally on the land, e.g. weeds or rocks.

2. The 'thing' must be likely to cause harm if it 'escapes'. The rule has been applied to a wide range of substances and items from oil, gas and explosives to flagpoles and fairground equipment. There is no requirement that the 'thing' must be intrinsically dangerous.

3. There must be an escape, i.e. the person injured must show that he or she was not on the defendant's land at the time when the harm occurred.

4. The escape must cause damage. This may be damage to property and was thought to include injury to the person. However, doubts have been expressed about whether the rule can be used in a claim for personal injuries.

5. The defendant must bring the 'thing' on to his or her land for their own purposes, and in doing so must be making a non-natural use of the land. This latter point is probably the most difficult and contentious. It appears that the defendant must be using his or her land in some extraordinary and unusual way. The idea of non-natural use is closely bound up with public benefit (or lack of it) and will tend to change with changing social needs. In the actual case *Rylands v. Fletcher*, the non-natural use was bulk storage of water in a reservoir. That was in 1868. Would the court still decide the case in the same way today? In *Rickards v. Lothian*, a non-natural use was defined as: 'Some special use bringing with it increased danger to others, and not merely the ordinary use of land, or such a use as is proper for the general benefit of the community.' It has been said that where planning permission has been granted for a particular use of land, it would be difficult to argue that such a use was non-natural. Recent House of Lords authority on the issue of what might constitute a non-natural use of land is *Transco plc v. Stockport Metropolitan Borough Council*. According to Lord Bingham, a cause of action under *Rylands v Fletcher* should involve establishing that the use of the land was extraordinary and unusual rather than non-natural. Lord Bingham also said that the same use of land might amount to being extraordinary and unusual in some circumstances but not others. In the same case, Lord Hoffman thought that a useful guide to determine whether there has been a 'non-natural' use of land is to consider whether the damage caused was 'something against which the occupier could reasonably be expected to have insured himself'.

Although liability is said to be strict, i.e. not dependent on whether the defendant took reasonable care, a number of defences may be available to the defendant. These include:

1   *Fault of the claimant* – if the damage is caused by the claimant's own wrongful act, he or she cannot recover.
2   *Consent by the claimant* – presumably tenants on different floors of a building impliedly consent to the presence of a water supply running through the building. If flooding occurs on an upper floor causing damage to a lower floor there will be no liability.
3   *Statutory authority* – a defendant may be able to point to an Act of Parliament which excuses his or her behaviour which would otherwise be tortious.

Finally, it must be remembered that the same set of facts could give rise to liability in both nuisance and *Rylands v. Fletcher*.


# Negligence

Negligence as a tort is not confined to protecting an interest in land or concerning liability arising out of the use of land, but can compensate for harm caused in a wide range of circumstances.

Negligence as a tort consists of the breach of a duty owed to the claimant to take reasonable care, which results in damage of the right type which is not too remote. Many more people are injured by careless acts than by acts which are intentional. As the law of tort developed, a number of specific situations in which liability would be imposed on a person for negligent behaviour were recognized by the law (e.g. where an employer fails to take reasonable care to ensure the safety of employees, or where an occupier fails to take reasonable care to ensure the safety of visitors), but there was no general principle established which could be applied to any and every set of circumstances.

Such a general principle began to emerge in 1932 from the famous 'snail in the bottle' case, *Donoghue v. Stevenson*. In that case, a manufacturer of ginger beer distributed the product in opaque bottles. One bottle was sold by a café owner to a customer who shared it with his friend. A decomposed snail slithered out from the bottle, causing the friend to become seriously ill. As she had not bought the ginger beer, the friend had no claim in contract. She successfully brought a claim against the manufacturer for negligence. The House of Lords decided that a manufacturer owes a duty to take reasonable care to see that its products are not contaminated. The prime importance of the case is that it tried to formulate some principles which would link together all the previously separate instances of negligence and by which new sets of circumstances could be tested.

Over the years, the separate tort of negligence has overtaken all other torts in importance: first, in the number of actions which are brought under this head, and second, in the way that the tort of negligence has begun to take over some of the liability which would previously have been covered by torts like trespass or nuisance.

The tort of negligence is usually analysed under three main headings:

- A duty of care owed to the claimant.
- Breach of that duty.
- Damage resulting of the right type which is not too remote.

## The duty of care

Liability in negligence can only arise if a duty to take care is owed. When a court is faced with a situation which has been considered previously, it may be able to follow a precedent from an earlier case in determining whether a duty exists. Alternatively, an Act of Parliament may impose a duty, e.g. the Occupiers' Liability Act 1957. In either case, establishing the existence of the duty is relatively straightforward.

Where the court has to consider an entirely novel situation, how does it decide whether or not a duty to take care exists? One of the judges in *Donoghue v. Stevenson* suggested the 'neighbour test' for answering this question. He said that it was impossible in law to have a rule which states that you must love your neighbour. But there could be a rule that you must not injure your neighbour – you must take reasonable care to avoid acts or omissions which you can reasonably foresee would be likely to injure your neighbour. Your 'neighbour' for this purpose is any person who is so closely and directly affected by your act or omission that you ought to have him or her in mind when directing your mind to the act or omission in question. This 'neighbour principle', as it is usually described, has often been thought to rest exclusively on reasonable foresight of harm. In a simple example, if you run a bath for a baby, you can reasonably foresee that the baby will be injured if you do not take reasonable care to test the temperature of the water before putting the baby into the bath. It is clear from the cases that mere foresight of harm is not now to be regarded as establishing a duty of care, without more. Emphasis nowadays is more on a test of proximity. Is there a sufficient degree of proximity between the parties so that the claimant should reasonably have contemplated injury to the defendant? The advantage of using a test of proximity is said to be that it can be more stringent where the circumstances require it. Certainly, the very wording of the 'neighbour principle' contains the test of proximity. A neighbour is a person 'closely and directly affected by the act or omission'.

The House of Lords subsequently refined the neighbour principle in *Anns v. Merton London Borough Council* in 1977, by the addition of a 'public policy' element. This consisted of the court weighing factors such as social conditions, a balance of interests and fear of inflicting excessive liability. This produced a result that in establishing whether a duty was owed, the test became a two-stage process. First, apply the neighbour principle, and if a duty appears to be owed, then ask whether there are any policy considerations which may justify the exclusion of, or restriction of, the duty.

By the mid-1980s, it had become clear from the cases that the judges were no longer happy with this two-stage formulation of the test for establishing duty. It left the introduction of policy considerations until too late a stage in the assessment of whether a duty existed. It would be more appropriate to take account of such factors when deciding whether the relationship between the parties was sufficiently proximate. A key factor to be taken into account by the court is whether it is just and reasonable to impose a duty. The use of this new approach is made much easier by the fact that the Anns case was overruled by *Murphy v. Brentwood District Council* in 1990.

In the light of recent cases, it is therefore better to state that before a duty of care can exist, it must be shown that there is foresight of harm, that there is a proximate relationship between the parties, and that it is just and reasonable to impose a duty. Although it is convenient to list the elements in this way, it should be borne in mind that they are inevitably interrelated. Moreover, in determining what is just and reasonable, account may be taken of factors which previously were considered under the 'policy' head.

An example of a recent case where what was just and reasonable was considered is *Commissioners of Customs & Excise v. Barclays Bank plc*. In this case, Barclays had failed to comply with orders freezing the bank accounts of two of its customers. The order had been obtained by Customs & Excise for unpaid VAT. Customs & Excise sued Barclays for the money it was unable to recover since the bank had allowed these customers to withdraw large sums of money from the supposedly frozen accounts. Although it was forseeable that Customs might be unable to recover sums owing to it if the accounts were not frozen, the House of Lords decided that it was not just and reasonable to impose liability, potentially of over £2 million, on Barclays for not complying with a court order which it had no opportunity to resist.

In essence, the purpose of requiring a claimant to establish a duty of care is simply to determine whether the defendant will have to pay for the harm caused. In cases where it is not considered appropriate for a defendant to have to pay, he or she can be insulated from the need to do so by manipulation of the duty concept, which is inherently flexible when expressed in the three-fold way seen above.

There are certain well-recognized situations where the courts have faced particular problems in establishing whether a duty should exist because of policy considerations. These include:

1   Cases involving pure economic loss.
2   Cases involving negligent misstatement.
3   Judicial immunity situations.
4   Nervous shock situations.
5   Negligent exercise of statutory powers.
6   Pure omissions to act.

It should be remembered that these areas constitute 'problem' areas of the law of negligence, and may well prove to be the fields where new developments occur.

## 1   Cases involving pure economic loss

As a general rule, no duty of care is owed where a claimant suffers only pure economic loss. An example can be found in the case where a research institute negligently allowed a virus to escape which caused foot and mouth disease, and as a result cattle needed to be slaughtered over a wide area. Markets were closed to stop the spread of the infection. The claimant was a cattle auctioneer who lost profits as a result, i.e. his only loss was pure economic loss. He could not recover damages for negligence. No doubt the court feared that if his claim was allowed, every cattle transporter, supplier of cattle food and dairyman would come forward to claim. Liability would have been excessive in terms of the numbers of claims and the possible amounts involved (*Weller v. Foot and Mouth Disease Research Institute*).

The judges are particularly anxious not to create liability 'in an indeterminate amount for an indefinite time and to an indeterminate class'. They also keep in mind that a person who suffers mere economic loss can frequently correct the situation, for example by accepting a smaller level of profit or by increasing charges. This may justify treating economic loss differently from personal injury, which is usually irreparable. They may also be taking a realistic account of which party could more easily have insured against the loss.

The rule can undoubtedly be very harsh, and sometimes almost illogical in application. Take, for example, the case of *Spartan Steel and Alloys Ltd v. Martin and Co. (Contractors) Ltd*. The defendants had negligently cut the electric power cable under a road. This resulted in power being cut off in the plaintiff's metal foundry. Work was in progress (a 'melt') which had to be aborted, resulting in damage to the metal. That was physical damage and the claimants could recover for that harm plus the resulting loss of profit on that 'melt'. As there was no power, they lost the possibility to carry out other 'melts' on which they lost profits. That was classed as pure economic loss for which the plaintiff could not recover damages.

Pure economic loss therefore means loss which is not injury to the person or damage to the claimant's property. However, there have inevitably been cases where the courts have sought to 'stretch' the definition of the phrase, sometimes with disastrous results. One such infamous case is *Dutton v. Bognor Regis UDC*. In that case, the defendant's negligence had caused a house to be built with defective foundations. The foundations were defective from the very moment of building. When a subsequent purchaser bought the house, could it be argued that the purchaser was suffering from physical damage to his property as a result of the negligence, or alternatively was he merely suffering loss which was purely economic. The house was damaged from the outset. The subsequent purchaser did not therefore 'suffer' damage to the property itself, but only to its value. Yet in the Dutton case, the damage was accepted as damage to property. No doubt the court in that case was anxious to allow the house owner to recover, but in achieving that result, the case created many problems and was clearly wrong. Fortunately, it too has been overruled by the House of Lords decision in Murphy.

During the period when Dutton was regarded as good law, it had to be seen as an exception to the usual rules about pure economic loss. The courts appeared to have created another exception in the very difficult case of *Junior Books Ltd v. Veitchi Co. Ltd*, which came before the House of Lords in 1982. The claimants had nominated the defendants as subcontractors to lay a new factory floor. The floor laid by the defendants was defective and the claimants claimed to have suffered losses totalling in excess of £200,000. The figure consisted largely of loss of profits while the factory had to be closed for the floor to be relaid. No one had been injured by the floor being defective and no harm had been caused to the property, i.e. the losses were pure economic loss. Nevertheless, the court thought that the claimants should recover damages. This decision seems to have been based largely on the 'close relationship' between the claimant and the defendant. There was no contract between them but the defendant was a nominated subcontractor. The court also took account of the fact that making the defendant liable would not open up liability to 'an indeterminate class'. It is certainly true that, in nominating the defendants, the claimants knew of them and relied on them, and although they had no direct contract with the defendants, the defendants were doing their work under the main contract so that was the next best thing. But it could be argued that by allowing the claimants to sue in tort in this case, when they could not sue in contract because of the privity of contract rules (see now the Contracts (Rights of Third Parties) Act 1999), the courts may be undermining the very rules of the law of contract – and that in itself may be a policy consideration for not allowing recovery for pure economic loss in a situation such as the Junior Books case!

Any idea that the Junior Books decision might see the beginning of the end of the economic loss rules has not been borne out by subsequent court decisions. In *Leigh and Sillavan v. Aliakmon Shipping* in 1986, where a shipper damaged goods through negligence, the buyer of the goods (to whom ownership of the goods had not yet passed) sued in respect of the damage. As the goods did not yet belong to the buyer, the harm was not damage to his property, but mere economic loss. The buyer's problem was that under his contract to buy the goods, he had to undertake to bear the risk of their loss or damage. The House of Lords would not allow the buyer to succeed against the shipper. Although the shipper had been negligent, he had caused the buyer only pure economic loss. Choosing not to try to bring this case within any exceptional principle, the court found that as an aspect of policy, it would be a bad thing to allow the buyer to recover, as that would effectively render useless the protection which the shipper enjoyed under his contract with the seller. In other words, allowing the buyer to succeed would undermine the rules of the law of contract.

In a trio of cases concerning building and building work (*Simaan General Contracting Co. v. Pilkington Glass Ltd (No. 2)*, *Greater Nottingham Co-op Society Ltd v. Cementation Piling and Foundations Ltd* and *D & F Estates Ltd v. Church Commissioners for England*), all decided by the Court of Appeal and the House of Lords, it seems clear that the possible trends suggested by Junior Books have been halted. Great emphasis in these cases is laid on the role of the contracts, and claims should be pursued down that route where possible. Within contracts, the parties have the opportunity to establish their responsibilities. Once they have done that, then it would be wrong to impose any further liability.

The difficulty created by Dutton's case, and perpetuated by the decision in Anns, can best be illustrated by reference back to the facts of *Donoghue v. Stevenson*. In that case, the negligent act (putting the snail into the ginger beer) caused harm to the claimant. What if, instead of complaining about harm she had suffered, the claimant were merely complaining about some harm to the ginger beer itself, e.g. that the presence of the snail had turned it sour? That would have been a complaint about the quality of the product, in which case the claimant would not have succeeded in her action in tort. Now translate that example into the building situation. A builder builds a house with defective foundations. When this is discovered by the owner, it has not caused him any physical injury, nor has the house damaged other property. The house merely suffers from a defect (defective foundations – snail in bottle) which damages the property itself (damaged house – damaged ginger beer). Yet in the Dutton and Anns cases, the courts chose to draw a distinction and to allow the claimants to recover.

The halt to this line of cases drawn by *Murphy v. Brentwood District Council* is greatly to be welcomed. In that case, Lord Keith said of the Anns case: 'It has engendered a vast spate of litigation, and each of the cases in the field which have reached this House [the House of Lords] has been distinguished. Others have been distinguished in the Court of Appeal. The result has been to keep the effect of the decision within reasonable bounds, but that has been achieved only by applying strictly the words of Lord Wilberforce and by refusing to accept the logical implications of the decision itself. These logical implications show that the case properly considered has potentiality for collision with long-established principles regarding liability in the tort of negligence for economic loss. There can be no doubt that to depart from the decision would re-establish a degree of certainty in this field of law which it has done a remarkable amount to upset.' The facts of the Murphy case are considered on page 260.

## 2   Negligent misstatements

The policy of excluding pure economic loss claims caused particular problems in those cases where the negligence consisted of a negligent statement rather than a negligent act. In such cases economic loss is the most likely form of harm to result. For example, if a firm wishes to do business with a new and unknown customer, it may have reservations about extending credit to that customer. To protect itself, the firm may ask permission to approach the customer's bank for a credit reference. If the reference is negligently given by the bank and the firm relies on it, extends credit and sustains heavy economic loss as a result, can the firm sue the bank for negligence when the loss it has suffered is purely financial?

In a case in 1964, *Hedley Byrne v. Heller*, where the facts were similar to the example given above, the courts drew a distinction between negligent acts and negligent statements. If the negligent statement results in physical damage, the normal rules of negligence operate. So, in one case where an architect negligently stated that a wall was safe to be left standing, he was successfully sued in negligence by a workman who was injured when the wall collapsed (*Clay v. Crump*). Where the negligent misstatement causes economic loss, liability may now arise either under the principles set out in the Hedley Byrne case or by virtue of the Misrepresentation Act 1967. (Liability for false statements in the law of tort had previously been covered only by the tort of deceit. In order to succeed in an action for deceit, it is necessary to prove that the false statement has been made knowingly, without belief in its truth, or recklessly, careless whether it be true or false. There has to be dishonesty, mere carelessness is not enough. Some relief can also be obtained from the law of contract in cases where the false statement does not amount to a term of the contract, but induces the plaintiff to enter the contract. In those circumstances, the plaintiff may be permitted to rescind the contract. It would be useful at this point to refer to the relevant material on misrepresentation in the law of contract.)

The general effect of the Hedley Byrne case is that certain negligent misstatements are now actionable, even if the only loss is pure economic loss. The judges in the case were no doubt cautious in view of the sizeable step they were taking to change the law. In consequence, they wanted to impose restrictions on those negligent statements causing economic loss which should be actionable. Not only should the maker of the statement reasonably foresee that it was likely to cause harm, there must also be a special relationship between the plaintiff and the defendant. The judges in the Hedley Byrne case were not at all clear exactly what would amount to a special relationship. It seemed it would exist if the party making the statement knew, or ought to have known, that the other was trusting him and was going to rely on what he said. Much of the need to establish a special relationship under the Hedley Byrne case is caused by the judges' desire not to open up too wide a potential liability.

The possible scope of the Hedley Byrne principle is very wide. It could apply, for example, to professional advice given by architects, surveyors, engineers, accountants, lawyers or doctors, amongst others. It is clear from the cases that some considerable overlap can occur between the rules of the law of tort and the law of contract. Most professional people are likely to have a contract with their client, so it is more likely that they will be sued for breach of contract in respect of any negligent service. However, two cases serve to illustrate how reliance on a negligent misstatement made outside the context of any contract can cause extensive loss. In *Smith v. Eric S. Bush*, a prospective purchaser of a house sought a mortgage. The building society instructed a surveyor to carry out a house valuation for them, and on the strength of that survey, the building society loaned money to the purchaser. Relying on the survey, the purchaser decided to go ahead with the purchase. There was no contract between the purchaser and the surveyor, who had been engaged by the building society. However, the court held that the surveyor was liable to the purchaser, as the situation was akin to contract, in that the surveyor knew that the consideration he received derived from the purchaser. He also knew that the purchaser would rely on his report in determining whether to buy the house. The valuation report was negligently prepared, and in consequence, the purchaser was successful in a claim for damages against the surveyor. The surveyor sought to rely on a disclaimer of liability, but the court further held that it would not be fair and reasonable to allow reliance on the disclaimer, as the surveyor must have known that the purchaser would be supplied with a copy of the valuation report and would be likely to rely on it.

In *Caparo Industries plc v. Dickman*, the auditors of a company issued inaccurate and misleading accounts for a company, Fidelity, in which Caparo held shares. Caparo bought further shares on the strength of the audited accounts, which were circulated to all shareholders. Later the same year, they mounted a takeover bid for Fidelity, but once they acquired the company, Caparo discovered the true state of its finances. They alleged that the auditors owed them a duty of care when making statements, both as a shareholder and as a potential bidder to take over the company. In the House of Lords, it was held that no duty was owed to them in either capacity. As a member of the general public, looking at the audited accounts with a view to takeover, there was no sufficiently proximate relationship between the auditors and Caparo. As a shareholder wishing to buy more shares in the company, no duty was owed as an individual shareholder should be in a position no different from that of a member of the general public. The accounts were prepared and circulated to members of the company to enable them to better manage the company, not to permit them to make a personal profit. To find otherwise would be to impose a virtually unlimited and unrestricted duty on the auditors, and would suggest that foresight of harm, without more, was sufficient. Such a finding would run contrary to all the other recent decisions. In Caparo the court indicated that the law in this area should be allowed to develop on a case-by-case basis.

Looking back to the Hedley Byrne case itself, the actual claim of the claimant failed because the defendants had expressly disclaimed responsibility for their reference. The possibility of using such a disclaimer successfully is limited now by the provisions of the Unfair Contract Terms Act 1977. Under s.2, where a person seeks by a notice to exclude or restrict his or her liability in negligence which results in loss or damage other than personal injury or death, he or she may only rely on that disclaimer if it satisfies the requirements of reasonableness. It should be noted, however, that the Unfair Contract Terms Act 1977 applies only to a business liability situation.

The law relating to negligent statements is complicated because of the overlap of the rules of the law of tort and the law of contract. Certainly the Hedley Byrne principles can be important where a false statement has been made which does not induce a contract. Where a contract has resulted it seems likely that a plaintiff would prefer to use the rules of the Misrepresentation Act 1967, to take advantage of a more favourable burden of proof which would then lie with the defendant. The law is probably much more confused than it need be, due in part to the fact that the Misrepresentation Act 1967 was passed on the basis of earlier Law Revision Committee reports and suggestions, made at a time before Hedley Byrne was decided in the courts. The legislators went ahead with the Act, making no attempt to blend in the developments which had occurred subsequently in the Hedley Byrne case.

## 3 Judicial immunity situations

As a result of Hedley Byrne, the courts have recognized that there are times when there will be liability in tort for negligent misstatements. Nevertheless, no liability attaches if the statement is made by a judge, or arbitrator in the course of legal proceedings. Quite simply, the reason for this is public policy. It is regarded as being against the public interest to have the constant possibility of litigation being reopened. The remedy, where it is believed that an error has been made, is to use the appeal system.

The immunity of arbitrators was considered in the case of *Sutcliffe v. Thackrah*, where an architect who had negligently certified work claimed to be acting as an arbitrator in the certification process. That claim was not accepted by the court but, had it been correct, no action in negligence would have lain against the architect. The immunity of arbitrators is dealt with by s.29 of the Arbitration Act 1996.

## 4 Nervous shock situations

Nervous shock in this context means actual illness like psychiatric disorder or depression. Nervous shock as a consequence of negligent behaviour causes no problems where the victim has also sustained physical injuries. So, a victim of a road accident, run over by a negligent motorist, may be claiming in respect of broken limbs, fractured skull, punctured lungs *and* nervous shock. The cases which cause problems are those where the only harm is the nervous shock. The problem is more acute in those cases where the victim of the nervous shock was never in any danger of suffering physical harm.

Take the example of the mother who looks out of her bedroom window and sees her small son run over by a reversing lorry. No doubt it is likely she will suffer nervous shock but she was in no personal physical danger. Or, take the example of the person who sees nothing of an accident to a loved one, but reads of it later, or hears of it from a third party. Are such people owed a duty of care by the defendant? Inevitably, these cases raise the usual policy issue, that liability may come to be owed to an 'indeterminate class'. Moreover, the courts may be anxious that these cases could give rise to false claims, as medical knowledge about shock and mental illness is not a particularly exact science.

Many of the issues in this area were clarified by the House of Lords in 1983, in the case of *McLoughlin v. O'Brien*. A horrendous road accident had occurred, in which the claimant's husband and children were badly injured, and one child was killed. A friend came to tell the claimant the news of the accident and took her to the hospital. There, the claimant learnt the full extent of the tragedy and saw her family in harrowing and distressing circumstances. It was held that her nervous shock was reasonably foreseeable, and she should be permitted to recover damages.

A significant number of the anxieties about the development of this area of negligence were aired in the McLoughlin case. Should claims be limited to those people who saw an accident or its immediate consequences? Or should claims be limited to those people who had some reasonably close relationship with the victims of the accident? (If this point is regarded as significant, that could eliminate claims from persons who suffer nervous shock as a result of assisting in an accident or disaster, where they might see and experience dreadful injuries and suffering.) Should claims be limited to those people who are reasonably close in time or place to the accident?

If damages for nervous shock are dependent on reasonable foresight of harm to the claimant in question, that in itself will act as a significant filter to ensure that only a proportion of cases are likely to succeed. In an action brought by relatives of spectators killed in the Hillsborough football stadium disaster, in respect of nervous shock suffered as a consequence of watching the events unfold on television, the House of Lords rejected their claim for damages. It seemed that watching edited pictures on TV could not satisfy the test of proximity, namely that the psychiatric disorder must have been the product of what the claimant saw or heard of the event or of its immediate aftermath. In a separate action, police officers on duty at Hillsborough were not awarded damages for nervous shock.

## 5   Negligent exercise of statutory powers

*Dutton v. Bognor Regis UDC* opened up a significant development in the law of negligence, when an owner of a building sued the local authority, alleging negligence with regard to the inspection of the foundations of the building. In many of these cases, the builder himself will not be worth suing if he has gone into liquidation or where there are doubts about his solvency. Local authorities look like richer pickings in these circumstances!

Where a local authority acts under statutory powers, as for example in the building inspection processes, there may be liability for negligence if, once having decided to inspect, the inspector fails to carry out the inspection properly. Of course, where there is merely a statutory *power* to act, as opposed to a statutory *duty*, the local authority has some discretion about whether to do anything at all. Its resources may be so scarce that it has to determine priorities about which inspections it will make. In these circumstances, if it decided to make no inspection, the question then arises whether the authority could be negligent for failing to inspect. It seems that there would be no liability in negligence, so long as the local authority has given proper consideration to the question of whether or not to inspect.

These issues were all considered in *Anns v. Merton London Borough Council* in 1977. This was another case involving defective foundations. Although the decision has now been overruled, some consideration of the approach taken by the House of Lords in Anns is vital to an understanding of the later developments. So far as local authorities are concerned, Anns appeared to decide that a local authority may be liable in tort to a building owner in relation to its statutory control over building operations. The extent of the liability would be the cost of remedying a dangerous defect in the building which had resulted from a negligent failure by the local authority to ensure that the building was constructed in conformity with the appropriate standards laid down in building regulations.

The difficulty created by this broad-ranging approach can best be illustrated by comparing a defective chattel with a defective building. If you make a chattel which contains some latent defect that causes damage or injury to persons or property, you may be liable in negligence if you made the chattel negligently (*Donoghue v. Stevenson*). But if the chattel is merely defective in quality, then your liability in that regard lies only in contract. Anyone acquiring the chattel who has no contract with you may suffer economic loss as a consequence of having to have the chattel repaired or having to simply abandon it but, of course, you would owe no liability in tort in respect of mere economic loss.

Now apply these same principles to a defective building. If a builder constructs a building with a latent defect (e.g. defective foundations), the builder may be liable in tort for injury to the person or damage to property arising from the dangerous defect. But if the owner of the house gets to know about the defect before it causes any injury or damage, then the owner would have rights in contract if the house had been built for him, but would have no rights in tort, because his loss would be mere economic loss – i.e. the cost of the repair, or the cost of abandoning the building if it cannot be repaired. To hold otherwise would be to impose on the builder a warranty as to the quality of his work which would be owed not only to the original owner but also to all who subsequently acquired an interest in the property.

The Anns case, following the earlier decision in Dutton, erred in failing to take account of this distinction between defects causing external harm and defects which merely harm the product itself. If, however, the distinction is a proper one, then it raises the question about liability of local authorities. If the builder's duty should be limited as suggested above, then it seems altogether wrong to make the local authority liable to a greater extent than the builder. The function of the local authority in inspecting building work is to ensure that there is compliance with building regulations. If the local authority were to negligently fail to ensure compliance with those regulations, it seems absurd that it would then incur a liability greater than that of the builder himself. In the Murphy case, overruling Anns, the Lord Chancellor, Lord Mackay, makes the point that it is not a proper exercise of judicial power for the courts to create large new areas of responsibility on local authorities in respect of defective buildings. That, properly, is the role of Parliament. He was strengthened in his view by the consideration he gave to the Defective Premises Act 1972 (see page 275).

During the period whilst the Anns case was still good law, it was clear in a number of other cases that the judges were anxious not to let the Anns principles in relation to the exercise of statutory powers run too far. In *Curran v. Northern Ireland Co-Ownership Housing Association Ltd*, a house extension on the plaintiff's home had been financed by a grant provided by one of the defendant organizations. It was allegedly so badly built that the work had to be completely redone. The plaintiff argued that he had bought the house in reliance on the fact that a grant had been given, assuming that a grant would only have been paid to the previous owner if the work had been done properly. The defendant organization had a statutory power to make grants to improve houses. A grant could only be made if the property was in stable structural condition on completion of the work, and if the work had been done to the satisfaction of the defendant organization. If the work was defective, could a subsequent owner of the property hold the defendant liable for negligence? Their Lordships found no negligence here, and indicated that caution should be exercised about extending duties of care to statutory bodies exercising their statutory powers to control third parties. Such an extension should only occur if:

1    the statutory power was directed to safeguarding the public, as would be the case in situations like Anns where the building regulations could be seen to have such a purpose;
2    it is clear that a proper exercise of the statutory power would have prevented the danger; and
3    if the negligent exercise of the statutory power created a latent defect which could not have been discovered and remedied before harm occurred. The whole object of the power in the Curran case was to protect the public purse, not to protect the health and safety of the general public. In consequence, the case could be easily distinguished from the Anns case.

Thankfully, the Anns case has now been overruled by *Murphy v. Brentwood District Council*. This case again arose out of defective foundations. A pair of semi-detached houses had been built on an infilled site on a concrete raft foundation to prevent damage from settlement. The local authority approved the plans. Some years after the house was completed, the owner noticed serious cracks, and he then discovered that the raft foundation was defective and had become distorted. Necessary repairs would have cost £45,000, which the plaintiff could not afford. Instead, he sold the house, and realized £35,000 less than the market value, had the house been in sound condition. The plaintiff sued the local authority to recover that sum. The House of Lords held that the plaintiff was unable to recover from the local authority, as his loss was mere economic loss.

Although the Murphy decision has been greeted with relief, in that it seems to take the law back to a point from which it should never have moved, the case does still leave some questions unanswered. For example, would the local authority be held liable in negligence if a defect in a building, which their negligence has failed to prevent, then caused actual physical injury to someone, or physical damage to some other property? This point was not decided in the Murphy case and must await a future decision. Cases are already emerging at first instance which are raising some of these questions.

Inevitably, these complex cases are often the result of an aggrieved person looking for someone who is financially worth suing. It has already been suggested that the erring builder may be insolvent, but the other difficulty which a claimant faces is the problem of limitation periods. In all tort actions, a claim must be commenced within the appropriate time limit laid down by the Limitation Act 1980. An action relating to damage to property must be commenced within six years of the cause of action accruing. Usually, this is the moment when damage occurs. That does not necessarily mean when the defective building was constructed. Since the decision of the House of Lords, in *Pirelli General Cable Works Ltd v. Oscar Faber and Partners* in 1983, the cause of action arises when physical damage actually occurs to the building, and it is matterless whether the claimant knew or could have known about it. This rule is undeniably harsh, especially as it fixes the date for the cause of action accruing for subsequent as well as present owners. It runs counter to an earlier test proposed by the courts, that time should only run from a point when the claimant could have discovered the defects. Some help with this problem has now been given under the Latent Damage Act 1986. In those instances where damage cannot be discovered immediately (i.e. defective foundations), the claimant must bring an action within three years from the time when damage was discovered, or when it should have been discovered, if using reasonable care, subject to an overriding rule that the action must be commenced within 15 years of the building being completed.

## 6    Omissions to act

As a broad general rule there can be no liability in negligence for a pure omission to act. If a road accident occurs where victims are trapped in a car, a passer-by commits no legal wrong if he or she makes no effort to save those trapped. This is described as pure omission, pure in the sense that there was no existing duty to act. If there is some existing duty, then an omission can be as wrongful as a negligent act. So, a parent of a small child has a duty to care for the child. If he or she omits to feed the child, such behaviour could be actionable.

If there is no duty to act, but someone decides to act as may be the case in the example above, where the passer-by decides to try to release the crash victims from the car, then if he or she behaves negligently, the law generally takes the view that there will be no liability for the negligence unless the passer-by makes matters worse. As there is no duty to act, then if you do act, there is no duty to effect an improvement.

This principle can be regarded as being subject to a most important exception in relation to the exercise of statutory powers. In such a case, although there is no duty to act, the defendant has a duty, if he or she chooses to exercise the power, to effect an improvement.

## Breach of the duty of care

If a duty to take reasonable care is owed to the particular claimant, the next stage is for the claimant to prove that the defendant was in breach of the duty, i.e. that the defendant failed to take reasonable care. The obvious question must then be asked – how much care should the defendant have taken? What would amount to reasonable care in the particular circumstances? If the defendant falls short of a reasonable standard, then he or she is in breach of duty.

Inevitably, there is no absolute precision in measuring what amounts to reasonable care in any given situation. A court will need to weigh up factors like: the degree of risk involved in the defendant's behaviour; how serious was the harm which was likely to be caused; was there any particular social value or utility in the defendant's activities; and how costly and inconvenient would it have been to take precautions to eliminate or reduce the risk? The standard of care expected is the care of a reasonable person, although no such person may exist in fact! This means that the defendant's behaviour must be assessed against an objective standard, and it is matterless that the defendant tried hard or did his or her best. Of course, where a defendant holds out as possessing a particular skill or capability, he or she will be expected to display reasonable amounts of that skill or capability. So a surgeon must operate with the care that a reasonable surgeon would take; a builder must build with the care a reasonable builder would take; and so on. If there is a suggestion that a child has been negligent it may be appropriate to ask whether the defendant child took the care that could be expected of a reasonable child of the same age.

It is important to remember that, in formulating a standard of reasonable care, everyone is assumed to have knowledge about common facts, e.g. fire burns, water can drown, knives cut. In particular fields, a person with expert knowledge will be expected to display a reasonable amount of such knowledge. When judging whether a defendant has taken reasonable care, some guidance may be given by looking at what is usual or common practice in the activity or industry or operation in question. So an employer who fails to provide protective clothing for employees may be doing exactly the same as other employers in the same line of business. Arguably, all the employers may be being negligent, so general practice is not an absolutely reliable guide.

In assessing how much risk is involved in the defendant's activities, one judge has said that 'people must guard against reasonable probabilities, not fantastic possibilities'. If you can only conceive of harm occurring in some obscure or bizarre turn of events, then the risk may be so small that you are justified in ignoring it. After all, the law only requires reasonable care, not an absolute guarantee of safety in all circumstances.

In practice, it may be very difficult to separate questions like degree of risk from seriousness of harm likely to be caused. Obviously, even in cases where a very small risk is involved, it may be necessary to take more elaborate precautions where the harm caused, if any, would be very serious. If an employer has huge tanks full of liquid, he may be able to foresee that employees may fall in, or objects may fall in causing the liquid to splash out. If the liquid is cold water, some barrier to prevent employees falling in may be all that is necessary in order to show reasonable care. If the liquid is some caustic solution, likely to burn on the slightest splash, then the employer may have to consider some protective cover for the tanks. The precautions which are reasonably necessary may also be determined by asking whether employees are meant to be in the vicinity of the tanks or not, and whether employees have been issued with protective clothing.

It will already be apparent that reasonable care will fluctuate according to the exact circumstances of the case. Failing to test the temperature of the bath water may well be negligent if the bath is intended for a baby or someone lacking the mental ability to perceive the dangers of scalding water for themselves; in other circumstances such a failure may not be negligent, because the user of the bath should have checked for themselves. A defendant engaging in some activity designed to save life and limb may well escape liability for negligence even though his or her behaviour may amount to a failure to take reasonable care in less socially useful circumstances. In *Watt v. Hertfordshire County Council*, firefighters rushed to the scene of an accident where a jack was needed in order to free a woman who was trapped. Because of the emergency nature of the call, the jack was transported on an unsuitable vehicle, and was not secured during the journey. It shifted on the lorry, and a fireman was injured. There was held to be no negligence here, but it was made clear that if the jack had been transported in a similar way in a commercial enterprise, there would have been a failure to take reasonable care.

Even in those cases where some minute risk is foreseen, it may be possible to eliminate that risk entirely, but only by massive expenditure. The law does not impose such an impossible standard. If the risk could be eliminated or reduced by reasonable expense or practicable precautions, then the defendant should have done what he or she could. But again, it must be emphasized that he or she is only required to behave reasonably. In *Latimer v. AEC Ltd*, oil normally ran away in channels in a factory floor. It had been spread over the surface of the floor when the factory was flooded. The factory occupier scattered some sawdust, mopped up as much of the oil as possible, and gave a warning to employees to take care as the floor was slippery. The plaintiff slipped and was injured but the defendants were held not liable as they had taken reasonable care. The risk was not particularly great, any harm they could foresee was not likely to be particularly serious, and short of closing down the factory (which would have been extremely costly), there was nothing more they could have done.

The discussion of reasonable foresight of harm may raise the idea that what the courts are actually concerned with is hindsight. Inevitably, when a court tries to consider what the defendant should have foreseen, the judges will be influenced by policy factors, as they are in so many aspects of the development of the rules of negligence. So, for example, in cases involving motoring accidents, the fact that the defendant is insured could well have a quite distorting influence on the judgment of what amounts to reasonable behaviour. In *Nettleship v. Weston*, it was held that a learner-driver must achieve the standard of a reasonably competent driver. Likewise, in medical negligence cases, policy factors like a fear of encouraging defensive medical practice may play a part in determining what is reasonable.

It has already been indicated that it is up to the claimant to show that the defendant was in breach of duty, i.e. that the defendant had failed to take reasonable care. The claimant may be assisted in proving negligence by s.11 of the Civil Evidence Act 1968. If the defendant has been found guilty of a criminal offence involving negligence, that fact may be used as evidence of negligence in subsequent civil proceedings. This is obviously a very useful rule where a person has already been convicted of certain driving offences, but it could also be relevant where an employer had been found guilty of offences under the health and safety legislation.

A claimant seeking to prove negligence may also rely on the maxim *res ipsa loquitur*: the facts speak for themselves. This could be pleaded in cases where the facts argue of no explanation other than negligence. In one case, a man was walking along past a warehouse when sacks of sugar fell on his head. What other explanation could there possibly be than that someone was being negligent? It is only appropriate to plead *res ipsa loquitur* if the defendant had management or control of the thing which caused the harm. This tends to depend on whether there was any possibility of some external interference. Where a claimant relies on the maxim, it is open for the defendant to rebut the finding of negligence, if he or she can do so.

## Damage resulting of the right type which is not too remote
The type of harm which must be suffered in order to succeed in a claim for negligence has already been considered. The two significant issues which

must now be addressed are the questions of causation and remoteness. Causation is concerned with the problem of whether the defendant's conduct caused the claimant's damage. Remoteness is concerned with the cut-off point at which the law regards the defendant as no longer liable to compensate the claimant. If such a cut-off point were not established, a defendant could owe limitless liability.

1   *Causation* – a defendant may behave negligently towards a claimant and yet still not be the cause of the injuries. In *Barnett v. Chelsea and Kensington Hospital Management Committee*, a doctor was negligent in failing to examine a patient who had been brought into casualty, complaining of feeling sick after drinking tea. In fact, the patient was dying from poison and nothing could have saved him, so the doctor's negligence did not cause his death. It is usual to approach the causation question by using a 'but for' test. But for the defendant's negligence would the claimant have died in the Barnett case? Unfortunately, this test is by no means conclusive, and it may be particularly ineffective in giving an answer where the harm caused to the claimant was inflicted by two or more people, in quite separate incidents. In *Chester v. Afshar*, it appeared that the House of Lords was stretching conventional causation principles. In this case, the claim related to a surgeon's failure to advise a patient on risks associated with an operation. The claimant suffered a known but unusual side-effect of the operation. Counsel for the defendant surgeon argued that the risk associated with the operation was the same whether or not the surgeon gave advice about potential side-effects – that is, it was not the lack of advice that caused the damage, it was the operation, which had not been carried out negligently and which the patient may have chosen to have at some later date even knowing the risks. The majority of the court held that the test of causation was met on grounds of policy. The risk of suffering side-effects as a result of the operation came within the scope of the surgeon's duty to warn the patient, and therefore the side-effects suffered could be regarded as having been caused by the surgeon's breach of that duty. Subsequently, in *White v. Paul Davidson & Taylor*, the Court of Appeal refused a similar type of claim relating to legal advice on the grounds of lack of causation. Chester was distinguished on its facts and the point made that this case clearly did not set out to establish new rules on causation.

2   *Remoteness* – assuming there is no difficulty in proving causation, the claimant must then show the harm he or she suffered was not too remote from the defendant's negligent act. Any act can give rise to endless consequences, and one can often remark with hindsight that if only a particular event in the past had not happened, none of the later events would have resulted.

As a feature of not imposing excessive liability, the law has created the rule that a claimant can only recover for harm which is not too remote a consequence of the negligent act. As soon as such a rule comes into existence, there must be principles which will assist in finding the 'dividing line'. The test used by the court is that a defendant is only liable for such consequences of a negligent act as could reasonably be foreseen. An objective test is used. Would a reasonable person reasonably foresee a particular consequence? As an example, if it is intended to light a bonfire, it is reasonably foreseeable that people may fall into it, or that a spark from the fire may set alight nearby fencing, trees or property. But is it reasonably foreseeable that fumes from the bonfire will choke to death battery chickens being reared on adjoining land?

In a case which is accepted as establishing the approach to issues of remoteness, a ship called the *Wagon Mound* negligently discharged oil into Sydney Harbour as it was sailing out. By the action of the wind and tide, the oil eventually floated all around the wharf area where the claimants were carrying out welding operations. The welding work was suspended until the claimants checked whether the oil would ignite. They discovered that this was unlikely as it was furnace oil with a high ignition point. Welding work recommenced and an extensive fire was caused when the oil did in fact ignite. The defendants, the owners of the *Wagon Mound*, were held not liable. Although they owed a duty to take care in relation to the claimants, and were in breach by negligently discharging the oil, the damage they had caused was not a reasonably foreseeable consequence of their negligence.

The damage caused must, therefore, be of a reasonably foreseeable type. In the *Wagon Mound*, no doubt, the ship owner would have been liable if the claimants had been complaining about oil fouling their slipways. This refinement can only be pursued so far. Take the example of *Bradford v. Robinson Rentals Ltd*, where a van driver was sent on a long journey in the depths of winter in a defective and unheated van. The employers owed him a duty, and were in breach. Could the employee recover for his injuries, which consisted of frostbite? No doubt that is quite an unusual complaint in the UK but nevertheless some harm to his health was reasonably foreseeable and frostbite was therefore in the same class of risk. Compare that example with *Doughty v. Turner Manufacturing Co. Ltd*, where the employer had a vat of acid maintained at a very high temperature. It was reasonably foreseeable that employees might be injured by falling in or by being splashed if any object fell in. The employer would certainly be liable if an accident had occurred in either of those ways. What actually happened was that the lid of the vat, made from asbestos, slid into the acid. A chemical reaction was caused which resulted in an explosion and employees were injured by the acid erupting. The employers were held not liable as the damage was too remote. It had occurred in a manner which was not reasonably foreseeable.

When deciding the remoteness question a court will bear in mind the following points:

1   *The defendant must take the claimant as he finds him* – if a defendant driver is unfortunate enough to knock down a claimant who suffers from a

weak heart and dies as a result, he or she may argue that, whereas the claimant's injury has proved fatal, any normal victim would only have been shocked and bruised. Again, he or she may crash into the back of another vehicle, which may be old and rusty or an expensive Rolls-Royce. But in each example the defendant will be liable to the full extent of the injury or damage caused, if injury or damage of that type could have been foreseen.

2   *The consequences of the defendant's breach of duty may have been 'overtaken' by some act which intervenes* – the defendant's breach may satisfy the 'but for' test, yet another event is regarded as the sole cause of the claimant's damage. For example, an employer's breach of duty towards an employee causes the employee a slight injury, and an ambulance is called to take the injured man to hospital. The ambulance driver is negligent, causes a crash and the employee is killed. The crash is then the new act intervening.

If the court takes the view that the harm which the employee has suffered is not too remote from the employer's act, it would consider the question of contribution, if more than one person is liable to the claimant. The situation is governed by the Civil Liability (Contribution) Act 1978. The Act provides that where two or more persons are liable in respect of the same damage, each may recover from the other, or others, a contribution of an amount regarded as 'just and equitable', taking into account his or her share of responsibility for the damage.

Even if the claimant has satisfied all the above aspects of proof, the claim may still fail if the defendant has a complete defence (e.g. *volenti*). More usually the claim may fail in part, if it could be shown that the claimant was partly to blame for his or her own injuries, i.e. contributorily negligent. The present rules are to be found in the Law Reform (Contributory Negligence) Act 1945. If a claimant suffers damage partly through his or her own negligence and partly through the negligence of the defendant, the compensation will be reduced to the extent which the court thinks just and equitable, taking into account the claimant's share of the responsibility.

To use the defence, the defendant must show that the claimant failed to take reasonable care for his or her own safety. As with negligence itself, the standard of care varies according to the circumstances. Thus, an experienced, mature workman can be expected to take more care when carrying out his work than a new apprentice. There has been a modern extension of the use of this defence in motoring cases when an injured claimant has failed to wear a seatbelt or crash helmet. In such cases, he may well find himself regarded as contributorily negligent (*Froom v. Butcher*).

If the claimant has consented to the risk, he or she should have no claim if injured. The scope of the defence is very limited and a defendant is more likely to plead contributory negligence. The defence can be relevant, however, when a defendant alleges that the claimant impliedly consented to a lower standard of care, for example in relation to sporting events, both for spectators voluntarily sitting there to watch and participants in the sport.

Any account of the rules of negligence shows that there are many obstacles in the way of a successful claim. Often, cases turn on their own individual facts and there are numbers of conflicting precedents. The present process of claims through the courts has been called 'a lottery'. The question was considered by a Royal Commission on Civil Liability for Death and Injury, which reported in 1978. Changes recommended by the Commission have not been made, but it was suggested in its report that motor vehicle injuries should be dealt with on a 'no-fault' basis. This would mean that an injured person could be compensated out of a central fund, regardless of whether he or she could prove anyone to be at fault. Naturally, such a system removes the vagaries and imponderables always likely to be present in a negligence action.

## Occupier's liability

Although the law can now be said to have general principles relating to negligence, there are some categories which pose their own special problems and for which particular rules have been created. An important example is the liability of occupiers of premises, whose duties are now laid out in the Occupiers' Liability Acts 1957 and 1984. The importance of the rules is obvious. Where a person enters premises and trips on a worn stair or collides with an obstacle because the lights are not working, liability can occur under the Acts. The main duty under the 1957 Act is a common duty of care owed by occupiers to their visitors. This requires the following analysis:

- Who is an occupier?
- What are premises?
- Who are visitors?
- What is a common duty of care?

### The occupier

This word is not defined by the Act, but is generally understood to mean the person who has occupational control of the premises. He or she is the person in the best position to know who is likely to be using the premises, and to know about, and be able to put right, anything wrong with the premises. The occupier does not need to be exclusively in occupation; it is possible for premises to have more than one occupier. For example, in *AMF International Ltd v. Magnet Bowling Ltd*, a building contractor was held to be a joint occupier.

### The premises

This word is wide enough to include not only buildings, but also land itself. What is more, by s.1(3) the Act extends its meaning to cover vessels, vehicles or aircraft. Premises will also include sporting stadia.

## The visitors

The range of people 'using' premises can be very wide, from a canvasser, a postman, an invited guest, a gasman, a deliveryman, even a burglar. The most important point is to distinguish between people lawfully on premises and trespassers. A person's presence on premises could be lawful because:

1    He has express or implied permission given by the occupier.
2    His entry is authorized by law (e.g. a policeman with a search warrant, an electricity official to turn off the supply, or factory inspector to investigate a work accident).
3    His entry is by virtue of a contract (e.g. going into a cinema after buying a ticket).

Even where entry on to premises is initially lawful, a visitor can subsequently become a trespasser if the permission to be there is withdrawn. For example, a shopper might enter a department store in a drunken state, create a disturbance and then be asked to leave. If refusing to go, he or she becomes a trespasser. What is more, a visitor must only use the premises for the purposes for which he or she is invited or permitted to be there.

It should be noted that persons who enter under a public or private right of way are not visitors for the purposes of the 1957 Act.

## The common duty of care

Under s.1 of the 1957 Act the occupier's duty is owed in respect of dangers due to the state of the premises themselves *and* due to things done or omitted to be done on the premises.

By s.2(2), the duty owed under the Act consists of taking such care as is reasonable in all the circumstances of the case to see that the visitor will be reasonably safe in using the premises for the purposes for which he or she is invited or permitted to be there.

The Act also sets out a number of factors to be taken into account when determining whether or not there has been a breach of duty:

1    When an occupier is considering how careful a visitor can be expected to be, he or she must be prepared for children to be less careful than adults (s.2(3)).
2    An occupier can, however, expect visitors who use the premises in the exercise of their calling (e.g. where a window-cleaner comes to the premises to clean the windows) to appreciate and guard against risks incidental to such calling (s.2(3)). For example, if the occupier asks a lift engineer to repair a defective lift, he or she will not be liable for damages caused to the engineer by that lift, but will be liable if the engineer cuts a hand on an already broken pane of glass in the front door when entering the premises.
3    If the occupier has given warning of a danger (e.g. by a notice stating 'Cliff edges crumbling – danger of collapse') this is not enough, in itself, to

protect him or her from liability (s.2(4)). But it may in all the circumstances be sufficient to enable the visitor to be reasonably safe, and by this means the occupier may therefore have discharged their duty.

4  The occupier may know of a state of affairs on the premises which he or she cannot personally remedy, which will involve asking a contractor to undertake work of construction, maintenance or repair. Section 2(4) provides that a visitor injured as a result of the faulty execution of such work by a contractor will not succeed in a claim against the occupier so long as:

(a)  the occupier acted reasonably in entrusting the work to a contractor, and

(b)  the occupier had taken reasonable steps to believe that the contractor was competent, and

(c)  the occupier has taken reasonable steps to believe that the work had been properly done. Note that this could require the occupier to employ someone to supervise or check the contractor's work, e.g. where building work is being carried out, by the employment of an architect.

This section was extensively considered by the House of Lords in *Ferguson v. Walsh and others*. The plaintiff had been injured while working for Walsh, the first defendant, on a demolition job. Walsh had been asked to undertake the work by X, who in turn had been given the job by the district council who owned the site in question where the demolition was to be carried out. By the terms of the contract between X and the district council, X was not allowed to subcontract the work without the council's permission. Nevertheless, that was exactly what X had done by his contract with Walsh. The first issue in the case was to decide if Ferguson was a visitor for the purpose of the Act. It was decided that he was, because he was effectively there by X's authority, and X was a joint occupier. Whether the council as joint occupiers were liable to Ferguson depended on s.2(4). The council had acted reasonably in entrusting the work to X and were held not to be liable. On its wording, s.2(4) appears to be limited to construction, maintenance or repair, but the House of Lords said that those words should be construed broadly and purposively, by which interpretation they were sure that 'construction' would embrace demolition.

5  The liability of the occupier may be limited or excluded, to the extent permitted by law. Section 2 provides that the occupier owes a common duty of care 'except in so far as he is free to and does ..., restrict ... or exclude his duty to any visitor or visitors by agreement or otherwise'. There are two principal limitations to an occupier's right to exclude liability:

(a)  Even where he can successfully limit his liability he can only cut it down to the level of responsibility he would owe a trespasser. Otherwise a person entering lawfully could have less protection legally than a trespasser.

(b) More important, any limitation of liability must be read subject to the Unfair Contract Terms Act 1977. In a 'business liability' situation an occupier cannot exclude or limit liability for death or personal injury caused by negligence. Nor can he limit liability for other damage caused by negligence unless the restriction satisfies the requirement of reasonableness.

6    When a person enters or uses premises by virtue of a contractual right, then unless the contract states otherwise, it is implied in it that the occupier owes a common duty of care. If the contract does contain express terms about the occupier's liability, these may be subject to the Unfair Contract Terms Act 1977.

A visitor to premises may suffer personal injury or damage to property, or both. When he or she claims against the occupier, the defences which may be pleaded are:

1    Consent – s.2(5) expressly preserves consent as a defence.
2    Contributory negligence.

## Occupier's liability to trespassers

As the 1957 Act extends protection only to visitors, it has no relevance when the person injured on premises is a trespasser or non-visitor. The rules which had been developed by the courts, culminating in the House of Lords' decision in *British Railways Board v. Herrington*, have now been overtaken by the Occupiers' Liability Act 1984. Cases like Herrington had determined that an occupier owed a lower standard of care to a trespasser, a standard described as a duty of common humanity. The lower standard took account of the fact that a trespasser was uninvited. Nevertheless, common humanity was a variable standard, depending on factors such as: whether the occupier knew of the likely presence of the trespasser; the kind of trespassers they were likely to be, e.g. children, burglars, poachers, squatters; the seriousness of the risk or danger; and whether or not the occupier knew of the risk.

The interplay of these factors can be seen in a case like Herrington. A child trespassed on the railway line and was electrocuted. The fence guarding the line was in a state of disrepair. The stationmaster knew that children were trespassing and had alerted the police. The line was something of an enticement, as the broken fence allowed children to take a short cut by that route to the nearby park. It was held that British Railways Board had failed to show common humanity. In view of the high level of danger, the likelihood of trespass, and the relative ease and cheapness of eliminating the danger for an organization with such resources, the child succeeded in an action for damages.

Under the 1984 Act, 'premises' and 'occupiers' are defined in the same way as for the 1957 Act. The duty under the Act is owed to persons 'other than visitors', and the duty consists of taking such care as is reasonable in all the circumstances of the case to see that the non-visitor does not suffer injury on the premises by reason of the danger concerned. For these purposes 'injury' is not to include loss or damage to property; it covers personal injury only. In this respect, the scope of the Act in relation to a non-visitor is narrower than the scope of the 1957 Act to a visitor.

The duty under s.1 is only owed by the occupier of premises if:

1    He or she is aware of danger or has reasonable grounds to believe it exists.
2    He or she knows, or has reasonable grounds to believe, that the non-visitor is in the vicinity of the danger or may come into that vicinity.
3    The risk is one against which, in all the circumstances of the case, he or she may reasonably be expected to offer the non-visitor some protection.

As the occupier owes a duty to take such care as is reasonable in all the circumstances, factors which were previously relevant, like the age of the trespasser, the nature of the premises, the character of the entry and the extent of the risk, will all still play a part in determining whether there is a breach of duty. One significant difference now is that the subjective element under the old common humanity rules may have gone. It used to be the case that the trespasser had to take the occupier as he found him or her, so if an occupier had limited financial resources, that might have been relevant in determining whether common humanity had been exercised. However, in determining what care is reasonable in all the circumstances, it is no doubt still open for the court to consider how burdensome it would be for the occupier to eliminate the risk.

Section 1(5) allows the occupier to discharge his or her obligations in appropriate cases merely by giving warning notices. It is doubtful if a notice saying 'Keep out' would suffice, as it would merely alert persons entering the premises to the fact that they were trespassers, and not to the fact that there was some danger. Warning notices may be totally ineffective if the trespassers are children. And of course, as the Act is concerned largely with trespassers, they may choose to enter premises at some unconventional place where no notice is displayed. The Act specifically retains the defence of *volenti* (consent).

One unsatisfactory feature of the Act is that it does not make clear whether it is possible to exclude or limit the liability it creates. If an occupier puts up a notice stating that he or she is excluding all liability under the 1984 Act, it would be normal to test such a notice against the rules of the Unfair Contract Terms Act 1977. That Act seems not to be relevant here, as it limits its definition of negligence so far as the statutory duty of care is concerned to the Occupiers' Liability Act of 1957. This aspect of the 1984 Act awaits clarification.

House of Lords authority on an occupier's responsibilities under the 1984 Act is found in *Tomlinson v. Congleton Borough Council*. In this case, the claimant was injured whilst diving into a lake. The lake was used for boating but signs warning that the water was dangerous and that swimming was forbidden were placed around the lake. The council were aware that, despite these notices, the lake was used for swimming and had plans to alter the edges of the lake to make it less accessible to swimmers. In the Court of Appeal it was held that the danger posed by swimmming in the lake was sufficient to have required the council to make the lake totally inaccessible. In failing to do so the council was therefore liable to compensate the claimant for the injuries that he suffered. When the case was heard in the House of Lords, the court held that the Occupiers' Liability Act 1984 required that the relevant danger was 'due to the state of the premises or to things done or omitted to be done on them'. The court decided that the lake was not unusually dangerous and that the claimant had taken the risk of swimming in, and diving into, the lake despite the obvious warning signs.

It is widely believed that the 1984 Act will make little difference to the sort of trespasser plaintiff who will succeed against an occupier. In cases like *Pannett v. McGuiness & Co.*, decided on the old 'common humanity rules', demolition contractors were occupiers of a site in a busy urban area where it was reasonably to be expected that there would be large numbers of children. It became necessary to burn large quantities of rubbish and, realizing that this would attract children, three workmen were employed to keep them away. Children, including the plaintiff aged five, were chased away from the site on a number of occasions. While the fire was still burning, the three men left the site and the plaintiff re-entered and was injured by the fire. The nature of the activity was hazardous, the likelihood of trespass was great and the attraction to children was obvious. Applying the principles of the 1984 Act, the defendants would be aware of the danger, would know that trespassers were in the vicinity of the danger and in all the circumstances the risk would surely be one against which the contractor could reasonably be expected to offer some protection. The defendants were held liable under the pre-Act rules and the same outcome would be likely to occur today.

## Liability of non-occupiers

For complicated reasons connected with the overlap of duties in contract and tort, the law has been slow to develop protection in tort for the person who suffers damage because of a defective building, where the defects are due to a non-occupier. A simple example of this is the case of a builder who built a house in 1970 for X. X sold the house to Y in 1980. The house has defective foundations and Y now discovers cracks appearing and distortion of door and window frames. Y has no contract with the builder – he bought the house from X. The builder had been negligent. Can Y succeed in a claim against the builder?

In 1972, the Defective Premises Act was passed, imposing a duty on persons who take on work for the provision of dwellings. That duty is owed to the person ordering the work and to any person who subsequently acquires an interest in the dwelling. The extent of the duty is to see that the work is done in a workmanlike and professional manner with proper materials so that the dwelling will be fit for habitation when completed.

Some very significant limits operate, however, to cut down the apparent effectiveness of the protection given by this Act:

- The protection does not apply to dwellings already protected by an approved scheme of the National House Building Council.
- The Act only applies to dwellings, not to all types of buildings.
- Any action based on the Act must be speedily brought as there is a limited period of only six years, which runs from the date of completion of the dwelling.

Although the protection afforded by this Act is useful, the very limited extent of it has encouraged claimants to seek alternative ways of obtaining redress. Earlier cases such as Dutton and Anns demonstrated a willingness of the courts to allow claims against local authorities to succeed, based on negligent inspections of defective foundations, despite the loss in these cases being properly classifiable as pure economic loss. The trend in those cases, already discredited as they were in *D & F Estates Ltd v. Church Commissioners for England*, has now been halted by the decision of the House of Lords in *Murphy v. Brentwood District Council*.

In that case, one of the judges, Lord Mackay, the Lord Chancellor, said: '… I am of the opinion that it is relevant to take into account that Parliament has made provisions in the Defective Premises Act 1972 imposing on builders and others undertaking work in the provision of dwellings obligations relating to the quality of their work and the fitness for habitation of the dwelling. For this House in its judicial capacity to create a large new area of responsibility on local authorities in respect of defective buildings would in my opinion not be a proper exercise of judicial power.' Lord Mackay was supported in this view by Lord Jauncey, who added: 'Parliament imposed a liability on builders by the Defective Premises Act 1972, a liability which falls far short of that which would be imposed on them by Anns. There can therefore be no policy reason for imposing a higher common law duty on builders, from which it follows that there is equally no policy reason for imposing such a high duty on local authorities.'

It seems now that a first purchaser of a defective building is therefore limited to contractual remedies and/or the protection provided by the 1972 Act. A subsequent purchaser of the property will have no contract with the original builder, and may therefore have to rely exclusively on the 1972 Act, subject to any rights acquired under the Contracts (Rights of Third Parties) Act 1999.

A practical application of the Act can be seen in *Andrews v. Schooling*, which was decided in 1991. An Edwardian house had been converted into flats, where the ground floor flat was bought by the plaintiff. Her flat also incorporated a basement. When the conversion work was undertaken by the builders, no damp-proof course was installed. Once the plaintiff took up occupation, she then found that her flat was affected by penetrating damp in the basement. She sued the builders under s.1 of the Act. The builders sought to argue that they would indeed be liable if they had put in a damp-proof course and done it badly (misfeasance) but that they could not be liable if they had done nothing (non-feasance). The court rejected that argument, indicating that it made no difference to the plaintiff whether it was misfeasance or non-feasance, as in either case the builders had not provided a house which was fit for habitation. Of course, in awarding damages to the plaintiff, we have to accept that this was to compensate her in respect of economic loss, and it is arguable whether that was ever the intention of the Act.

To date, there is no sign of Parliament tackling the issues raised by the Murphy case, and the limitations indicated on the 1972 Act.

# Breach of statutory duty

Many Acts of Parliament create obligations or duties, where a breach may result in the imposition of a penalty, very often a fine. Sometimes the courts have allowed the Act of Parliament to be used as the basis for a civil action in tort. Such an action is called breach of statutory duty. The difficulty lies in knowing which Acts of Parliament can be used in this dual way. Most industrial safety legislation has been so used, yet by contrast the Health and Safety at Work etc. Act 1974 makes it clear that its duties give rise to no civil liability (s.47). However, breach of regulations made by the Secretary of State under the 1974 Act does give rise to civil liability (s.47(2)), unless the regulations provide otherwise. To succeed in an action for breach of statutory duty a person must, therefore, show:

1    The Act in question was intended to create a civil action. This is the biggest hurdle unless a plaintiff can rely on a precedent showing that the Act has been used before as the basis of a civil claim.
2    The Act imposes a duty on the defendant which he or she has broken.
3    The breach of duty has caused harm which is not too remote.

In general, this action is most used by employees in relation to Acts of Parliament concerning industrial safety, when the result is often to give an injured employee two heads of claim – an action based on negligence and an action based on breach of statutory duty. The attraction of this tort for the employee is that the standard of behaviour expected from the employer is prescribed by the particular statute.

An example of breach of statutory duty of particular relevance to the construction industry is to be found in the Building Act 1984. This is the Act under which the Building Regulations are made. These regulations relate to the design and construction of buildings, and are designed to secure the health, safety and welfare of people in the building. Section 38 specifically provides that where a breach of any duty imposed by the Building Regulations causes damage, including death or personal injury, a civil claim for breach of duty will lie. This is in addition to any action in negligence which may be available in relation to the construction of the building. (Note that this section is not yet operative.)

# Defective products

Where a person is injured by using a defective product and is looking for some legal redress, there can be considerable overlap between the law of contract and the law of tort. If the injured person purchased the product, then he or she may sue the seller under the Sale of Goods Act 1979 and Sale and Supply of Goods Act 1994 if there has been a breach of any of the implied terms relating to reasonable fitness for purpose and satisfactory quality. Moreover, in a consumer transaction, the seller cannot exclude or limit liability for breach of these implied terms, because of the provisions of the Unfair Contract Terms Act 1977. In exceptional cases a buyer of defective goods may also have some limited contractual rights against the manufacturer, where there is a guarantee in operation. Such rights have been strengthened by the Sale and Supply of Goods to Consumers Regulations 2002. This 'contract route' to gain legal redress is limited to persons who can prove that they bought the goods, so it cannot assist a person who receives the goods as a present, or who is injured when using borrowed goods. It is rather inefficient, in the sense that it may set in motion a chain of legal actions – buyer v. seller, seller v. wholesaler, wholesaler v. importer, importer v. manufacturer. Several costly actions may thus be necessary to lay liability at the door of the manufacturer. The 'contract route' relies on suing a retailer who owes a strict liability, who will be liable whether at fault or not. This does nothing to ensnare the manufacturer, who is the person best placed to remedy such faults for the future, and who could redistribute any losses much more widely if made liable for the defective product.

The 'tort route' for damages in respect of defective products finds its origins in *Donoghue v. Stevenson*, where the defective product was contaminated ginger beer. It was consumed by a friend of the purchaser. The friend had no contractual rights. The House of Lords recognized in that case that a manufacturer owed a duty to take reasonable care to the ultimate consumer of the product where injury could be foreseen and it was unlikely that there would be any intermediate examination of the goods. Of course, under this rule, the range of people who can sue is potentially much wider and the manufacturer's liability is not strict. The manufacturer owes a duty to take reasonable care. The plaintiff will bear the burden of proving negligence against the manufacturer. That may not prove easy, although the plaintiff may be able to rely on the *res ipsa loquitur* principle. A plaintiff's claim may be defeated or reduced if the manufacturer proves that the plaintiff was contributorily negligent, e.g. by failing to follow the instructions issued with the product. And of course, a plaintiff needs to show that he or she suffered harm of the right type. In an action for negligence, we have already seen that the harm must normally be physical harm, such as personal injury or damage to property. A plaintiff will not usually succeed if he or she has suffered mere economic loss. But earlier discussions in the area of negligence have already focused on the fine dividing line between physical injury and economic loss. In some cases, there has been a suggestion that a plaintiff can recover for economic loss, e.g. *Junior Books Ltd v. Veitchi Co. Ltd*, but that case turned on a peculiarly proximate relationship between the parties which is not usually likely to exist between a manufacturer and a consumer.

Thus, the tort and contract routes can both be seen to be problematical. This area is now also governed by legislative controls. This is particularly important as there are inevitable policy decisions to be taken into account. English law has had to come into line with the European Community, which issued a Directive in 1985 requiring member states to harmonize their rules on product liability. This has been achieved by the Consumer Protection Act 1987 and the General Product Safety Regulations 2005, which impose strict liabilities that cannot be excluded or limited, contractually or otherwise (s.7 of the Act). The Act provides that a producer of defective products is liable for any damage caused by the defect (s.2). The terms 'producer', 'product' and 'defect' are all defined by the Act.

The producer is defined by s.1 as the person who manufactured the product or, in the case of a substance which has been won or extracted, the producer is the person who won or extracted it, or in the case of a product where its essential characteristics are due to an industrial or other process being carried out, the person who carried out the process.

A product, defined by s.1 (2), means any goods or electricity, and includes a product which is comprised in another product as a component part.

The Act recognizes that it may be easy to identify the person who sold goods but not so easy to establish who is the producer. This problem is met by s.2(3), under which a supplier of a product is liable if the plaintiff asks the supplier within a reasonable time of damage occurring to identify the producer, in circumstances where it is not reasonably practicable for the plaintiff to identify the producer, and the supplier fails to satisfy the request within a reasonable time.

A defect in a product is defined by s.3. There is a defect if the safety of the product is not such as persons generally are entitled to expect. In determining what persons generally are entitled to expect, a court could take account of:

1    The manner in which, and purposes for which, the product has been marketed, and any instructions for use, or warnings in relation to the use of the product.
2    What might reasonably be expected to be done with, or in relation to, the product.
3    The time when the product was supplied by its producer to another.

Nothing in this section is to be taken to infer that a product is defective merely because a later issued product adheres to an even higher safety standard. The factors specifically listed in s.3 are not necessarily exhaustive.

In *Tesco Stores Ltd v. Pollard*, a childproof cap that could be opened with far less degree of force than that required by the relevant British Standard was nontheless not defective under the terms of the Act. The Court of Appeal held that the expectation with this type of cap was only that it be more difficult to open than a non-childproof cap, not that it meet the relevant British Standard.

Section 4 provides that it is a defence for a producer to show (amongst other things) that at the relevant time (usually when the producer supplied the product to another) the state of scientific and technical knowledge was not such that producers could have been expected to discover the defect. This is often described as 'the state of the art' defence. It is thought to be vital if technical and scientific innovation is not to be discouraged. It was one aspect of the Directive where member states had some choice about whether to incorporate such a defence.

In order to sustain a claim under the Act, a plaintiff would need to prove damage. This is defined by s.5 as including death or personal injury or any loss or damage to any property including land. It expressly excludes loss or damage to the product itself. The property lost or damaged must be property intended for private use or consumption. Where the harm to property is valued at less than £275 no claim lies, but this restriction does not operate where the damage is personal injury.

The Act provides rights which are in addition to existing common law rights. Section 2 specifically states that it is without prejudice to any liability arising otherwise than by virtue of the Act. This could be relevant in cases involving pure economic loss. The Act places an embargo on recovery for such loss but cases on the common law rules of negligence may develop differently (see *Junior Books v. Veitchi*). An action under the 1987 Act is obviously advantageous to a plaintiff because the burden of proof is primarily on the manufacturer.

## Vicarious liability

The word 'vicarious' means 'in place of another person', or 'in substitution for the proper person'. A rule has emerged in tort that an employer is vicariously liable for torts of employees committed in the course of their employment. The rule is one of policy. Its importance is that it ensures that an injured person can sue the employer instead of suing the employee. The employer is likely to have greater financial resources, and may have taken out insurance against such claims. There is no particularly satisfactory justification for the existence of the rule, although it is often said that an employer profits from an employee's work and should, therefore, bear the risk of it. The two main criteria for the operation of the rule are:

1   The person who commits the tort must be an employee. This, however, can be misleading, as there are a number of important situations in which an employer will also be vicariously liable for the torts of an independent contractor.

2   The employee must commit the tort in the course of employment.

## The course of the employment

There is no single test to determine when an employee is in the course of his or her employment. The decisions of the courts tend to turn on the individual facts of the case. Certain situations seem definitely to be covered:

1   Torts committed while the employee is doing what he or she is employed to do. For example, a man employed to drive a bus does so negligently and injures a pedestrian.

2   Torts committed while the employee is doing what he or she is employed to do, but doing it in a manner forbidden by the employers. For example, a petrol tanker driver is forbidden by his employers to smoke while making deliveries. He disobeys this order and causes an explosion which injures a passer-by.

3   Torts committed while the employee is doing what he or she is employed to do, but doing this in a criminal manner. For example, a solicitor's clerk employed to advise clients fraudulently advises a client to transfer property to him.

4   Torts committed while the employee is doing acts which are reasonably incidental to his or her work. For example, an employee opens his office window for ventilation and in doing so negligently knocks a potted plant out of the window on to a passer-by below.

5   Torts committed while the employee is acting in an emergency for the protection of the employer's person or property, when his or her actions would normally be outside the scope of employment. For example, an employee negligently moves a vehicle which he has no authority to drive, but which he reasonably fears is about to be damaged in a fire.

Where the employee's behaviour has amounted to a 'frolic of his own', the employer will not be vicariously liable. Reported cases cover the bus conductress who decides to drive the bus, the delivery driver who varies his proper route to visit a friend, and the garage attendant who assaults a customer with whom he has had an argument. In all of these cases, there is (at least temporarily) no course of employment.

The rule of vicarious liability is often said to be one of the distinguishing features of an employer–employee relationship. Generally, an employer is not vicariously liable for the torts of an independent contractor. If I engage a builder to build me a house, he is liable for any acts of trespass or negligence he may commit, but I am not. However, in some cases either the liability imposed is strict, or the duties owed by a person are so onerous that there can be no delegation to someone else. Then the employer is liable in addition to the contractor. Although these are often spoken of as examples of vicarious liability, it is more appropriate to regard them as examples of an employer continuing to be personally liable. This can arise in the following situations:

1   When the employer is negligent in choosing the contractor, he remains personally liable. For example, an occupier might need repair

work of a technical nature undertaken on his premises. He knows that the work involves electrical rewiring and yet he appoints an 'odd-job-man' with no training or expertise in this sort of work. The job is not properly done and causes a lawful visitor to the premises to receive a severe electric shock. The occupier would be in breach of his duty under the Occupiers' Liability Act 1957 and could not plead the benefit of s.2(4) of the Act, because he has not taken reasonable steps to satisfy himself that the contractor was competent.

2   When the liability of the employer is strict (e.g. as in the tort of *Rylands v. Fletcher*) and can occur even if reasonable care has been taken, the employer is liable whether the the tort has been committed personally or through an independent contractor.

3   When the employer undertakes a particularly hazardous activity, especially when it is on or near the highway, he is liable for torts committed personally or by an independent contractor. In one case, a heavy lamp suspended over a footpath fell into disrepair. The employer was held liable when a passer-by was injured by the activities of the contractor called in to repair it. In another case, a contractor was engaged to thaw out frozen pipes and chose to use a blow-lamp. The employer was held vicariously liable when a fire broke out as a result of the contractor's negligence.

4   When the employer authorizes or instructs the contractor to do something involving the commission of a tort, he remains liable. When a gas company engaged a contractor to dig up part of a street over which they had no such authority, the gas company was held liable when a passer-by fell over a heap of earth left by the contractor.

## Contractors' acts of collateral negligence

Even in the above cases where both employer and contractor are liable, the former will not be held responsible for the contractor's collateral acts of negligence. In other words, an employer will only be liable if the risk of harm arises from the work itself, rather that the negligent performance of the work. For example, if an employer engages a competent contractor to install replacement frames on windows overlooking the highway, he will be liable if injury to a passer-by is caused by the glass being knocked out of the old windows (injury due to the work itself) but he will not be liable if the injury is caused by a workman negligently dropping his hammer (injury caused by the performance of the work).

# 11
# Health and safety

## The background to legislation

The following comment from a report published by the Health and Safety Executive on safety in the construction industry sets the scene for the problems of accident prevention in the industry:

> Construction management has to contend with a number of problems which vary enormously from site to site during the life of the site – climate, regional attitudes, geology, the time of year, the type of contract, the scale of the job, the methods of payment, the type of employment, the rapid turnover of labour. There is also the very complex problem of the relationship between a main contractor and an ever-changing group of subcontractors. All these factors affect management in all its aspects, including the management of safety.

For a long time the law has played a part in controlling how employers must 'manage' safety. Acts of Parliament laying down duties and standards in specific working places date back to 1802. The earliest rules related to the employment of the very young in the mills, where the particular evil was the importing of large numbers of pauper children and engaging them in appalling working conditions. The procedure of making rules to control specific workplaces was established, and it will be seen that this became one of the worst features of the safety legislation.

With the pattern set, whenever a new loophole, crisis or new hazardous work area was discovered, special rules would be made to cover the situation. The end result was piecemeal legislation covering mines, factories, offices and shops, agriculture and so on, culminating in a series of specific workplaces each governed by its own Act of Parliament. The Acts were not sufficiently detailed in themselves to cover every conceivable situation, so they all had various sets of regulations (delegated legislation) made under them. The Factories Act 1961 alone had more than 200 sets of regulations made by virtue of its authority.

The main aim of the safety legislation was to establish standards and duties for the workplaces covered. Failure to abide by the prescribed standards and duties would lead to employers being prosecuted and fined. By operating in this way the rules could, therefore, be seen to fulfil

Galbraith's Building and Land Management Law for Students. DOI: 10.1016/B978-0-08-096692-2.00011-7
Copyright © 2010 by Elsevier Ltd

an accident prevention role. When an employer's breach of duty led to a workman being injured the various Acts were interpreted by the courts as giving the injured workman the right to sue for compensation.

The health and safety question can be seen to have three dimensions:

1   *The point of view of an injured employee* – he will want to know the answer to several questions. Should he report the accident? Can he get compensation? Is his employer insured? Will he get state benefits? Does it matter that he was partly to blame?

2   *The point of view of the employer* – he will want to know whether he will be prosecuted. Will his insurance premiums be increased? Will the employee claim damages? Should he alter existing safety arrangements?

3   *The point of view of the state* – here the issues will include the cost of health care for the injured person, lost productivity, the cost of state benefits and whether there are any lessons to be learnt for future safety policy.

The number (180) of fatal injuries to workers reported in 2008–9 fell from the previous year, thus continuing a trend that has seen fatalities at work reduced to a level which is less than one-third of that recorded in 1981. On the other hand, there is little room for complacency as the number of reported major injuries to employees has only decreased marginally in recent years. Estimates suggest that 29.3 million days are lost each year due to work-related injury and ill health, the latter involving 1.2 million people and contributing 24.6 million working days lost. In part the improvement in fatal and non-fatal injuries is attributable to shifts from manufacturing and heavy industries to service sectors, where risks are generally lower.

Dissatisfaction with the existing system of piecemeal legislation led to a review by the Robens Committee on Safety and Health at Work. Robens recommended a major revision of the approach to health and safety with a move away from legislation which applied to specific types of workplace to a single Act which would provide comprehensive coverage to workplaces in general. It also stated that the promotion of safety and health at work was the joint responsibility of both management and employees. The implementation of this new thinking on health and safety took the shape of the Health and Safety at Work etc. Act (HSWA) 1974. This was a piece of framework legislation which established certain general duties on employers, workers and others, introduced new requirements for safety representatives and committees, and amended the existing regulatory and enforcement machinery with the establishment of the Health and Safety Commission and the Health and Safety Executive. It was never intended that the existing legislation be substituted overnight, rather the aim was that over time the Acts which related to specific places of work would be replaced by regulations and codes of practice of more general application. However, although some important changes were made, overall progress on this process of substitution proved to be slower than anticipated and towards the end of the 1980s much of the old legislation was still in place. But new impetus was brought to the process of reform in the shape of the Single European Act of 1987, which introduced Article 118A, thereby enabling improvements to the work environment and to health and safety to be adopted by qualified majority voting rather than by the unanimous assent of member states. Since that time a programme of safety, hygiene and health reforms has been put in place and, starting with the European Framework Directive in 1989, a series of directives has been passed which has had a profound effect on the working environment in the UK and across Europe.

On 1 January 1993 new regulations were introduced intended to implement the UK's obligations under the Framework Directive and its subsequent measures. These became known as the 'six pack' but before turning to these we must consider in more detail the structures and bodies established by the Health and Safety at Work Act itself.

## The Health and Safety Commission and Executive

It is worth remembering that safety is not just an issue affecting those in employment. The general public is awakened to safety issues from time to time when a major disaster occurs, such as the King's Cross tube station fire in 1987 or the Paddington rail crash in 1999. At such a time, there are often demands for safety standards to be improved, and a public inquiry (held by virtue of s.14 of HSWA 1974) may highlight changes that are necessary. The government has overall responsibility for introducing legislation to promote health and safety, and several of its departments, such as the Department of Transport, will see this as part of their work. Other organizations (e.g. the Royal Society for the Prevention of Accidents, RoSPA) exist to promote health and safety, and although their work may be more general, it can have useful implications for health and safety in employment.

The Health and Safety Commission (HSC) was established by the HSWA in 1974. It had up to nine members, who were required by the Act to be representative of employers, trade unions, local authorities and other bodies, such as RoSPA, concerned with health and safety. The role of the Commission was to undertake, by whatever means it considered appropriate, to secure the health, safety and welfare of people at work, and to protect other people from risk to their health and safety created by work activities. The Commission carried out research, provided training, disseminated information and advice, and submitted proposals to the Minister for new regulations.

The Commission's accident investigation or inquiry role was significant, as it tended to have a high public profile and created significant pressure for change or some government action. An investigation would be less formal than an inquiry, so whereas the latter was the proper method to adopt in the King's Cross situation, an investigation would be more appropriate for, say, a serious accident at a fairground. The person appointed to conduct an inquiry would be given formal powers to compel the attendance of those required to be present. Inquiries would usually be held in public and might well involve visits to the site.

One major initiative of the Health and Safety Commission was to establish advisory committees on issues such as major hazards, nuclear safety and toxic substances, as well as industry-based advisory committees designed to promote the health and safety of workers in a particular industry. One example is the Construction Industry Advisory Committee (CONIAC). This kind of specialist committee has played an important role in assisting the Commission in tasks such as drawing up guidance and promoting awareness of health and safety issues. CONIAC also plays a part in identifying training and education needs and promoting improvements in competence through continuing professional development.

The most significant development which resulted from the establishment of the Health and Safety Commission was the new unified focus for safety issues and policy-making, which had previously been lacking. The Commission operated on a day-to-day basis through the Health and Safety Executive (HSE), which was the enforcing authority. Some enforcement duties were allocated to local authorities, and to avoid duplication between HSE inspection and local authority inspection, some types of premises, e.g. shops, catering establishments and offices, were specifically allocated as the responsibility of the local authorities. However, local authorities did *not* inspect their own offices! That task was performed by health and safety inspectors.

Between 1999 and 2000 the HSC published a consultation document and strategic statement, 'Revitalising Health and Safety', setting out the government's plans for reform. Subsequently the HSC issued in 2004 its strategic vision for 2010 and beyond: 'To gain recognition of health and safety as a cornerstone of a civilized society and, with that, to achieve a record of workplace health and safety that leads the world.' The key challenge for this strategy was seen to be the need to ensure that appropriate risk management was made relevant to the modern and changing world of work. It saw a role not only for the HSC, HSE and local authorities, but also for employers, workers and their representatives.

In April 2008, after consultation on the roles of the two bodies, the HSC and the HSE merged to become a single regulatory body under the title of the Health and Safety Executive. In its first annual report the objectives of this merged body were stated to be to:

- influence people and organizations – duty holders and stakeholders – to embrace high standards of health and safety;
- promote the benefits of employers and workers working together to manage health and safety sensibly; and
- investigate incidents, enquire into complaints about health and safety prectices and enforce the law.

More immediately an employer, landowner or occupier with contractors on his land must pay particular regard to the following legislation:

- the Health and Safety at Work etc. Act 1974;
- the 'six pack' and other regulations stimulated by EU Directives.

# The Health and Safety at Work etc. Act 1974

The 1974 Act adopted the recommendations of the Robens Report, placing emphasis on greater involvement and self-regulation rather than reliance on state regulation. The Act provided the framework of a new system, and under this 'umbrella' new codes of practice have been created. As this happened, the intention was that old Acts and Regulations would gradually be repealed.

The main objectives of the Act are set out in s.1:

1   To secure the health, safety and welfare of persons at work.
2   To protect persons other than persons at work against risks to health and safety arising out of work activities.
3   To control the keeping and use of dangerous substances.
4   To control the emission of noxious and offensive substances.

To achieve these objectives, the Act lays general duties on employers (s.2), employees (s.7), the self-employed (s.3), occupiers of workplaces (s.4), and manufacturers and suppliers (s.6).

The Offshore Safety Act 1992 now extends the 1974 Act to cover the health, safety and welfare of persons on offshore installations.

## *Duties owed by the employer – sections 2 and 3*

So far as is reasonably practicable, an employer must ensure the health, safety and welfare of employees at work. This broad generalist approach is certainly a remarkable contrast with the detailed format previously adopted by safety Acts and Regulations. The employer must (so far as is reasonably practicable):

1   Provide and maintain plant and systems which are safe and without risk to health.
2   Make proper arrangements for the safe handling, storage, use and transport of articles and substances.
3   Provide necessary information, training, instruction and supervision.
4   Maintain the place of work and access to it so that it is safe and without risk to health.
5   Provide and maintain a safe and adequate working environment.

The Act does not define the terms 'health', 'safety' or 'welfare' nor the phrase 'so far as is reasonably practicable'. The latter phrase is familiar to the lawyer, because it was also used to qualify certain duties created by the Factories Act 1961. For example, s.28 of that Act provided that floors and stairs must, so far as was reasonably practicable, be kept free from obstruction. The meaning of this phrase has been judicially considered and has a narrower meaning than 'physically possible'. It implies that: 'A computation must be made in which the quantum of risk is placed on one scale and the sacrifice involved in the measures necessary in averting the risk is placed on the other. If the risk is insignificant in relation to the sacrifice, the defendants discharge the onus on them.' The following case provides an illustration. As a result of an accident at work, the Health and Safety Inspector issued an Improvement Notice, requiring a dairy to provide protective footwear free of charge to all employees involved in operating hydraulic trolley-jacks. The employer successfully appealed against the notice. The expense of providing the footwear free of charge was disproportionate to the risk to the employees. The dairy's existing arrangements, whereby employees could buy the footwear at cost price and pay by instalments, were adequate to satisfy the firm's duty to secure its employees' safety so far as was reasonably practicable (*Associated Dairies v. Hartley*).

Nothing in the Act allows these duties to be used as the basis for a claim for compensation. The sanctions for their enforcement are criminal. However, an injured workman would be able to proceed against the employer under ordinary negligence principles.

Section 2(3) provides that an employer must prepare and, when necessary, revise a written statement of general policy with regard to health and safety, and must bring the statement to the notice of employees. The introduction of such written policy statements was seen as an important application of the ideas about greater involvement and self-regulation expressed in the Robens Report. If employees are to be involved, they must know what the safety policy is. The Act gives no guidance on the form or content of the safety policy. This will naturally vary with the size and type of organization and the specific hazards involved. The lack of a model form is sensible, as it compels employers to think positively about safety problems. This approach to employer and employee awareness is also encouraged by some of the 'six pack' regulations. Failure to abide by the rule about issuing a safety policy is an offence usually punishable by a fine.

Section 3 imposes on employers and self-employed persons a duty to conduct their undertakings in such a way as to ensure, as far as is reasonably practicable, that persons not in their employment who may be affected are not exposed to risks to their health and safety. This section is a valuable protection for members of the public but an unusual application of it can be seen in *R. v. Swan Hunter Shipbuilders Ltd.* In that case a fire broke out on board a ship being built by Swan Hunter. In one part of the ship the fire became particularly intense because a subcontractor had left an oxygen hose there. Swan Hunter was aware of the dangers of using oxygen equipment particularly because oxygen enrichment could occur in confined spaces. In order to alert its own employees to this danger, Swan's had distributed a booklet to them, setting out practical safety rules. These booklets were not given to all subcontractors as a matter of course. As a result of the fire, eight men died. Swan Hunter was convicted of a breach of its duties to provide a safe system of work and to provide information and instruction (s.2). It was *also* convicted of a breach of its duty under s.3. By not providing information to non-employees, Swan Hunter had failed to ensure that persons not in its employment were not exposed to risk to their health.

## Duties laid on controllers of premises (s.4)

Section 4 imposes duties on controllers of premises, where the premises are used by persons who are not employees, and where those persons may use plant or substances provided there for their use. A frequently quoted example of such premises is a coin-operated dry cleaners. Here, the controller must take such measures as are reasonable to ensure that the premises, plant and substances are safe and without risk to health so far as is reasonably practicable. (This is the criminal law provision corresponding with an occupier's duties at civil law under the Occupiers' Liability Act 1957.)

Section 4 can also be significant where a site is in the control of the main contractor. That site will be used by subcontractors and their employees, i.e. people who are not employees of the main contractor. In one case a prosecution was brought in London under s.4. The employee of a subcontractor was driving a small dumper truck out of a service lift when the lift suddenly rose as he was halfway out. The driver was killed when he became trapped between the dumper truck and the lift shaft. The tragedy could have been avoided if the interlock on the lift had not been jammed with pieces of timber, thus tricking the machine into believing that the lift gates were shut. The main contractor was fined £1000.

## Duties laid on manufacturers and suppliers (s.6)

These duties are regarded as among the most important methods of securing an improvement in safety standards at work. Generally, manufacturers or suppliers of any article or substance for use at work must, so far as is reasonably practicable:

1    Ensure that the article or substance is safe and without risk to health when properly used.
2    Carry out such tests and examinations as are necessary to achieve (1) above.
3    Take such steps as are necessary to ensure that adequate information is available about how to use the article or substance, so that when properly used, it is safe and without risk to health.

These duties allow faults and defects to be traced back to their source, and prosecutions can be brought against persons with the ability to remedy them. The effect of s.6 is not to absolve the employer from its duties, but it is recognized that the power of the employer to do much is often limited. These duties do not create any new civil liability. Any civil proceedings must be brought under ordinary negligence rules. However, the possibility to prosecute under s.6 is a significant increase in the scope of the inspectorate's power. Note also that similar duties are imposed on designers, importers, erectors and installers.

## Duties laid on employees (s.7)

Under the old Acts and Regulations duties were placed directly on employees, and this section merely extends an old principle. The duties which an employee owes while at work are:

1    To take reasonable care for his or her own health and safety.
2    To take reasonable care for the health and safety of others affected by his or her acts or omissions at work.
3    To cooperate with the employer where necessary to enable the employer to fulfil its duties.

As with the other duties under the Act, s.7 cannot give rise to a civil claim for compensation but an offending employee can be prosecuted. Employees are most likely to be prosecuted in those cases where the employer has fully complied with its duties. The most notable feature about prosecutions brought against employees is that fines tend to be smaller than those imposed on employers, although the maximum fine in the magistrates' court is £5000 and/or 12 months' imprisonment. In a case brought against two nurses under s.7, where they had put an elderly patient into a scalding bath without testing the water, and the patient had died as a result of the accident, both were fined what was then the maximum amount of £1000.

Although s.7 relates specifically to employees, they could also, with others, be prosecuted under s.8, which provides that no person shall intentionally or recklessly interfere with, or misuse, anything provided under the Act for health and safety purposes. Those guilty of an offence under s.8 could face fines up to £20,000 or 12 months' imprisonment, or both, on summary conviction.

Under the Management of Health and Safety at Work Regulations 1999, employees are under a duty to use machinery and equipment as they have been trained and instructed to do so, and they must inform the employer of any situation which is an imminent danger, or where there is noted to be a gap in the health and safety arrangements.

## Safety representatives

In safety matters, the Robens Report considered that management and employees had an 'identity of interest' which should lead to 'participation in working out solutions'. Indeed, they found evidence in some working situations of existing participation by employees, e.g. the representatives appointed by coalminers to inspect mines. Many firms and organizations had also established voluntary safety committees before 1974. This voluntary effort needed statutory support, and the 1974 Act permitted regulations to be made providing for the appointment of safety representatives and for the establishment of safety committees. Such regulations have been in force since 1978 and provide an accompanying code of practice for recognized trade unions to appoint safety representatives from amongst the employees. Generally, representatives must have been employed for two years by the employer or have two years' experience in similar employment. The main tasks of the representative are:

1    To be consulted on safety matters.
2    To inspect plant and premises, usually at three-monthly intervals but more often if there has been a change in work methods, or an accident.
3    To investigate complaints from employees with regard to health, safety and welfare.
4    To investigate potential hazards and dangerous occurrences.

To perform these functions effectively, the representative needs time and training. The Regulations provide for time off with pay for carrying out their responsibilities.

The appointment of safety representatives was a step designed to overcome the apathy on safety matters which had been identified by Robens. This can be regarded as enforcement through cooperation. Two points may be noted, however:

1    The representatives have no obvious 'teeth' for enforcement. Given that they are appointed by recognized trade unions, it could be argued that any necessary 'teeth' would be the usual forms of industrial pressure or action. If this is what is envisaged, it seems to contradict the idea of a 'common interest' in safety matters.
2    By limiting the power to appoint safety representatives to recognized trade unions, no statutory machinery exists to appoint representatives in non-unionized areas of employment. Presumably, employees in these areas were to be left to 'flourish' in the state of apathy which the Robens Report identified. The initiative in these cases had to come from management.

With the decline in union recognition in the 1980s and 1990s, the part played by safety representatives decreased correspondingly. However, EU law once again came to the rescue. Based on a decision of the ECJ in a case involving collective redundancies and transfers of undertakings, it was believed that the 1989 Framework and the UK regulations implementing it imposed a similar requirement, namely that, in the absence of a recognized trade union, an employer should consult with elected employee representatives. The Health and Safety (Consultation with Employees) Regulations 1996 introduced just such a requirement and an obligation on employers to give elected employee representatives such information as is 'necessary to enable them to participate fully and effectively in the consultation'. Employee representatives receive many but not all of the rights and protections given to their trade union-appointed counterparts. Thus they are protected against detrimental treatment or unfair dismissal based on their role as safety representatives and have the right to paid time off, but have no right to inspect plant and premises. The reintroduction of a statutory recognition procedure for trade unions (see Chapter 9) may reduce the number of elected representatives but they clearly have an important part to play in non-unionized workplaces.

## Safety committees

Regulations made under the Act provide for the establishment of safety committees if the employer receives a request in writing from two safety representatives. However, employee-elected representatives have no right to request that a safety committee be set up. Once a committee is established, s.2 lays down that it should keep health and safety measures under review. The recommendation about health and safety committees is that they should not be given any other functions or responsibilities. There are no prescribed rules about membership, but the make-up of the committee, and the seniority of the management representatives chosen to sit on it, may well give a good indication of a firm's general attitude to safety questions!

One useful function which a safety committee can fulfil is to monitor the advice available which is disseminated by bodies like the Health and Safety Executive. That advice can then be pursued and applied more specifically to the particular work situation. For example, in demolition work, an area notorious for the number and seriousness of accidents, the committee might consider the Executive's guidance notes, and then recommend to the employer that it should adopt a policy of drawing up a 'written methods statement' for each demolition job. This would focus the employer's attention on the hazards of the particular job to be undertaken (e.g. the presence of asbestos, unusual structures, toxic chemicals, safe access for work to be carried out at height, dust levels, and the need for protective clothing). If this is prepared after a site visit and at tender stage, it will allow the contractor to tender at a price level which takes account of doing the job safely.

By monitoring all the reports and advice available, the safety committee can also play a useful role in alerting employers to wider health and safety issues. Again, taking demolition as the example, it may be thought that the hazards are all too obvious. But this type of work may be even more hazardous than conventional construction work because of hidden dangers. It may seemingly have been fungal spores which caused severe pulmonary illness in Lord Caernarvon and some of his team who had excavated the tomb of Tutankhamun. Indeed, the illness was referred to as the 'curse of the Pharaohs'! Some old buildings could present similar hazards, calling for breathing equipment or protective clothing.

It should be remembered that the inspectors employed by the Health and Safety Executive have always seen advice giving as one of their key functions and a safety committee can act as an impetus to the employer to take the benefit of their advice.

## Enforcement of the Act

A system of inspectors for the enforcement of safety legislation has existed for more than 150 years. Indeed, it is the oldest established inspectorate in the world. With the establishment of the Health and Safety Executive, all inspectors were brought together under one umbrella, but there are still different branches of the inspectorate. The factory inspectorate, which has responsibility for construction, is by far the largest branch. Its inspectors are responsible for almost half a million sets of premises, and the focus of their work has been dramatically broadened by the 1974 Act. In one of their publications for young people who may be contemplating a career with the inspectorate, they say:

> As well as gaining an awareness and understanding of the practical problems that industry faces, a Factory Inspector must be able to deal firmly and diplomatically with a wide range of people. Many situations will require delicate handling and the successful inspector is one who is able to adapt his approach and meet the needs of the situation without losing sight of his main objective – an improvement in health and safety conditions. The Factory Inspector has to be conversant with a wide range of technical issues, ranging from the nature of toxic chemicals to the techniques for guarding a whole range of machinery. He must also have a thorough understanding of health and safety law, as well as having the tact and diplomacy to deal with the sensitive issues that arise from time to time.

Training for new inspectors is partly 'on the job' and partly by formal training courses. A warrant is issued immediately to new inspectors. This is their authority to act. It specifies the powers conferred on them and they must produce it on request when seeking to exercise those powers.

Although inspectors have always had power to prosecute, there has been internal conflict over their role, which they have seen to be the improvement of standards of safety, rather than the mere enforcement of the law. Indeed, the time and effort needed to mount a prosecution was often felt to be wasted because of the low level of fines imposed. The inspectorate was fragmented with separate inspectors for factories, mines and quarries, etc. Valuable time was often wasted making routine inspections of places where there was no great danger, leaving insufficient time to concentrate on places where there were real hazards. In its analysis of the problem of enforcement and the inspectorate, the Robens Report found that:

1    There should be a unified inspectorate.
2    The principal objective of the inspectorate should be to give impartial advice to industry.
3    The services of the inspectorate should be more selectively used, foregoing the system of routine visits.
4    Where sanctions need to be used by inspectors, the penalties imposed should be more severe.
5    A range of alternative procedures should be available to inspectors, allowing them to use sanctions which were immediate and constructive.

These proposals were implemented by the Act, and the scope of an inspector's powers is now set out in s.20. To enable them to perform their duties effectively, they may:

1    Enter premises (if necessary with a constable) taking necessary equipment with them.
2    Make investigations and examinations.
3    If necessary, order the premises to be left undisturbed.
4    Take photographs and measurements or recordings, and if necessary take away samples.
5    Question persons whom they believe can give them information relating to their investigation.
6    Inspect records, accident books, test certificates, etc.

When establishing how their time can best be spent, inspectors involved in the construction industry might prioritize as follows:

1    Investigating fatal accidents and large incidents.
2    Visiting demolition sites and sites where steel erection is taking place. This is seen as particularly important, not only because of the high level of accidents, but also because these may involve examples of employers using untrained employees.
3    Investigating any site involving asbestos. This takes a high priority because of the high incidence of asbestos related deaths.
4    Investigating non-fatal accidents, where there has been a breach of law.
5    Investigating complaints from the general public. Very few of these come from trade unions. Some come from groups such as parents who are worried about an adjacent building site. Others come from rival contractors, possibly disgruntled that they failed to get the job!

Each year, the inspectorate focuses on one particular type of inspection, for instance roofing or transport. This is seen as an important part of their preventative role. The inspectors also lay considerable stress on their instruction role and point to guidance notes issued, e.g. on use of ladders, or demolition practice. In 2003, for example, the HSE issued a pamphlet entitled 'The High Five', aimed at small construction businesses. Under the image of a raised open hand it promoted five ways to reduce risks on sites, including steps to deal with falls from height and with asbestos. Guidance notes are less formal than codes of practice, of which there are very few. Ironically, codes and guidance are not always popular with 'customers', who actually seem to prefer strict regulations!

Most inspectors emphasize that, by and large, they get a good reception from employers, who accept how wide the powers of the inspector are. This may be due in part to their training in the skills necessary to present new ideas and persuade people to make changes in their working practices. The inspectors deal with people from the shop floor and people at director level, with back-street garages and multinational corporations. In all these dealings, they hope to advise and persuade. Often the threat of action by an inspector is the only 'enforcement' necessary.

If that fails, or the situation otherwise merits it, an inspector may use the powers granted by ss.21 and 22 to serve either an improvement notice or a prohibition notice. The advantage of both of these notices is that they can be issued by the inspector immediately.

### Improvement notices (s.21)

The inspector may discover a breach of HSWA 1974 or the Regulations and decide to serve an improvement notice on the person in breach. This notice will set out the breach of duty which the inspector thinks is occurring, specify the Act or Regulation in question, and give the inspector's reasons for this opinion. The notice will also set out the time limit for correcting the breach, the matters to be put right and the way in which this is to be done.

The earliest date which an inspector can set for compliance is 21 days after service of the notice. This is to allow time for an appeal to be lodged, but it creates the biggest drawback about improvement notices – it means that it is impossible to get something done immediately. This can sometimes undermine the inspector's authority. Take the example of a site with no welfare hut. This is certainly not likely to be a situation warranting the use of a prohibition notice, but if the inspector serves an improvement notice, the employer can continue to ignore the notice for the next 21 days.

There is a right of appeal against the notice:

- Against the time limit imposed.
- Against the substance of the notice.

The effect of lodging an appeal is to suspend the operation of the improvement notice. On appeal, which is heard by the Employment Tribunal, the notice may be affirmed, cancelled or modified. The most frequent modification is an extension of the time limit. There must be proper grounds for an appeal. Mere lack of finance to carry out the improvements is doomed to failure as an excuse. Similarly, arguments that there has never been an accident, that the employees have not complained, or that the breach is trivial, have not found favour with tribunals.

## Prohibition notices (s.22)

This type of notice can be served when activities being carried on involve a risk of serious personal injury. Unlike an improvement notice, no specific breach of the Acts or Regulations need be specified, but the inspector must give reasons to support the decision. The notice directs that the activities in question must stop either immediately or within a specified period until matters have been remedied. One significant difference about a prohibition notice is that lodging an appeal does not suspend its operation. The repercussions of a prohibition notice are serious for an employer, as business may be brought to a standstill. In consequence, these appeals are usually heard as a matter of urgency.

Anyone served with either variety of notice who fails to comply with it commits an offence for which he or she may be prosecuted. As well as imposing a fine, the court may order him or her to attend to the matters specified in the notice.

## Prosecutions (ss.33–42)

The two types of notices already mentioned may not be considered appropriate by the inspector. Knowledge of an employer's past record may convince the inspector that immediate prosecution is necessary. Prior to 1974, proceedings were always summary in the magistrates' courts, and the levels of fines tended to be quite low. The Act has now created the possibility of summary proceedings in the magistrates' court or proceedings on indictment in the Crown Court. The Health and Safety (Offences) Act 2008 has modified the penalties available to the courts.

Prosecutions in the magistrates' courts must be brought within six months of the date of the offence, but there is no time limit if a case is tried on indictment. In magistrates' courts, inspectors will conduct prosecutions themselves. Commonly, the charge may be failing to comply with s.2 of the 1974 Act. Breach of ss.2–6 could lead to a fine in either the magistrates' court (maximum £20,000 and/or 12 months' imprisonment) or Crown Court (unlimited fine and/or two years' imprisonment). Failing to comply with an improvement or prohibition notice, or a court remedy, could result in a fine within the same limits and/or a period of imprisonment (maximum of 12 months or two years respectively). Breaches of all health and safety regulations are subject to similar sanctions as well as giving rise to civil liability. Sometimes fines may amount to hundreds of thousands of pounds and in the case of the Southall rail crash, Great Western Trains were fined £1.5 million. However, often the fines imposed may seem to those seriously injured or to bereaved families insultingly small. In the Southall case the company was also charged with manslaughter by gross negligence but the Court of Appeal held in *A-G's Reference (No. 2 of 1999)* that such an offence in the case of a non-human defendant required evidence establishing the culpability of an identified human defendant for the same offence. In the absence of such a person the charge was dismissed. The Law Commission had previously recommended the establishment of an offence of 'corporate killing' based on 'management failure' without the need to identify any specific individual responsibility. The offence of corporate manslaughter has been belatedly introduced by Parliament and is outlined below.

In any proceedings for an offence under the Act or its regulations the onus of proving that everything reasonably practicable has been done to satisfy a particular duty rests upon the accused. Despite the imposition of criminal liability by HSWA the Court of Appeal has held in *Davies v. Health and Safety Executive* that the reverse burden of proof in health and safety cases is justified, necessary and proportionate, and does not infringe the principle in Article 6(2) of the European Convention of Human Rights that anyone charged with a criminal offence must be presumed innocent until proven guilty.

Prosecutions are only brought in extreme cases. In 2008–9 there were 1245 offences prosecuted and 14,427 prohibition and improvement notices issued by the HSE and local authorities. In the construction industry, with its relatively high rate of fatalities and other injuries, weapons such as notices and prosecution form a useful part of the inspector's tools.

On 1 January 1993 the government introduced six new sets of regulations under the Health and Safety at Work etc. Act 1974. These were intended to implement the EU Directive on Health and Safety. The regulations were commonly referred to as 'the six pack' and covered the following areas:

- Management of Health and Safety at Work Regulations 1992
- Provision and Use of Work Equipment Regulations 1992
- Personal Protective Equipment (EC Directive) Regulations 1992
- Manual Handling Operations Regulations 1992
- Workplace (Health Safety and Welfare) Regulations 1992
- Health and Safety (Display Screen Equipment) Regulations 1992.

In fact all of the 'six pack' have subsequently been amended to a greater or lesser extent. There are extensive guidance notes attached to the regulations. They are cross-industrial and apply to all places of work. Originating from the EU, they go beyond the Health and Safety at Work etc. Act 1974 and are designed to harmonize health and safety legislation. The central theme is, however, that there must be a continuing proactive approach to health and safety. An example is that where traditionally British legislation refers to employers doing as much as 'reasonably practicable', the new regulations are in a sense more prescriptive, and terms such as 'suitable and sufficient' or 'effective and suitable' are used. These imply a direct application to the job or site in question and require an employer to prepare individual risk assessments by competent persons.

The Manual Handling Operations Regulations provide a good example:

> Previous legislation applied to specific occupations whereas the new regulations apply to employees whatever their occupation, including the self-employed, and apply to the movement of loads by hand or other

bodily force, and their application is therefore extremely wide, involving the preparation of individual risk assessments for any task involving an element of manual handling and the reduction of risk wherever possible.

## Provision and Use of Work Equipment Regulations 1998

The 1998 Regulations impose on employers various obligations with respect to the provision, use, maintenance and inspection of 'work equipment', namely 'any machinery, appliance, apparatus, tool or installation for use at work'. The breadth of this definition was recently illustrated in *Spencer-Franks v. Kellogg Brown and Root Ltd*, where a door closer was held to be work equipment and this was so even though at the time it was being repaired, rather than being used as a door closer. In the same way a ramp giving access for carers to a wheelchair-confined user of care facilities has been held in *Smith v. Northamptonshire County Council* to be work equipment, although on the facts the employers were held not to be liable to a care assistant who was injured when the ramp collapsed as her employer had not constructed or installed the ramp. An employer may also escape liability under these Regulations where an employee improvises equipment and is injured as a result. In *Couzens v. T McGee & Co Ltd*, a lorry driver's claim failed when a piece of metal which he had stored in the cab of his lorry had caused a crash and resulted in injury to the driver. His employer had neither expressly nor impliedly consented to the use of the piece of iron by the driver for scraping out the lorry and it was thus not 'work equipment' for the purposes of the Regulations. A similar outcome was reached in *Ammah v. Kuehne & Nagal Logistics Ltd*, where an employee had used a plastic box to stand on to reach a high shelf despite the provision of suitable equipment by his employer and warnings and instructions on proper working methods. The employee's claim for compensation for breach of duty at common law and under statute was rejected.

## Control of Substances Hazardous to Health (COSHH) Regulations 2002 (as amended)

These rules were originally enacted in 1988 in response to an EU Directive on hazardous agents. Substances are defined as 'any natural or artificial substance whether solid, liquid, gas or vapour'. Harm may be caused by inhaling the substance, or ingesting or absorbing it.

Employers are required to assess the risk presented by the substances, both to employees and others. They must prevent or control exposure to substances, and monitor it, and carry out health surveillance of employees exposed to risk. The regulations impose obligations on employers in regard to training and information for employees.

According to the HSE 'an estimated 1.3 million businesses in Great Britain are engaged in activities that involve the use or production of substances hazardous to health'. These businesses are thus subject to the COSHH Regulations 2002 and the subsequent amendments. COSHH does not apply to asbestos or lead, which have their own regulations.

Employers may even be liable under COSHH when they have provided protective equipment such as latex gloves where those gloves themselves cause a harmful reaction to an employee. In *Dugmore v. Swansea NHS Trust* the Court of Appeal held that employers are subject to an absolute duty which requires them to seek out risks and prevent them. There was no reason to limit liability to situations where a risk was reasonably foreseeable or could only have been revealed by a risk assessment.

In June 2007, a European Union regulation operating alongside COSHH came into force. REACH concerns the Registration, Evaluation, Authorization and restriction of CHemicals. Its aims include the provision of a high level of protection of human health and the environment from the use of chemicals and the imposition of responsibility on manufacturers and importers of chemicals such that they understand and manage the risks associated with their use. A system of registration of relevant information applies.

## The Factories Act 1961

The main body of this Act has been replaced by the Health and Safety at Work etc. Act 1974 and the Regulations and Codes of Practice made under it.

## Construction regulations

The vast range of workplaces and working situations covered by the Factories Act 1961 meant that the rules, though detailed, could not be very specific. Each working situation will tend to have hazards peculiar to it. In the construction industry statistics show that many accidents are caused by falls – from ladders, scaffolds, platforms or roofs – or by materials falling. Another significant cause of accidents, sometimes fatal, is the use of lifting equipment and machinery. Significant numbers of less serious accidents occur when employees step on, or strike against, objects. There is extra danger for workers in the industry when excavation and tunnelling takes place. To take account of these special risks, there were several sets of Regulations in the construction industry:

- Construction (General Provisions) Regulations 1961
- Construction (Lifting Operations) Regulations 1961
- Construction (Working Places) Regulations 1966
- Construction (Health and Welfare) Regulations 1966.

These Regulations were made by the Secretary of State for Employment under powers granted by the Factories Act 1961. In place of these old legislative measures there were more forward-looking regulations such as the Construction (Design and Management) Regulations 1994, the Construction (Health, Safety and Welfare) Regulations 1996 and the Control of Asbestos at Work Regulations 2002. Like the 'six pack' these owed much to a European influence.

## Construction (Design and Management) Regulations 2007

These Regulations combine two earlier sets of regulations, the Construction (Design and Management) Regulations 1994 and the Construction (Health, Safety and Welfare) Regulations 1996. The merged regulations seek to ensure that all duty holders work together in order to improve the planning, preparation and conduct of projects with the aim of ensuring the health, safety and welfare of all construction, maintenance and repair workers. Duty holders include the client, the CDM coordinator, the designer, the principal and any other contractors, and the workers. The CDM coordinator has a central role as the person who provides advice and assistance to the client on how to comply with the regulations, and who coordinates health and safety measures through the planning, preparation and construction stages of a project. A CDM coordinator is only required for a notifiable contract, namely one likely to involve more than 30 working days, or 500 person days, of construction work. There must be notification of the HSE. A further duty imposed on CDM coordinators is the preparation and updating of a health and safety file containing information that those involved in carrying out construction or cleaning work on a structure in the future are likely to need, but could not be expected to know. This file is passed to the client at the completion phase. The 2007 Regulations are supported by an approved code of practice, Managing Health and Safety in Construction. Both the regulations and the code of practice emphasize the need to involve workers and their representatives in planning, training and discussions about health and safety on a project.

## Control of Asbestos Regulations 2006

The HSE estimates that about 4000 people in Great Britain die each year from mesothelioma and asbestos-related lung cancer as a result of past exposure to asbestos. It is predicted that these figures could continue to rise in the future. Historically, exposure to asbestos was particularly high in industries like shipbuilding and railway engineering, and asbestos was a popular building material during the second half of the last century. Those presently at high risk include, perhaps obviously, asbestos removal workers but also plumbers, carpenters and electricians. According to the HSC even computer installers and cabling engineers are at risk of exposure.

The 2006 Regulations combine three earlier sets of regulations which covered prohibition of the use of asbestos, asbestos licensing and the control of asbestos at work. The importation, supply and use of any form of asbestos is banned, and a duty to manage asbestos in non-domestic premises is imposed. Guidance on the latter duty can be found in an approved code of practice, the Management of Asbestos in Non-Domestic Premises.

The duty under the 2006 Regulations requires those in control of premises to take reasonable steps to determine the location and condition of materials containing asbestos and to keep up-to-date records of that information. They should presume materials contain asbestos unless strong evidence indicates the contrary and should assess the likelihood of anyone being exposed to asbestos fibres. A plan to manage these risks should be prepared. These duties fall on all those with responsibility for maintenance and/or repair of non-domestic premises. This could apparently include landlords, tenants or a managing agent. Where employers or the self-employed are working with asbestos or doing work that may disturb asbestos, they should take steps to prevent exposure to asbestos fibres or ensure that exposure is kept as low as reasonably practicable. These steps should not rely on the use of respiratory protective equipment but should prevent the spread of fibres.

The introduction of the previous Asbestos Regulations in 2002 provided a certain symmetry with the decision of the House of Lords in *Fairchild v. Glenhaven Funeral Services Ltd* in the same year. Here their Lordships overruled a decision of the Court of Appeal that victims of mesothelioma were not entitled to recover compensation from former employers because they had had more than one employer and it could not be scientifically established which specific employer was responsible for causing their exposure to the disease.

## Work at Height Regulations 2005 (as amended)

Several years ago the HSE made the prevention of falls from height in construction one of its priorities. Falls from height are the most common cause of fatal and major injury. Poor management control rather than equipment failure is said to be the most likely cause. Thus the HSE campaign has resulted in its inspectors visiting workplaces to ensure that, where working at a height cannot be eliminated, fall prevention measures (e.g. guardrails, scaffolding and safe working platforms) are in place or that fall arrest systems such as safety harnesses are being used. Inspectors also check that equipment is maintained and inspected, and that workers have been properly trained and are supervised. The Regulations require employers to ensure that all work at height is properly planned and organized, that those working are trained and competent, and that risks from fragile surfaces and falling objects are properly controlled. Account should be taken of weather conditions when they might endanger health and safety.

# Corporate Manslaughter and Corporate Homicide Act 2007

This long-called-for piece of legislation finally came into force in April 2008 after much debate, delays and procrastinations. At the time of writing there has been only one prosecution brought under the Act and that case has itself been adjourned. As noted above, accidents at work (or amongst the travelling public) are generally brought to public attention only when a major disaster occurs, such as the Alpha Piper explosion in 1988 or the capsizing of the *Herald of Free Enterprise* off Zeebrugge in 1987. As a result of the latter incident it was finally accepted that a corporation could be charged with the offence of gross negligence manslaughter, although on the facts of the particular case the court was unable to identify the necessary 'controlling mind and will' to impose liability on the non-human owners of the vessel. This was so despite a finding that the shipping company's organization was described as being infected with the disease of sloppiness from top to bottom. Although charges of gross negligence manslaughter have subsequently been successfully brought against corporations, these cases have involved small private companies where it has been easy to identify the non-human corporation with its human managers and owners. Where a corporation has a complex hierarchical structure there has been no comparable outcome (see the Southall case and Great Western Trains above).

With the enactment of the Corporate Manslaughter and Corporate Homicide Act 2007 more than 10 years since a proposed offence of corporate killing was recommended by the Law Commission, Parliament has eventually filled a significant gap in the sanctions available to deal with the deaths that result from not only high-profile disasters but from the all too frequent accidents at work. The Act provides that an organization (including a corporation, public body, police force, partnership or trade union) is guilty of an offence if the way in which its activities are managed causes a person's death and amounts to a gross breach of a relevant duty of care owed by the organization to the deceased. Relevant duties of care would include those owed by employers to their workers or to those employed by their subcontractors, as well as duties owed as occupiers of premises and the carrying on by the organization of any construction or maintenance operations. However, organizations are only liable for gross breaches, namely those where the conduct alleged to amount to a breach of duty falls far below what can reasonably be expected of the organization in the circumstances. It is the jury in any trial that must determine this question and in doing so the jury must consider whether the organization failed to comply with any relevant health and safety legislation, and if so how serious that failure was and how much of a risk of death it posed. The jury may also consider the organization's attitude and approach to health and safety. Did it encourage or tolerate a lax regard for health and safety? A further qualification on an organization's liability is that it is guilty of an offence only if the way in which its activities are managed or organized by its senior management is a substantial element in the breach of duty. Under the Act senior management means the persons who play significant roles in the making of decisions about how the whole or a substantial part of an organization's activities are to be managed or organized, or the actual managing or organizing of the whole or a substantial part of those activities. As noted earlier, a major stumbling block in securing the conviction of a company for gross negligence manslaughter at common was the need to identify a person sufficiently senior in the company whose guilty mind (*mens rea*) and actions or omissions (*actus reus*) could be attributed to the company. Commentators are undecided as to whether the new senior management test has removed this hurdle.

The Act abolishes gross negligence manslaughter as far as organizations are concerned and imposes liability only on the organization itself and not on its directors or managers. There is no offence of aiding, abetting, counselling or procuring the commission of an offence of corporate manslaughter. However, an individual might still be charged and convicted of gross negligence manslaughter at common law. The sanction for breach of the Act is a fine, which the Sentencing Guidelines Council has suggested should seldom be less than £500,000 and might be measured in millions of pounds. The court also has power to order remedial action and to make a publicity order. The latter may involve publication of the fact of conviction, the particulars of the offence, the amount of any fine and the terms of any remedial order.

## Compensation for injuries at work

Where a workman is injured or is killed in an accident at work, he or his dependants may wish to sue for damages. Damages are only available where an employee can show that the employer was in some way at fault. Generally, this means that the employee must prove the tort of negligence. Although negligence is still developing as a tort, the example of the duty owed by an employer to take reasonable care for the safety of his employees was recognized very early. Not only may the injured employee be able to sue in negligence, he may also have a claim for damages for breach of statutory duty. Negligence and breach of statutory duty are considered in detail in the chapter dealing with tort (Chapter 10). Apart from claims for damages, the injured employee may also be entitled to disablement benefits under the Industrial Injuries Scheme. He may be entitled to benefits under his own private insurance arrangements, or from his trade union.

When the court is considering a claim for damages from an employee, it will expect the employer to have taken reasonable care, given the state of knowledge and recognized practice in particular industries and trades. An employer will be expected to keep abreast of developing knowledge. The duties owed are not absolute so the employer must weigh up the risks involved and set these against the effectiveness of precautions that could be taken, bearing in mind the cost and inconvenience of taking those precautions.

An employer owes the duty to take reasonable care of all employees as individuals. Where some employees are inexperienced or do not easily understand English, or are newly recruited, the employer may have to take more precautions in their case than for other workers. However, the courts have said that they do not see the relationship of employer and employee as that of nurse and imbecile child!

The employer's duty to take reasonable care has been analysed regularly by the courts and they have been inclined to break down the duty into a number of broad heads:

- A duty to provide reasonably safe plant, equipment and premises.
- A duty to provide reasonably safe fellow workers.
- A duty to provide a reasonably safe system of work.

In all of these aspects the duty is ongoing. A safe system of work, once introduced, must be regularly checked and monitored. It may need to be changed to meet new standards created by new knowledge.

Employers are in breach of duty if they fail to take reasonable care. The duty is not absolute; they do not guarantee an employee's safety. It will be for the injured employee to prove negligence. In some cases this will be admitted by the employer and the subsequent court action will concern itself only with the measure of damages. Employers may seek to defend themselves by pleading that the employee was partly to blame – contributory negligence. If that plea is successful, it will operate to reduce the damages payable by that proportion which the employee is found to be at fault.

Recent years have seen a growth in actions by employees on the grounds that they have suffered bullying or stress at work. HSE research indicates that nearly half a million people suffer work-related stress at a level which they believe makes them ill. In one of the earliest stress-related cases, *Walker v. Northumberland County Council*, a social worker successfully claimed damages in respect of mental stress which resulted in psychiatric injury as a consequence of ever-increasing workloads and the employer's failure to provide help despite the employee suffering an earlier breakdown. Subsequent appeal court decisions seem to take a contradictory approach to the extent of an employer's duty, but best practice would seem to suggest that employers should put into place arrangements to make work less stressful and to ensure that those arrangements are implemented, at least where an employee has suffered an earlier psychiatric illness. An employer may be expected to provide counselling or to have in place an occupational health department to deal with stress-related injuries at work. In at least one case it was suggested that breach of the Working Time Regulations may be a factor in a stress claim. Failure to deal with claims of bullying may similarly result in civil claims either for breach of contract or for harassment under discrimination legislation. Indeed, it may even be possible to pursue a claim under the Protection from Harassment Act 1997 according to the House of Lords in *Majrowski v. Guy's and St Thomas's NHS Trust*, although it has subsequently been suggested that the relevant conduct must be of an order which would sustain criminal liability.

Employers invariably must have Employers' Liability Insurance in accordance with the Employer's Liability (Compulsory Insurance) Act 1969 to ensure that funds are available to meet any claim for compensation. Individuals also have rights under the Third Party Rights Against Insurers Act 2010 to obtain satisfaction of a judgment directly against an insurer, which is particularly important if the company employing the individual has gone out of business. Insurance companies currently take in hundreds of millions in premiums to meet potential liabilities for accidents or disease suffered as a result of employment. Insurers will also often undertake regular inspections of buildings and work processes to assess a company's compliance and knowledge of current health and safety legislation. This, together with their claims experience, i.e. how much they have to pay out in claims, will determine future premiums; bearing in mind that certain larger companies undertaking hazardous work will have premiums running into six figures, it is clear that businesses must give serious consideration to health and safety from an economic point of view as well as a legal and moral one.

# 12
# Land law

## Classification of property

Land is one of a number of kinds of property which can be owned, so it is useful to establish the meaning of the word 'property' and then to attempt to classify types of property. When considering sale of goods we saw that the word 'property' was used in the special sense of 'ownership'. The more general meaning of the word is, simply, anything capable of being owned.

The various categories of property are so diverse (e.g. land, jewellery, cars, stocks and shares, patents, money) that it is not surprising that the law has needed to develop different rules for each type to cover buying, selling or other transfer. Of course, if you buy a loaf of bread, it can be handed over to you. The same is not true when you buy land. A transaction involving land may be far more complex, especially when several people may have different rights over the same property at the same time. A person may own a house in which a tenant is currently living. The house may be mortgaged to a building society, and the owner may have leased shooting and fishing rights on the land and granted rights of way. This is a simple example but it serves to illustrate why the rules relating to land can be complex.

Obvious ways to divide property of all types are into things which are movable or immovable, or things which are tangible or intangible. Unfortunately, English law has not used either of these classifications, but instead divides all property into that which is real and that which is personal. This might be simple if real property consisted of land and personal property was everything else. But real property consists only of freehold interests in land, so (for historical reasons) leasehold interests are classed as personal property. As the terms 'real' and 'personal' do not conjure up as clear a picture of what they cover as the other classifications, it is interesting to note the derivation of the names. Historically, English law has always given preferential treatment to land and interests in land. If a person had his freehold land wrongfully taken from him the law would ensure that he could recover it by means of a 'real' action. If any other type of property was wrongfully taken from him the law merely allowed him to make a claim for damages against the wrongdoer, by means of 'personal' action. So property came to be classified according to the type of court action needed to protect it, which accounts for the anomalous classification of leasehold interests in land.

Galbraith's Building and Land Management Law for Students. DOI: 10.1016/B978-0-08-096692-2.000012-9
Copyright © 2010 by Elsevier Ltd

# What is land?

Land is the surface of the earth, the airspace above it and the ground below. An owner of land could expect the law to protect his or her interests if someone unlawfully tunnelled under the land. However, Acts of Parliament limit a landowner's rights in this respect. Certain mining and mineral rights have been granted by statute to public corporations, e.g. coal mining rights to the Coal Authority.

As for the airspace above the surface of the earth, it used to be assumed that the landowner owned the space up to the heavens. This rather romantic idea has now been restricted by cases such as *Bernstein v. Skyviews and General Ltd*, where it was held that an owner owns only as much of the airspace above the land as is reasonably necessary for ordinary use or enjoyment of the land. In that case, a firm specializing in aerial photography flew over the plaintiff's house to photograph it. This act was held not to constitute a trespass, given the height at which the aeroplane flew. Most normal aircraft activity is excused under the Civil Aviation Act 1982, which precludes actions for nuisance or trespass if the aircraft flies over at a reasonable height.

It is difficult to think of airspace as being a valuable commodity but it may have significant commercial worth. For example, it may be possible to rent out space for erection of an advertising hoarding. The commercial value of airspace was at stake in a case, *Woollerton and Wilson Ltd v. Richard Costain Ltd*, where a builder was using a tower crane. The arm of the crane swung over adjacent land, trespassing to the owner's airspace. The builder had not sought permission for this activity. When it became clear that he was trespassing, he offered to pay a weekly sum to the landowner. The landowner obviously thought he could hold out for more. However, his greed was 'rewarded' when the court granted him an injunction to prevent the builder using the crane, but its operation was postponed for several months, by which time the need for the crane had disappeared! An injunction was granted with immediate effect in similar circumstances in *Anchor Brewhouse Developments Ltd v. Berkley House Ltd.*

It is clear that the law now sees some limit on the extent to which land extends into the air. The precise limit will depend on the type of land and the sort of use to which it could be reasonably expected to be put. An owner of a grouse moor may need a considerably greater height of air than the owner of a suburban bungalow, where the only likely use for airspace will be for the erection of a TV aerial.

On the surface of land there are likely to be structures such as houses or factories. Buildings form part of the land, and so do fixtures. A fixture is something on or attached to the land for its improvement, which is regarded as forming part of it, e.g. a garden gate, rose bushes or a dry-stone wall. A chattel is an item of personal property which does not form part of land, e.g. a watch or a pen. The same items may, in different contexts, be either fixtures or chattels. Thus, a pile of stones in a builder's yard is personal property, but the same stones in the form of a dry-stone wall would be a fixture.

It may be important to determine whether an object is a fixture or a chattel. When land is sold, all fixtures pass with the sale unless otherwise agreed; when a tenant installs fixtures in a landlord's property, the fixtures become the property of the landlord; and, where land is mortgaged, all fixtures form part of the mortgage security, even those installed after the mortgage was created. Further, where a testator in a will leaves all real property to one person and all personal property to another, fixtures will pass with the real property and chattels with the personal property.

The question of whether an item is a fixture can also be relevant where a landowner becomes insolvent. In the case of *Lyons Co. v. London City Bank*, the owners of chairs had hired them out to the owner of a hall. In order to comply with local bye-laws on safety, the hall owner had fixed the rows of chairs to the floor. He then became insolvent, and the receiver was keen to show that the chairs formed part of the hall owner's land, as fixtures. The chair owner was equally keen to show that they were still his chattels, as he would be limited to a fairly hopeless claim for their value in the insolvency proceedings unless he could take the chairs back. The court found that the chairs were chattels.

In determining whether an item is a fixture or a chattel, the basic question is whether the item is physically attached to the land. If an item is attached to the land, this raises a presumption that it is a fixture. A statue which is cemented in place would thus be presumed to be a fixture. If an item is not attached to the land, the presumption is that it is a chattel. Thus, a free-standing statue, held in place only by its own weight, would be presumed to be a chattel.

It is possible to rebut the presumption that things attached to the land are fixtures, by showing that the item in question was fixed only to allow it to be enjoyed as a chattel and not for the improvement of the land as land. Thus, in *Leigh v. Taylor*, where a valuable tapestry was fixed to battens on a wall, the tapestry was held to be a chattel because, in order to display and enjoy the tapestry, it was necessary to fix it in some reasonably robust way appropriate to its considerable weight.

Equally, it is possible to rebut the presumption that an item which is not attached to the land is a chattel, by showing that the item does form part of the land. Thus, a dry-stone wall would clearly be regarded as a fixture, because it is there to improve the land, although unattached other than by its own weight. Similarly, a free-standing statue which formed part of an overall landscape garden design might be a fixture.

Clearly, it may at times be difficult to determine whether an object is a fixture or a chattel. Thus, in relation to a sale of land, it may be sensible for the parties to reach an express agreement about those items to be included in the purchase price and those which the vendor can remove. If such an agreement were always reached, there would be fewer disputes over greenhouses and sheds, TV aerials, dishwashers, carpets and garden ornaments!

In relation to tenancies, whilst items installed by a tenant in a landlord's property become the property of the landlord, the tenant may be entitled to remove 'tenant's fixtures' before or within a reasonable time of the termination of the tenancy. Tenant's fixtures are fixtures installed by the tenant which are domestic and ornamental (such as stoves and paintings), trade fixtures (such as petrol pumps and industrial machines), or agricultural fixtures. In removing tenant's fixtures, the tenant must do as little damage as possible, and must repair any damage done.

As ownership of land includes the surface and the ground below, a landowner might imagine that he or she owns everything in or on the land. The special position of mineral rights has already been mentioned, but it is also worth noting that any treasure as defined in the Treasure Act 1996 found on the land belongs to the Crown. If necessary, the coroner may hold an enquiry into findings of treasure.

Wild animals living on a person's land are incapable of being owned. Once the animals are dead, they belong to the landowner, so a poacher should hand over the fruits of his labours! Percolating and flowing water cannot be owned, but a landowner is permitted to abstract quantities of it for uses connected with the land. Similarly, if the land is adjacent to a river, he may take water for domestic purposes but if he takes it for other purposes connected with the land, he must be careful to observe the rule that landowners further downstream are also entitled to the flow of the river unaltered in quantity and quality. If a landowner pollutes the water which he returns to the river he can be sued in nuisance.

## Freehold and leasehold estates

People are often surprised to learn that in England individuals do not own land. All land belongs to the Crown, and the greatest interest which a landholder can own is the freehold interest or estate. This peculiarity of English law is explained by reference to history. When William conquered England in 1066 he claimed all the land for himself, and then parcelled it out to his barons as a reward for their support. In return, they owed allegiance to him and undertook various services, some merely ceremonial, some valuable. This was the feudal system. It was not until 1925 that our legal system was rid of most of the consequences of feudalism. Modern land law is based on the Law of Property Act 1925.

As every landowner is technically a tenant of the Crown, it is common to refer to his or her interest in land as an 'estate'. The law recognizes two estates – freehold and leasehold. The major difference between a freehold and a leasehold is the length of time for which the land will be held.

Freehold is the nearest equivalent to absolute ownership of land, and is more technically described as a 'fee simple absolute in possession'. Basically, a fee simple continues to exist as long as there are persons entitled to take it on the owner's death either under a will or on intestacy. (Intestacy is the name used to describe a situation where there is no will to govern the disposal of property of a deceased person. Statutory rules then operate to determine who will inherit.)

By contrast, a leasehold estate is described as a 'term of years absolute'. This means that a landlord (who may be the owner of the freehold or may himself be a tenant) grants to a tenant exclusive possession of property which the tenant is to enjoy for a fixed period. If the period is quite short it is usual to call it a tenancy (for example, a monthly tenancy). Longer terms are usually referred to as leases (for example, a 99-year lease), but there is no precise or rigid rule.

Both freehold and leasehold estates are property which can be sold and can be inherited under a will or on intestacy, and are the only two legal estates which can exist in land.

A new way of owning land called 'commonhold' has been introduced by the Commonhold and Leasehold Reform Act 2002. This Act provides that a freehold estate can be registered as a freehold estate in commonhold land. Commonhold is designed for blocks of flats and commercial developments. A commonhold is made up of 'commonhold units' and the 'common parts'. There must be at least two commonhold units in a commonhold and these will usually consist of flats or commercial units. The freeholder of a commonhold unit is called the 'unit holder'. The rest of the commonhold is designated as the common parts. The unit holders must form a company limited by guarantee, called the 'commonhold association', which will manage the commonhold. Each unit holder will own the freehold in his or her unit whilst the commonhold association will own the freehold in the common parts so that, for example, it can insure and maintain the common parts. The Act contains detailed provisions about the setting up and running of a commonhold.

## Interests in land

Rights in land other than the two legal estates (freehold and leasehold) are called interests in land. Basically, there are two types of interests in land, legal interests and equitable interests. Section 1 of the Law of Property Act 1925 provides that a limited number of rights in land may, if created by deed, be legal interests. The most common of these are easements and profits *a prendre* (provided that their duration is either that of a fee simple absolute in possession or that of a term of years absolute), and charges by way of legal mortgage. The nature of each of these interests is considered below.

Any interest in land which is not a legal interest must be an equitable interest. Examples of equitable interests are: interests under restrictive covenants; interests (including fee simples, leases, easements or profits *a prendre*) granted for life; interests (including fee simples, leases, easements and profits *a prendre*) created without the use of a deed; interests under contracts to create legal estates or interests; and interests under trusts.

Traditionally, the essential difference between a legal interest and an equitable interest is that a purchaser of land is bound by all legal interests in the land but will not necessarily be bound by all equitable interests in it. The position has been modified by statute, however (Land Charges Act 1972; Land Registration Act 2002).

The extent to which a purchaser of land can discover and is bound by interests in the land which he or she purchases, and the extent to which the owner of an interest in land can ensure that his or her interest will bind a purchaser of the land, are considered in the course of the following section.

# The conveyancing process

The conveyancing process is the process of transferring ownership of land, i.e. selling land. At present there are two systems of conveyancing in England and Wales. Unregistered conveyancing is the transfer of unregistered land, and registered conveyancing is the transfer of registered land. The traditional system is unregistered conveyancing, but as all land in England and Wales is now subject to compulsory registration, under the provisions of the Land Registration Act 2002, registered conveyancing is now replacing it.

In order to determine which system of conveyancing is appropriate, it is necessary to determine whether the land in question is registered or unregistered. This requires a search of the Register, which is made by sending a form to the District Land Registry in whose area the relevant land is situated.

## Home information packs and Energy efficiency

A vendor must supply a purchaser with an energy performance certificate, free of charge, at the earliest opportunity. An energy performance certificate must be prepared by an approved energy assessor and it will set out information about the energy efficiency of the property.

From 1 October 2008, occupiers of buildings with a total useful floor area over 1000 square metres occupied by public authorities and by institutions providing public services to a large number of persons and therefore frequently visited by those persons have been required to show a display energy certificate in a prominent place. The idea behind this is to enable the public to compare the energy performance of public buildings and to promote improved energy use. There are also obligations on those who have control of air-conditioning systems (with a maximum calorific output of more than 12 kW) to ensure that the system is inspected at least every five years by an energy assessor.

## Unregistered conveyancing

(For the purposes of this section the terms purchaser and vendor will normally imply the solicitors acting on behalf of the purchaser and vendor.)

The stages of the process of conveying unregistered land are as follows:

1 The contract is drafted by the vendor and two copies are sent to the purchaser. The vendor is required to show that he or she has the necessary title to the land which he or she has undertaken by the contract to transfer. The vendor must produce an abstract of title and send it to the purchaser within a reasonable time, unless the contract lays down a specific time limit. The abstract of title is a summary of the documents which prove the vendor's title to the land. In practice, a chronological list of the relevant documents together with photocopies of them are provided by the vendor. Traditionally title was sent to the purchaser after exchange of contracts but the modern practise is for it to be sent with the draft contract.

2 The purchaser searches the local land charges register, which is kept by the local authority, in order to determine whether any local land charges are registered in relation to the land. It is usual to send in a search form, requesting the Registrar's staff to perform the search on behalf of the purchaser.

There are two types of local land charges: restrictions imposed on the use of land, such as planning permission restricting the number of houses to be built on the land; and financial charges imposed on the land in respect of work carried out by the local authority, such as the imposition amongst frontagers of the cost of making up a road. If a local land charge is correctly registered then the purchaser is bound by it even if he or she does not search the register. If a search by the Registrar's staff fails to reveal a local land charge which has been correctly registered, or the local authority has failed to register an existing local land charge, the purchaser is still bound by the local land charge but is entitled to compensation from the local authority.

Using the standard search form, to which additional enquiries may be added, the purchaser makes a variety of other enquiries of the local authority in relation to matters not registerable as local land charges. These may include issues relating to planning, schemes for new roads and slum clearance, and whether the local authority is obliged to maintain roads adjacent to the property. If the local authority's answers are incorrect, the authority may be liable in negligence or for breach of contract.

The purchaser will also make a search with the local water company to make sure that the property enjoys mains water and drainage. Additional searches may also be carried out depending upon the property and its location, for example a coal mining search where the property is located in an area of past, present or future coal mining.

3   The purchaser makes a variety of enquiries of the vendor, usually by sending a standard form, which may be customized by adding or deleting enquiries as appropriate. Enquiries normally relate to matters such as easements and covenants, services enjoyed, fixtures and boundaries. The vendor is not obliged to reply, but if he or she does, incorrect information may amount to a misrepresentation or a criminal offence of fraud under the Fraud Act 2006.

4   The purchaser may wish to have a survey of the property carried out. This is not essential, but may be a prudent step. If the purchaser needs to raise money by mortgage, the lender will usually conduct its own survey. Where this is the case, it may still be sensible for the purchaser to have his or her own survey carried out as the lender is only concerned to know whether the property is good security for the mortgage, not whether it is worth the purchase price.

5   The purchaser examines the abstract of title. If he or she is not satisfied that the vendor has the necessary title, a written requisition is sent to the vendor within a reasonable time (unless the contract specifies a specific time limit). The vendor must reply within a reasonable time (unless the contract specifies a specific time limit). If the vendor does not reply in time or fails to show that he or she has the necessary title, the purchaser may be entitled to rescind the contract.

6   If the survey and the replies to all searches and enquiries are satisfactory, and raising any necessary finance presents no problems, it is then safe for the purchaser to make a legally binding contract with the vendor.

The purchaser may agree the draft which has been sent by the vendor, or may want amendments. When both parties agree to a final draft of the contract, the purchaser sends one copy to the vendor and keeps the other. The vendor then produces two final copies of the contract and sends them to the purchaser. The purchaser checks them and returns one to the vendor. The vendor and purchaser each sign their copy and then exchange copies (exchange of contracts), the purchaser giving a deposit if required. The contract becomes binding at the time of exchange.

Section 2 of the Law of Property (Miscellaneous Provisions) Act 1989 provides that a contract for the sale or other disposition of any interest in land is only valid if made in writing. The writing must embody all the terms agreed, though it may incorporate terms by reference to another document. The contract must be signed by or on behalf of both parties. If there is an exchange of contracts (see above), at least one of the documents, but not necessarily the same one, must be signed by or on behalf of each of them. (These rules are considered in more detail on page 114. See also 'Subject to contract' on page 117.)

The existence of the contract does not prevent the vendor selling the land to a third party, unless the contract is registered (see registration of land charges and registered land below). However, if this is done, he or she will be acting in breach of contract. Exchange of contracts does not transfer legal title to the property from the purchaser to the vendor.

7   Immediately before completion (the transfer of legal title in the property from purchaser to vendor), it is necessary for the purchaser to search the registers kept by the Land Charges Department of the Land Registry. In practice these searches are normally carried out by officials of the department on the purchaser's behalf.

Of the five registers kept by the Department, the most significant is the register of land charges. Certain interests in land are registrable as land charges. Examples are: mortgages not protected by deposit of title deeds; contracts to convey or create legal estates; options to purchase land; restrictive covenants created after 1925 and not in a lease; equitable easements or profits *a prendre*; and equitable interests not protected by deposit of title deeds, not arising under a trust and not otherwise being registrable as land charges.

Where a valid registrable interest is registered then, subject to certain exceptions, a subsequent purchaser of an estate or interest in the land will be bound by it, whether he or she searched in the register or not.

Where such an interest is not registered then, basically, a purchaser for value of a legal estate in the land will not be bound by it. This rule is well illustrated by *Midland Bank v. Green*. In that case, a farmer had made a legally binding contract to sell his farm to his son for a very favourable price. He then fell out with his son, and sold the farm instead to his wife. She knew about the earlier contract between father and son, but the son had never registered it as an estate contract. She paid a very low price for the farm. The son sued on the binding contract he had made with his father, but the court held that his mother could keep the farm, as she was a purchaser for value of a legal estate, and she was not bound by the son's interest because he had never registered it. The case is particularly powerful in reinforcing the rules about registration because the mother had actual knowledge of the estate contract, despite its lack of registration. Nevertheless, the non-registration was the paramount factor.

However, in some cases of non-registration, including contracts to convey or create legal estates, options to purchase land, restrictive covenants and equitable easements and profits, purchasers of interests in land, or purchasers of legal estates other than for money or money's worth, may be bound by the interest even though it was not registered.

Where a legal interest is not registrable as a land charge (for example, a legal easement or profit *a prendre*), a purchaser of an estate or interest in the land will automatically be bound by it. Where an equitable interest is not registrable as a land charge (for example, the interest of a beneficiary under a trust) a bona fide purchaser for value of a legal estate without notice of the equitable interest will not be bound by it, but other purchasers of estates or interests in the land will be. Thus, a purchaser of a legal estate who knows of the existence of an equitable interest will be bound by it. Further, a purchaser of a legal estate who would have discovered the existence of an equitable interest if he or she had properly examined both the title deeds and the vendor's land is deemed to have notice of it and is therefore bound by it. Moreover, a purchaser is deemed to have notice of, and is therefore bound by, interests which his agent either has or is deemed to have notice of.

8    If the purchaser is satisfied that the vendor has the necessary title, the transfer of legal title takes place by deed, the deed being called a conveyance. (Note: due to compulsory registration, a Land Registry form of transfer is usually used in practice.) The requirement to use a deed is contained in s.52 of the Law of Property Act 1925, but s.1 of the 1989 Act created some new rules about the nature of a deed. Most significantly, not only must the deed be signed, but also the signature must be witnessed and the witness must attest the signature, i.e. sign as witness to it. Two witnesses are required in any case where the executor of the deed needs someone to sign on their behalf, e.g. because the executor is too ill to sign it. Such formalities are imposed by law in circumstances where it may be necessary to prove the validity of documents.

Two copies of the draft conveyance are prepared by the purchaser and sent to the vendor. The vendor returns one copy to the purchaser, either approved or with amendments marked on it. When the final form of the conveyance is agreed, the purchaser makes a final copy, signs it and returns it to the vendor. On the date specified by the contract for completion, the purchase money is paid, the remaining legal work is completed (including handing over of the conveyance, which has been signed by the vendor, to the purchaser) and legal title to the property passes from the vendor to the purchaser. (Note: due to compulsory registration, title must be registered within two months of completion, otherwise the legal estate will not pass.)

## Registered conveyancing

(For the purposes of this section the terms purchaser and vendor will normally imply the solicitors acting on behalf of the purchaser and the vendor.)

The stages in the process of conveying registered land are as follows.

1   Local land charges and enquiries of local authorities and a search with the local water company (see unregistered conveyancing).
2   The vendor is required to show that he or she has the necessary title to the land which he or she has undertaken by the contract to transfer. In relation to registered land the vendor sends the purchaser an official copy of the entries on the register kept by the appropriate District Land Registry.

The register comprises:

(a)  The property register, which contains a description of the property.
(b)  The proprietorship register, which contains the names of the registered proprietors, states the class of title, and indicates whether the registered proprietor's ability to transfer the land is limited, for example because he or she holds the property in trust for a beneficiary.

The registered estate will either be freehold or leasehold for a term exceeding seven years. Different classes of title may be registered, the best and most common of which is absolute title. Absolute freehold title is equivalent to a fee simple absolute in possession. Absolute leasehold title guarantees that the lease was validly granted.

Where the Registrar is not satisfied that the registered proprietor owns the freehold or possesses a validly granted lease, he or she may register an inferior class of title which does not guarantee that the registered proprietor owns, respectively, the freehold or a validly granted lease. The inferior classes of registered title are good leasehold title (leases only), possessory title (freehold or leasehold) and qualified title (freehold or leasehold).

(c)  The charges register, which contains details of certain interests in the property which are owned by third parties. Important examples are 'registrable interests', such as mortgages (legal charges), and third party interests such as contracts to convey or create legal estates, restrictive covenants, equitable easements and profits *a prendre*, and interests of beneficiaries under trusts.

Third party interests may be entered on the register by the Registrar as a notice or restriction. As a general rule, a purchaser is bound by interests which are on the register, and takes the land free of any interests which are not registered, except those which are overriding.

Certain interests are 'overriding interests'. These bind a person who purchases or otherwise acquires registered title to property even though they are not entered on the register. Important examples are leases for a term not exceeding seven years, local land charges, and legal easements and profits *a prendre*.

3   Enquiries of the vendor/surveys (see unregistered conveyancing).
4   Formation and exchange of contracts (see unregistered conveyancing).
5   Completion takes place by deed, the deed being called a transfer. Two copies of the draft transfer are prepared by the purchaser and sent to the vendor (a standard form is used). The vendor returns one copy to the purchaser, either approved or with amendments marked on it. When the final form of the transfer is agreed, the purchaser makes a final copy, signs it and returns it to the vendor. On the date specified by the contract for completion, the purchase money is paid and the remaining legal work is completed (including transfer of the deed, which has been signed by the vendor, to the purchaser). Legal title to the property does not, however, pass from the vendor to the purchaser until the transfer has been registered.

## Trusts

Basically, a trust exists when one person, the trustee, holds property for the benefit of another person, the beneficiary. Thus, for example, X may hold the freehold of a property on trust for Y. X holds the legal estate of the property but Y has an equitable interest which entitles him to the benefits of the property.

Trusts may be created in a number of ways and for a number of reasons. A landowner may expressly create a trust of his or her legal estate, either during his or her lifetime or in a will. For example, assume that X, a widow who owns a house, has a child, Y, who is eight years old. The terms of X's will may provide that if X dies the house is to be held on trust for Y by two trustees, A and B, until Y reaches the age of 21.

Sometimes, when a landowner does not expressly create a trust of his or her legal estate, the courts may presume or imply that a trust has been created. For example, suppose that X, the purchaser of a house, agrees that Y, who lives with him, will acquire an interest in the house if Y contributes to the cost of purchasing the property by assisting with the mortgage repayments. Whilst a trust is not expressly created, the courts may be prepared to rule that X holds the legal estate on trust for himself and Y, their relative equitable interests being commensurate with their relative contributions.

In *Stack v. Dowden*, an unmarried couple had lived together for many years. Each had contributed financially to the purchase of the property but the transfer of the property to them did not contain an express declaration of their respective beneficial interests in their home. The House of Lords decided that the parties were entitled to joint and equal shares in the property, unless a clear contrary intention could be shown. The House of Lords established some key principles for determining the respective beneficial interests in this, and in similar, cases. These were as follows:

- A conveyance into joint names will result in a legal and beneficial joint tenancy, unless the contrary is shown.
- The burden of proof is on the owner seeking to show that they intended to hold their beneficial interests as tenants in common.
- The court must ascertain the parties' shared intentions in the context of the whole course of their conduct relating to the property. Factors to be considered include:
  - any advice or discussions at the time of the transfer, that would indicate their intentions at that time;
  - the reasons why they purchased the house jointly;
  - the reasons why the survivor of the couple was authorized to give a good receipt for capital monies;
  - the purpose for which the house was acquired;
  - the nature of the parties' relationship;
  - whether the couple had children for whom they both had responsibility to provide a home;
  - how the purchase was financed, both initially and subsequently;
  - how the parties arranged their finances, for example whether their accounts were held separately, together or a combination of both; and
  - how the couple discharged their outgoings on the house and other household expenses.

These principles have since been applied to a mother and daughter who purchased an investment property (*Laskar v. Laskar*).

## Co-ownership

It is quite common for one person to share the ownership of land with another. Obvious examples are a husband and wife who buy a house, or parties who buy premises from which to run their business. The law recognizes two types of co-ownership:

- Joint tenancy
- Tenancy in common.

With a joint tenancy, the co-owners are regarded as an indivisible whole, owning the entire property as one, whereas in a tenancy in common, the persons involved each own a share. This can be important if one of the co-owners dies. In a joint tenancy, the effect of death is to vest the deceased person's rights in the land in the surviving owners. If a house is owned as a joint tenancy by a husband and wife (i.e. as joint tenants), and the husband dies, the wife is then absolutely entitled to the house. If a property is owned equally by four partners (i.e. as tenants in common), each is at liberty to dispose of his or her share, and if one dies that share may be inherited by someone other than the surviving partners.

One problem associated with co-ownership is that there could potentially be a great many co-owners of one piece of land, which could greatly complicate transfers of the land or other related transactions. Thus, co-ownership, whether in the form of joint tenancy or tenancy in common, will always give rise to a trust. Such a trust is known as a 'Trust of Land'. The legal estate in the land must be vested in trustees, of whom there can be no more than four. Each of the persons entitled to a share will then hold an equitable interest. Essentially, a purchaser of land who pays two trustees for the land acquires the legal estate and is not bound by the interests of the beneficiaries. The trustees then hold the proceeds of sale on trust for the beneficiaries, whose interests in the land have been 'overreached'.

It should be noted that, in practice, in the case of co-ownership, the same persons are often both trustees and beneficiaries (e.g. where two persons have jointly purchased a house). Where one party desires a sale but the other does not, it may be necessary to obtain a court order before a sale can take place. The court may refuse to make such an order where the purpose for which the property was bought has not come to an end.

Special rules apply in the context of matrimonial proceedings. For example, upon divorce the court may adjust the relative interests of the spouses in the matrimonial home (Matrimonial Causes Act 1973). Equally, a spouse, cohabitant or former cohabitant (not including one of a homosexual couple) who is not a co-owner may, nevertheless, possess rights under the Family Law Act 1996.

If co-owners have not made it clear exactly how they intend to hold the land, rules exist to determine whether there is joint tenancy or a tenancy in common.

1    For a joint tenancy, the co-owners must be granted the same interest in the land. So if the land is given to A and B, they could be joint tenants, but if the land is given to A and B in shares of two-thirds and one-third respectively, they cannot be joint tenants.

2    For a joint tenancy, all co-owners must acquire their title under the same document.

3    For a joint tenancy, all co-owners must acquire their interest at the same time.

4    For either type of co-ownership, each co-owner must be as much entitled as the others to possession of any part of the land.

5    If no answer is provided by these various tests, then it becomes necessary to look at the deed creating the co-ownership, to see if it makes clear which type of co-ownership is intended. A grant to A and B jointly is construed as a joint tenancy; a grant to A and B in equal

shares is construed as a tenancy in common. Words such as 'in equal shares', which indicate that owners are to take distinct shares in the property, are said to be words of severance.

6    Where none of the tests is conclusive there are some factual circumstances in which equity will favour a tenancy in common, because it is fairer. These include cases where land is purchased by co-owners who have contributed in unequal shares, or as partnership property.

Even where these tests indicate that land is held as a joint tenancy it is possible for the parties to alter the arrangement in equity, by severing the joint tenancy. The simplest way to do this is for one tenant to serve notice in writing on the other joint tenants indicating that the server of the notice wishes to sever his or her share (s.36 Law of Property Act 1925).

## Rights and duties of owners of land

It is difficult to summarize the rights of a landowner. They certainly include the right to occupy the land without interference from others. (Interferences which the law will control include trespass and nuisance.) One important right of every landowner is a right of support for his or her land from adjacent pieces of land. Any withdrawal of such support, which causes the adjacent land to slip, e.g. excavation during building work, may result in a claim for damages. The right of support is a natural right, existing because of the nature of the land itself. A right of support may, however, be acquired, in which case the right is called an easement.

An owner's duties in relation to land include not only using it in a way which will not constitute a nuisance, but also complying with all the various statutory controls which now exist concerning land use. Significant amongst these are the planning laws contained principally in the Town and Country Planning Act 1990.

The controls on a landowner by statute range from restrictions on killing certain kinds of birds and animals on the land to positive obligations to repair dilapidated property. If an authority wishes to take an owner's land from him or her, it may be possible to do so under the rules of compulsory purchase. And rights of access may be reserved by statute, e.g. for the creation of a long-distance footpath. When the range of controls on a landowner is fully examined it is clear that the number of restrictions is far greater then the number of rights.

## Access to Neighbouring Land Act 1992

This Act enables a person to apply to the court for an order giving him or her access to neighbouring land (the servient land) in order to carry out work reasonably necessary for the preservation of the land (the dominant land). Examples of such work include repairing buildings, clearing drains and cutting hedges.

An order may be necessary where the owner or a tenant of the neighbouring land refuses consent to entry, the order being obtained against the person who refuses consent. The court will not make an order if this would interfere with or disturb the use or enjoyment of the servient land, or cause hardship to an occupier of it.

An order specifies a variety of matters, including the nature of the work which can be done, the areas of the neighbouring land which can be visited, the date of entry, terms to avoid or restrict damage, loss or injury and loss of privacy, the manner of the work, the days and hours during which the work may be carried out, and the persons who may enter the land to carry out the work. An order may impose compensation requirements and/or provision for payment to the person against whom the order is obtained, in consideration of entering the land. Breach of the terms of an order may give rise to an action for damages.

An order binds the person against whom it was made, and may be protected against successors and purchasers by registration. This may be necessary where an order permits entry during the term of a lease or in fee simple. Either party to an order may apply to the court for a variation.

## Party Walls etc. Act 1996

This Act applies to 'party walls' and 'party structures' (e.g. walls between terraced or semi-detached houses and structures separating upper and lower flats in the same house). Essentially, the Act gives the owner of one house the right (provided that certain conditions are satisfied) to enter the other house to carry out certain works (e.g. to carry out repairs). Before entering, however, it will usually be necessary to notify the owner of the other building, as required by the Act. Further, it may be necessary to carry out works to protect the property entered and/or to pay compensation to its owner.

## Easements

An easement is an interest in land which gives the owner of one piece of land the right to use or restrict the use of land belonging to another. Common examples of easements are:

- Easement of way, which gives a landowner the right to cross land belonging to another.
- Easement of storage, which gives a landowner the right to store goods on land belonging to another.
- Easement of light, which gives a landowner the right to the flow of light to a building on his or her land over land owned by another.
- Easement of support, which gives one landowner the right to have his or her land supported by land belonging to another.

However, the courts are not prepared to recognize every right as capable in law of being an easement. For example, a right to a view is incapable of being an easement. Equally, a right of total occupation or control of land cannot be an easement.

An easement must burden one piece of land (servient land) for the benefit of another piece of land (dominant land). For example, in the case of an easement of way, the owner of the dominant land has the right to cross the servient land for the benefit of the dominant land. An easement cannot exist unless it is created for the benefit of dominant land. Moreover, the easement must be of actual benefit to the dominant land, such as increasing its value. A personal benefit to the owner of the dominant land which does not benefit the land itself is not sufficient. For example, a right to cross land in Durham for the benefit of land in Somerset could not be an easement as the right could not be of benefit to the dominant land.

An easement cannot exist if the same person owns and occupies both the dominant land and the servient land. However, where both pieces of land are owned by the same person but one is occupied by a tenant, an easement can exist. Further, an easement cannot be created, whether expressly, impliedly or by prescription, unless at the time of its creation there is a person capable of creating it and a person capable of receiving it.

Easements are normally created expressly. A legal easement is created by deed, its duration being that of a fee simple absolute in possession or a term of years absolute. Landowners may grant an easement over their land to a neighbouring landowner (express grant). Alternatively, when landowners sell part of their land they may reserve an easement over the land which they sell (express reservation) or may grant the purchaser an easement over the land which they retain (express grant).

Further, a conveyance, transfer or lease of dominant land may automatically convert a licence enjoyed with the land into an easement, as s.62 of the Law of Property Act 1925 deems this to be an express grant. For example, if X, a tenant, has permission (a licence – see page 321) to store coal in a shed on land retained by her landlord, the grant of a new lease to X may, unless the new lease specifies otherwise, automatically convert the permission into an easement of storage (*Wright v. Macadam*).

Where easements are not created expressly, they may be created by implication (implied grant or implied reservation). Thus, where landowners sell part of their land and do not grant the purchaser an easement which is necessary for its enjoyment, or fail to grant to the purchaser an easement which both they and the purchaser intended would exist, the court may recognize as implied into the sale, respectively, the grant of an easement of necessity or the grant of an intended easement. For example, the court might be prepared to recognize the implied grant of an easement of way over a vendor's land where, following a sale, the purchaser discovers that he or she has no means of access to the land that has been purchased.

The courts are more reluctant to recognize implied reservation than they are to recognize implied grant. Implied grant may also occur where a landowner, prior to selling part of his or her land, enjoyed over it a right which could have been an easement, except that the dominant and servient lands were both owned and occupied by him or her (a quasi easement). If he or she sells the dominant land and retains the servient land, the effect of the rule in *Wheeldon v. Burrows* is that the court may recognize the implied grant of the easement to the purchaser provided that the easement was in use when the land was sold and either was necessary to its reasonable enjoyment or was used regularly and could be detected by visual examination of the land.

As well as being created by express or implied grant or reservation, easements may also be acquired by prescription. Basically, this means that if landowners, for the benefit of their land, use or derive a benefit from land belonging to another for 20 years or more, they may acquire an easement. Thus, if X crosses a path over Y's land for 20 years or more, or if the windows of X's house derive a flow of light across Y's land for 20 years or more, X may acquire, respectively, an easement of way or an easement of light. An easement can be acquired by prescription by a freehold owner or tenant against a freehold owner or tenant of other land, but cannot be acquired by a tenant against his or her own landlord.

In order to acquire an easement in this way, however, it is necessary that X used Y's land as though having a right to do so. Thus, if X had Y's permission to cross the land, then if X crossed Y's land in secret, so that Y would not discover him, or if X used force in order to cross Y's land, such as damaging fences, he would not acquire an easement in this way. It is also necessary that the use of the servient land be regular. (It is an oversimplification to suggest that 20 years' use will necessarily be sufficient to support a claim of an easement by prescription. In practice it is necessary to satisfy the requirements of at least one of three methods of prescription, the technical details of which fall outside the scope of this book.)

Basically, an easement may come to an end in any of three ways. First, if the owner of the dominant land agrees to 'release' it, normally by deed (express release). Secondly, if the owner of the dominant land abandons the easement (implied release). Thirdly, if the dominant and servient lands become owned and occupied by the same person.

# Profits *a prendre*

A profit *a prendre* is an interest in land in the form of a right to remove property from someone else's land. Common examples are rights to fish and to graze cattle. Basically, the legal rules which relate to profits *a prendre* are identical to those which relate to easements (see above), with two major differences. First, in relation to a profit *a prendre* there is no requirement of dominant land. Thus, whilst the benefit of a profit *a prendre* may be attached to dominant land, this is not required. Secondly, the only form of implied grant or reservation of a profit *a prendre* is the implied grant or reservation of an intended profit *a prendre* (equivalent to the acquisition of an intended easement).

## Licences

Basically, a licence is a permission to use land. It confers no legal estate or interest in land but merely prevents the licensee (the person to whom permission is given) from being a trespasser. An example is provided by permission to cross land belonging to another which either is not intended to create an easement or which cannot create an easement (for example, because there is no dominant land).

Since a licence is not an interest in land, it may be revoked by the licensor (the person who gave permission). What is more, it will not bind subsequent owners of the licensor's land. In contrast, the grant of a legal interest (such as a legal easement of way) confers a legal right to use land belonging to the grantor. Being a legal interest in land, it cannot be revoked by the grantor and will bind subsequent owners of the servient land. Even an equitable easement, which is an equitable interest in land, will bind all subsequent owners of the servient land if it has been appropriately registered.

A *contractual licence* is a licence in the form of a contract. A mere permission which is not contractual is called a *bare licence*. The courts may be prepared to enforce a contractual licence against the licensor, preventing him or her from revoking it until the contract comes to an end. They will not normally be prepared to enforce a contractual licence against a subsequent owner of the licensor's land. However, in exceptional circumstances, such as where the subsequent owner has behaved unfairly in relation to the licensee, they may be prepared to accept that a contractual licence has given rise to an equitable interest in land.

Sometimes a licence may be coupled to the grant of a legal interest in land, such as a licence to cross land in order to reach a stream and exercise a right to fish (a profit *a prendre*). Where this is the case, as long as the interest (the profit *a prendre*) continues to bind owners of the servient land, so will the licence.

## Restrictive covenants

A restrictive covenant is an undertaking which a landowner (the covenantor) makes, usually by deed, not to use land in certain ways. For example, X might covenant with Y (the covenantee) only to use his house as a single private dwelling. Thus, X would be contractually bound not to use his house for business purposes and not to divide it into flats. However, as a consequence of the principle of privity of contract (see page 328), if the covenantor X sells his land to Z, Y cannot enforce it against the new owner Z.

In certain circumstances, however, a restrictive covenant may give rise to an equitable interest in land, making its burden pass to a purchaser of the covenantor's land, and thus enforceable against such a purchaser. This can only happen if a number of requirements are satisfied.

First, the covenant must be negative in nature. This means that it must not require the covenantor to spend money. Thus, a covenant only to use premises as a single private dwelling would be negative.

Secondly, as in the case of easements (see above), the covenant must have been made for the benefit of the covenantee's land and not merely for the covenantee's personal benefit. (Certain bodies, such as the National Trust, are not required to own benefited land in order to enforce covenants against purchasers of the covenantor's land.) Thus, the land must receive a benefit from the covenant. This could be an increase in or preservation of its value in consequence of the restriction. Or it might be that the covenant prevents undesirable development or activities on the covenantor's neighbouring land.

Thirdly, the covenantor and covenantee must have intended that the burden of the covenant would pass to subsequent owners of the covenantor's land. The effect of s.79 of the Law of Property Act 1925 is that the parties are deemed to intend the burden of the covenant to pass in this way unless the agreement expressly indicates otherwise.

Finally, to be enforceable against all purchasers and subsequent owners of the covenantor's land, it is necessary for the covenantor to register the covenant as a land charge (unregistered land) or have it entered on the register as an interest (registered land).

Sometimes, when it becomes necessary to enforce a covenant, both the covenantor and the covenantee will have sold their land. Where this is the case, as well as establishing that the covenant has given rise to an equitable interest in land which makes its burden enforceable against purchasers of the covenantor's land, it will also be necessary to satisfy equitable requirements relating to the passing of the benefit of the covenant to the purchasers of the covenantee's land. The wording of the covenant may itself 'annex' the benefit of the covenant to the covenantee's land (express annexation), by indicating that the covenantor and covenantee intended the benefit of the covenant to pass to subsequent owners of the covenantee's land. In the absence of express wording to this effect, s.78 of the Law of Property Act 1925 appears to produce the same result provided that it is clear which land was intended to receive the benefit of the covenant. Alternatively, if the covenant was not annexed to the covenantee's land, its benefit may be transferred each time the covenantee's land is sold by the owner of the land to the purchaser (express assignment).

In relation to the enforcement of covenants, special rules apply to 'building schemes'. Basically, a building scheme exists when a developer of land intends to sell the land in plots, each plot being subject to the same covenants for the benefit of all the plots and all purchasers of plots appreciating that this is so. Where a building scheme exists, the benefit of the covenants automatically passes to purchasers of plots. Further, the burden of the covenants will pass to purchasers of plots even though the covenants were not made for the benefit of land possessed by the covenantee (the developer, who presumably will not retain any land when the final plot is sold).

In some cases, when it becomes necessary to enforce a covenant, the original covenantee may have sold the benefited land, but the covenantor has retained the burdened land. In this case, in order to enforce the covenant, it is only necessary to show that the benefit of the covenant has passed to the purchaser of the covenantee's land. This may be achieved in either of two ways. First, by showing that the benefit of the covenant has passed in equity, by annexation or assignment or under a building scheme. Secondly, by showing that the benefit of the covenant has passed at common law. In order to do this, the person who has purchased from the covenantee must satisfy a number of common law requirements.

A covenant may come to an end in a number of ways. First, an application may be made to the Lands Tribunal, who may discharge or modify restrictive covenants. They have power to do so if the covenants have become obsolete, or if they prevent reasonable use of the burdened land and are either against the public interest or provide no practical benefit of substantial value to anyone. The Lands Tribunal can also intervene if the owner of the benefited land has expressly or impliedly agreed that the covenant should be modified or discharged, or in circumstances where discharge or modification of the covenant would not injure the owner of the benefited land.

Secondly, the owners of the land benefited and burdened by a covenant may agree to terminate it (discharge it), usually by deed.

Thirdly, if the two pieces of land become owned by the same person, the covenant will come to an end (extinguish) unless the covenant exists under a building scheme.

## Mortgages of land

The enduring quality of land makes it a valuable asset and a superior form of security when needing to raise money. The owner of the land who borrows against it will naturally want to continue to use and enjoy it, but the person lending the money must be given certain rights over the land in the event of the borrower defaulting on payment. These requirements are provided for in the rules governing legal mortgages.

In registered land a mortgage is created by a charge by deed, expressed to be by way of legal mortgage.

The rules ensure that a mortgagee has the necessary interest in the land for it to constitute effective security for the loan. They are reinforced by the rights granted to a lender if the borrower defaults. In extreme cases, a mortgagee may need to sell the property in order to realize this security. To enable this to be done, he or she may rely on:

1   *The right to take possession* – naturally, if a mortgagee has to sell, obtaining vacant possession of the property is a crucial first stage. He or she is effectively entitled to possession by virtue of the interest in the land granted by the mortgage itself, but can only exercise this right through the court, i.e. a mortgagee must apply to the court for an order for possession.

2    *The right to foreclose the mortgage* – again this right is exercisable through the court. If an order is granted, its effect is somewhat drastic. It removes the right of a mortgagor to redeem a mortgage and vests the legal interest in the land in the mortgagee. In other words, the mortgagor loses all rights to the land. This is so extreme a result that it is unlikely that a court would make such an order. A mortgagee would normally be satisfied to exercise his or her right to sell.

3    *The right to sell* – the duty of a mortgagee when selling is to obtain the best possible price. The mortgagee then pays him- or herself what is due and any surplus is used to discharge subsequent mortgages. If there is any money left, it is paid over to the mortgagor.

4    *The right to appoint a receiver* – a receiver may be appointed to manage or sell the mortgaged property, paying off the loan and paying any surplus profits over to the mortgagor. This is common in relation to mortgages of commercial property.

Naturally a purchaser of land must be careful to ensure that what he or she is buying is not the subject of an unredeemed mortgage. In registered land it will be apparent on the register if there is a mortgage which needs to be redeemed. However, in unregistered land a legal mortgage, protected by depositing the title deeds to the land with the mortgagee, is not a registrable interest under the land charges system. However, the purchaser always needs to see the title deeds, and if they are not available, this will obviously suggest that they are being held by someone as security to protect a loan.

# Adverse possession

Basically, if a squatter (or a succession of squatters) occupies land for 12 years or more, the owner of the land loses his or her estate in the land and cannot bring an action to recover the land (Limitation Act 1980). Thus, the squatter may become the owner of the land. However, where land is let, whilst the squatter may acquire the tenant's lease, time does not begin to run against the landlord (i.e. the 12-year period does not commence in relation to the landlord's estate in the land) until the lease comes to an end.

A landowner can only lose an estate in land in this way if the squatter's possession of the land is adverse, in the sense that it is inconsistent with the landowner's enjoyment of the property. This will be so if the squatter exercises a sufficient degree of physical control over the land. What is sufficient may vary with the individual facts of each case. For example, fencing a piece of wasteland and cultivating it as a garden, or renovating a derelict building and occupying it as a house, might well be sufficient to amount to adverse possession.

# 13
# Landlord and tenant

The relationship of landlord and tenant arises from a lease, which is a grant of exclusive possession of land for a fixed period. Exclusive possession is the most important feature of a lease, as 'The tenant armed with exclusive possession can keep out strangers and keep out the landlord' (*Street v. Mountford*). Leases can be legal estates under s.1 of the Law of Property Act 1925, in which they are described as 'terms of years absolute'.

There are two types of lease: fixed-term leases and periodic tenancies. A fixed-term lease, such as a seven-year lease or a 99-year lease, expires when the term comes to an end (i.e. after respectively seven or 99 years). A periodic tenancy, such as a weekly, monthly or yearly tenancy, may be terminated at the end of a period. If not terminated, the tenancy continues for another period. Thus, a weekly tenancy might end after one week or might continue for several years.

Normally, a legal lease must be created by deed (s.52 LPA 1925), but if made in writing, the lease may take effect as an equitable interest. However, a legal lease for three years or less can be made in writing, and a deed is not required, provided that the lease commences immediately and the rent is the best that can reasonably be obtained without taking a premium (s.54).

Once a lease expires, and is not renewed, the leaseholder (sometimes called the lessee or, more commonly, the tenant) is left with nothing. This may raise the question why anyone would want to take a lease rather than buy a freehold estate. Commercial and personal considerations will inevitably play a part in such a decision. A businessperson may need offices or premises in a specific area where the only property available is leasehold. The need for the property may be very short term, or a person may lack the funds or the means of raising the funds to buy a freehold property.

From the point of view of the freeholder, there may be significant advantages in granting leases rather than selling land outright. Take the example of a person developing land as a shopping arcade. If he sells all the units within the arcade as freeholds, it will be quite difficult to impose positive obligations on freeholders so that the properties can be maintained in a uniform way and common areas kept in order. If he grants leases, he can impose conditions more easily, and maintain a greater measure of control.

Galbraith's Building and Land Management Law for Students. DOI: 10.1016/B978-0-08-096692-2.00013-0

# Leases and licences

As is seen below on a number of Acts of Parliament (such as the Rent Act 1977, the Housing Act 1988, and the Landlord and Tenant Act 1954, Part II) give security of tenure to tenants. That is, they protect tenants by imposing restrictions on the ability of landlords to recover possession of property which they have let. For present purposes, the significant feature of these provisions is that they apply to leases but not to licences.

A licence is a permission to do something in relation to land which would otherwise be a trespass. A licence does not create any interest in the land itself. So, for example, if you visit the cinema, you have a contractual licence to be on the premises. A contractor taking over a site is usually there by virtue of a licence, and not because he has any interest in the land itself (for a further discussion on licences).

A landlord may wish to retain greater flexibility in the management of his property by creating licences rather than leases, and thus avoiding the application of legislation which might give his tenants security of tenure. Traditionally, an agreement for occupancy of premises which was called a licence and/or which purported not to give the occupiers exclusive possession of the premises would be classified by the courts as a licence. For example, in *Somma v. Hazelhurst*, a single room was let to a man and woman who were living together. It was let under two separate agreements which gave the landlady the right to share the room with them and to nominate replacement occupiers to share with them. The Court of Appeal held that the agreement could not be a lease, as it did not give the occupiers exclusive possession. Rather, it was a licence, to which the protection given to tenants by the Rent Act 1977 did not apply.

In *Street v. Mountford*, however, the House of Lords held that the decision in *Somma v. Hazelhurst* was incorrect. *Street v. Mountford* concerned an agreement, described as a licence, under which two rooms of a house were let, the occupier agreeing that the agreement did not give her a tenancy protected by the Rent Act 1977. The House of Lords held that the agreement gave the occupier exclusive possession of the property and, consequently, it was a lease to which the Rent Act 1977 applied.

In relation to *Somma v. Hazelhurst*, their Lordships held that the true effect of the agreements in that case was to grant exclusive possession of the premises to the occupiers. The fact that two separate agreements were given rather than one and that the landlady reserved the right to share the room or nominate replacement occupiers was a mere sham, an attempt to disguise the true nature of the agreement. Thus, in reality, the agreement was a lease, not a licence, the two agreements being read together as one.

Basically, the effect of the decision of the House of Lords in *Street v. Mountford* is as follows. If the courts believe that the true effect of an agreement is to give the occupier exclusive possession of premises then, in the absence of special circumstances, the agreement will be treated as a lease. This will be so even though the agreement is described as a licence and even though the agreement purports not to give the occupier exclusive possession of the premises.

The House of Lords in *Street v. Mountford* accepted that there may be special circumstances in which an agreement which gives or appears to give exclusive possession is only a licence. Some examples which occur commonly include:

- Lodgers and hotel guests normally only have a licence. They do not have exclusive possession of the premises as they receive services which require licensors or their agents to have unrestricted access to the room.
- An employee who is required to occupy premises as a necessary part of the proper performance of contractual duties only has a licence, even though he or she has exclusive possession of the premises.
- An agreement made between relatives or friends or with charitable intent which gives exclusive possession may only amount to a licence if creation of a lease was not intended.

A significant number of cases involving the distinction between a lease and a licence have come before the courts in recent years. Each tends to turn on its own individual set of facts, and may have limited value as a precedent. In every case, the court is seeking to determine the intention of the parties, particularly whether the tenant truly has exclusive possession, however the landlord may have tried to disguise it.

## Assignment and subletting

Under a lease, both the landlord and the tenant possess interests in land. The tenant possesses a lease. The landlord possesses a reversion, the right to resume occupation of the premises when the tenancy ends.

A landlord may own the freehold of the premises let or may himself be a tenant. For example, L, who owns the freehold of a property, may grant a 99-year lease to T, who grants a seven-year lease to S. L is the head landlord, T is L's tenant and S's landlord, and S is T's tenant and L's subtenant.

A landlord who owns the freehold of the let premises may sell his interest. The purchaser becomes the owner of the reversion and thus becomes the tenant's new landlord. A tenant (including a landlord who does not own the freehold) may sell his lease. This is technically called an assignment. The purchaser becomes the new tenant (and also becomes a landlord if the vendor had a subtenant).

It may be, however, that the lease prohibits assignment by the tenant, either absolutely or without the landlord's consent. Where the lease prohibits assignment without the landlord's consent, s.19 of the Landlord and Tenant Act 1927 provides that consent must not be unreasonably refused. Further, undue delay in giving consent or notifying the tenant of a refusal of consent may render the landlord liable in damages to the tenant (Landlord and Tenant Act 1988). If the tenant assigns without consent, he has a defence to an action brought against him by the landlord, if the landlord's refusal was unreasonable.

In relation to tenancies granted on or after 1 January 1996, other than residential tenancies, the lease may specify circumstances in which the immediate landlord may refuse to consent to an assignment of the lease and may also specify conditions to which the giving of consent will be subject (Landlord and Tenant (Covenants) Act 1995). Refusing consent in the specified circumstances or giving consent subject to such conditions will not be unreasonable.

As an alternative to assigning the lease, the tenant may choose to sublet and receive rent from his own tenant. The lease may prohibit subletting, either absolutely or without the landlord's consent. Again, where subletting is prohibited without the landlord's consent, consent must not be unreasonably refused (s.19 Landlord and Tenant Act 1927). Likewise, the landlord may be liable in damages to the tenant where there is undue delay in giving consent, or notice of refusal.

# Enforcing covenants

The rules which govern the enforcement of covenants differ depending upon whether the lease was granted before 1 January 1996 or on or after that date.

## *Leases granted before 1 January 1996*

The original parties to a lease can enforce its covenants against each other, because the lease is a contract, and both landlord and tenant are privy to the contract. Where either or both the landlord and tenant are not the original parties to the lease, because the original landlord sold his reversion and/or the original tenant assigned his lease, there is no privity of contract between them. There is, however, privity of estate, due to the existence between them of the relationship of landlord and tenant.

Where there is privity of estate but no privity of contract between a landlord and tenant, only covenants which 'touch and concern the land' are enforceable between the parties. A covenant does not touch and concern the land if it benefits only the parties to the lease and not the land itself. Examples of covenants which touch and concern the land are covenants to repair, covenants to pay rent and covenants for quiet enjoyment. Examples of covenants which do not touch and concern the land are options to purchase and obligations to repair other premises.

Regardless of whether a covenant touches and concerns the land, the original tenant may remain liable for its breach under the contract. Thus, when a tenant assigns, it may be advisable for him or her to require on the part of the assignee a covenant indemnifying the assignor in respect of any breaches of covenant by the assignee. Such a covenant is implied by s.77 of the Law of Property Act 1925 into assignments of leases.

Where a tenant sublets, there is neither privity of contract nor privity of estate between the head landlord and the subtenant. Thus, covenants of the head lease are only enforceable between the head landlord and the subtenant if the requirements for enforcement of restrictive covenants (see page 321) are satisfied. In order to enforce a restrictive covenant against a subtenant, the landlord is not required to possess neighbouring land which is benefited by the covenant, however, as his reversion counts as land which can receive the benefit of a restrictive covenant.

Regardless of whether a covenant of the head lease is enforceable between head landlord and subtenant, the original tenant may remain liable under the contract for its breach. Thus, when a tenant sublets it may be advisable to require the subtenant to covenant to comply with the covenants of the head lease and to indemnify the tenant in respect of any breaches of those covenants.

## Leases granted on or after 1 January 1996

The Landlord and Tenant (Covenants) Act 1995 has modified the rules of law regulating the enforcement of covenants when the landlord sells the reversion or the tenant assigns the lease. The position concerning the enforcement of covenants between head landlord and subtenant essentially remains unchanged.

The major effects of the 1995 Act are that the enforcement of covenants is no longer dependent upon the existence of privity of contract or privity of estate or upon whether a covenant touches and concerns the land. Essentially, once the tenant assigns the lease, the landlord can no longer enforce the tenant covenants against the tenant and the former tenant can no longer enforce the landlord covenants against the landlord. The landlord can, however, enforce the tenant covenants against the new tenant and the new tenant can enforce them against the landlord. Moreover, if the former tenant required the consent of the landlord to assign the lease, he may have entered into an 'authorized guarantee agreement', requiring him to guarantee that the new tenant will perform some or all of the tenant covenants.

Where the landlord sells the reversion, the new landlord can enforce the tenant covenants against the tenant and the tenant can enforce the landlord covenants against the new landlord. Moreover, the former landlord can still enforce the tenant covenants against the tenant and the tenant can still enforce the landlord covenants against the former landlord. The former landlord may, however, notify the tenant that he wishes to be relieved from the landlord covenants. If the tenant does not object then the former landlord is relieved from the landlord covenants and can no longer enforce the tenant covenants. If the tenant objects, by serving a counter-notice, the county court must decide whether it is reasonable for the former landlord to be released.

# Covenants in leases

Covenants in a lease lay down the obligations of the landlord and tenant. There are two types of covenants: express covenants and implied covenants. Express covenants are expressly agreed upon by the parties to the lease. Implied covenants do not result from express agreement by the parties to the lease but are implied into it by common law or statute, unless the parties agree otherwise. However, certain covenants implied by statute into a lease cannot be excluded by agreement of the parties.

## Implied covenants (other than repairing covenants)

### Tenant's implied covenants

Other than repairing covenants (which are examined below), two major examples of tenant's implied covenants (i.e. implied covenants imposing an obligation upon the tenant) are the implied covenant to pay rent and the implied covenant to pay taxes relating to the property. In relation to rent, in the absence of an express covenant or contrary agreement, a covenant may be implied if the wording of the lease implies that the payment of rent is intended.

In relation to taxes, in the absence of an express covenant or contrary agreement, the tenant is under an implied obligation to pay rates and taxes relating to the property for which the landlord is not personally liable.

### Landlord's implied covenants

Other than repairing covenants, two major examples of landlord's implied covenants are the implied covenant not to derogate from his grant and the implied covenant for quiet enjoyment. In relation to non-derogation, in the absence of an express covenant or contrary agreement, the landlord is under an implied obligation not to do things which make the leased premises less suitable for the purpose for which they were let, assuming that the landlord knows the purpose, e.g. blocking the flow of air to timber-drying sheds.

Quiet enjoyment, despite its name, is not really about noise or lack of it. It covers any actions of the landlord which would stop the tenant from being able to use the premises. In the absence of an express covenant or contrary agreement, the landlord is under an implied obligation not to interfere with the tenant's quiet enjoyment of the premises (for example, by unlawful entry or eviction). With regard to both of these implied covenants, the landlord may also be liable in respect of acts done in breach of the covenants by persons claiming under him, such as tenants of other neighbouring property which he owns.

## Express covenants (other than repairing covenants)

### Tenant's express covenants

Examples of express covenants which may be made by a tenant are a covenant to insure, a covenant to pay taxes relating to the property, a covenant only to use the leased premises for certain purposes, a covenant not to assign or sublet, and a covenant to pay rent. In relation to insurance, a tenant may agree to insure the premises, the extent and nature of the policy being determined by the covenant.

In relation to taxes, an express covenant may impose upon the tenant an obligation more or less extensive than, or identical to, that which would otherwise be implied.

In relation to use of the premises, the tenant may agree to restrictions on his or her use of the premises. For example, he or she may agree only to use the premises as a single private dwelling, which would prevent both use for business purposes and subletting as flats. The significance of covenants not to assign or sublet is considered on page 327.

In relation to rent, the tenant will normally expressly covenant to pay a specified amount of rent on a specified day. Rent is normally payable in arrears unless the parties have agreed to payment in advance. A tenant may be entitled to deduct sums from the rent before paying it to the landlord. The lease may specifically permit or prevent the making of such deductions. Where the lease is silent, the basic principle is that the landlord is entitled to the entire rent. The lease may contain a term which suspends the tenant's obligation to pay rent in circumstances such as damage or destruction of the premises by fire.

The lease may contain a rent review clause, which permits regular rent reviews throughout the course of the tenancy, say once every three years. The lease will normally contain a formula to be used in determining the revised rent, perhaps an open market rent to be fixed by agreement or arbitration or a form of index linking, possibly to the retail prices index.

### Landlord's express covenants

Examples of express covenants which may be made by a landlord are a covenant to insure, a covenant to pay taxes relating to the property and a covenant for quiet enjoyment. In relation to insurance, if the landlord covenants to take out an insurance policy, it may be that the tenant agrees to pay the insurance premiums.

In relation to such taxes, the landlord may covenant to undertake liability which would otherwise fall within the tenant's implied obligations.

In relation to quiet enjoyment, the requirements of an express covenant may be wider in some respects but more limited in others than those of an implied covenant. For example, an express covenant for quiet enjoyment continues to bind the party who made it throughout the duration of the lease, whereas an implied covenant for quiet enjoyment only binds a landlord whilst he remains landlord.

### Provisos and options

A lease may contain a number of provisos and options. Examples of options are options to purchase and options to renew. An option to purchase may entitle the tenant to purchase the landlord's reversion, which may be the freehold or may be a lease depending upon whether the landlord owns the freehold or is himself a tenant. The tenant may be entitled to purchase at a fixed price or, alternatively, the lease may lay down a mechanism and formula to be applied in determining the purchase price.

An option to renew may entitle the tenant to renew the lease by serving notice on the landlord on a specified day. The tenant will be entitled to require a new lease for a term specified by the option. The lease may specify a fixed rent or, alternatively, may lay down a mechanism and formula to be applied in determining the new rent. Options should be registered in order to bind purchasers of the landlord's estate.

Examples of provisos are provisos for forfeiture and break clauses. A landlord cannot forfeit a lease unless the lease contains a proviso which gives the landlord a right of re-entry on breach by the tenant of conditions or covenants of the lease.

A break clause is a proviso which empowers one or both parties to a lease to terminate the lease at certain specified dates by serving notice on the other.

# The obligation to repair

When negotiations for a lease are taking place, it is important to establish the responsibilities with regard to repair of the property. The parties may expressly agree on responsibility for repairs, such agreement forming the 'express repairing covenants' of the lease. Where no such express agreement has been reached, the parties may find that there are repairing obligations implied by common law or by statute – 'implied repairing covenants'.

## Express repairing covenants

A lease may expressly impose repairing obligations upon landlord or tenant in the form of express repairing covenants. For example, a lease might require the tenant:

(a)  to put premises in repair at or within a reasonable time of the commencement of the tenancy; or

(b)  to leave premises in repair (i.e. to hand them over in repair) at the end of the tenancy; or

(c)  to keep premises in repair throughout the tenancy (i.e. to put the premises in repair at the commencement of the tenancy, keep the premises in repair throughout the tenancy and hand them over in repair at its end).

Where the landlord covenants to repair, he is not obliged to repair until he has notice of the disrepair, i.e. from the tenant or from his own inspection of the premises.

Once repairs are undertaken, the standard required is 'such repair as, having regard to the age, character and locality of the premises, would make them reasonably fit for the occupation of a reasonably minded tenant of the class who would be likely to take them'. This was established by Lord Esher in *Proudfoot v. Hart*. The criteria are applied by looking at the age, locality, etc. as they would have been at the start of the tenancy. This can be important where, for example, the locality becomes less desirable during the course of a long lease. The standard of repair required is not then reduced because the locality would no longer attract the same sort of tenant.

A covenant to repair requires painting and decorating in order to prevent the property falling into disrepair. A lease may also contain an express covenant to paint, which may specify the materials to be used and the frequency of the work.

Express repairing covenants may specifically exclude liability to repair in respect of disrepair resulting from 'fair wear and tear', whether caused by normal and reasonable use by the tenant for proper purposes or by the normal action of time and weather. The tenant must, however, undertake such repairs as are necessary to prevent the premises falling into further disrepair.

A covenant to repair does not normally impose an obligation to make improvements, or to renew the premises in their entirety. However, such a covenant will normally impose an obligation to rebuild in the case of destruction of premises by fire, unless the lease expressly excludes such liability.

## Remedies for breach of repairing covenants

1   *Damages*. Where a party to a lease fails to perform his repairing obligations he may be liable in damages.

Where the landlord is in breach of his repairing obligations, the tenant cannot recover damages in respect of the period prior to the date when the landlord was notified of the disrepair. If the landlord fails to repair within a reasonable time of receiving such notice, the tenant may sue for damages or carry out the repairs and recover their cost from the landlord.

Where the tenant is in breach of a covenant to put, keep or leave the premises in repair, s.18(1) of the Landlord and Tenant Act 1927 provides that the damages cannot exceed the amount by which the value of the landlord's reversion was diminished in consequence of the breach. The section also provides that, where the tenant is obliged to leave the premises in repair, no damages are recoverable if, at or shortly after the termination of the tenancy, the premises are to be demolished or structural alterations which would render the repairs valueless are to be made.

In the case of a lease for a fixed term of or exceeding seven years, of which at least three years remain unexpired, s.1 of the Leasehold Property (Repairs) Act 1938 provides that, before bringing an action for damages or forfeiting the lease in respect of the breach by the tenant of a covenant to keep or leave the premises in repair, the landlord must first serve notice on the tenant under s.146 of the Law of Property Act 1925. This notice must, amongst other matters, inform the tenant of his or her right to serve a counter-notice within 28 days of the service of the s.146 notice. If the tenant serves a counter-notice, the landlord cannot sue for damages or forfeit the lease without the leave of the court. The court may give leave if:

(a)    the value of the landlord's reversion has been substantially diminished by the breach or will be so diminished unless the breach is immediately remedied; or

(b)    the breach must be immediately remedied in order to give effect to the provisions of a statute, a bye-law or of delegated legislation; or

(c)    in those cases where the tenant is not in occupation of the whole of the premises, the immediate remedying of the breach is required in the interests of the occupier; or

(d)    the expense required in order to remedy the breach is small compared to the much greater expense, which would probably be required if the work required was postponed; or

(e)    there are special circumstances which make the giving of leave just and equitable.

Section 1 of the 1938 Act does not apply in respect of the breach of a covenant to put premises in repair upon or within a reasonable time of taking possession (s.3). Further, s.1 does not apply to breach of a covenant to carry out internal decorative repairs, though s.147 of the Law of Property Act 1925 empowers the court to grant the tenant relief from such repairs.

2    *Forfeiture.* A landlord can forfeit a lease if the tenant is in breach of the repairing obligations and the lease contains a proviso which gives the landlord a right of re-entry on breach by the tenant of those obligations (see page 338).

3    *Landlord's right to repair at the tenant's expense.* Where the lease empowers him to do so, the landlord may enter the premises, carry out repairs and recover their cost from the tenant.

4    *Specific performance.* A tenant may obtain an order of specific performance against the landlord compelling the landlord to carry out his repairing obligations. It appears that a landlord would only be able to obtain such an order against his tenant in very exceptional circumstances.

5    *Tenant's set-off of repair costs against rent.* If the landlord fails to repair within a reasonable time of receiving notice of disrepair and the tenant carries out the necessary repairs, the tenant may be entitled to deduct their cost from rent payments.

6    *Receiver.* Upon application by the tenant, the court may appoint a receiver to manage the property if it is just and convenient to do so. This

may be the case where the landlord refuses to perform his repairing obligations. (The general power of the High Court to appoint a receiver is contained in s.37 of the Supreme Court Act 1981. A specific power relating to buildings containing two or more residential flats is given to the county court by Part II of the Landlord and Tenant Act 1987.)

## Implied repairing covenants

Where a lease fails to lay down express repairing obligations, repairing obligations may be implied at common law. Further, repairing obligations may be imposed by statute, regardless of the terms of a lease.

### Relating to the landlord

1   *Fitness for habitation of houses* (s.8 Landlord and Tenant Act 1985). Where s.8 applies, it implies, regardless of any agreement or term of the lease to the contrary, a condition that the house is fit for human habitation at the commencement of the tenancy and an undertaking that the landlord will keep the house so fit throughout the tenancy.

Section 9 provides that a house is only unfit for human habitation for the purposes of s.8 if it is not reasonably suitable for occupation due to defects in its condition with regard to one or more of the following circumstances: state of repair; stability; freedom from damp; internal arrangement; natural lighting; ventilation; water supply; drainage and sanitary conveniences; facilities for preparation and cooking of food and for disposal of waste water.

It appears that s.8 only requires the landlord to repair where it is possible to make the premises fit for habitation at reasonable expense. The section entitles the landlord to enter the premises at reasonable times of the day to view their state and condition on giving 24 hours' notice to the tenant or occupier. The landlord's liability under s.8 does not arise until he has received notice of the want of repair. Breach by a landlord of the requirements of s.8 may entitle the tenant to vacate the premises without paying rent and to recover damages.

Section 8 applies to lettings of houses for human habitation provided that the rent does not exceed £80 per year if the house is located in London or £52 per year if the house is located elsewhere. Whilst these financial limits remain so low, the section remains of extremely limited application.

Section 8 does not apply to leases for a term of three years or more which require the tenant to put the premises into a condition reasonably fit for human habitation provided that neither party has the option to determine the lease within three years.

2   *Fitness for habitation of furnished houses*. Subject to express provision to the contrary in the lease, there is an implied condition at common law that a furnished house will be fit for habitation when the tenancy

begins. If this is not the case, for example because of defective drains, the tenant can either repudiate the lease or sue for damages. The landlord is not required to keep the premises fit for habitation throughout the tenancy as long as they are fit for habitation at its start.

3   *Repair of the structure of dwelling-houses* (s.11 Landlord and Tenant Act 1985). Where s.11 applies to a lease of a dwelling-house, its effect is to imply into the lease covenants on the part of the landlord:

(a)   to keep in repair the structure and exterior (including drains, gutters and external pipes);

(b)   to keep in repair and proper working order the installations for supply of water, gas and electricity and for sanitation (including basins, sinks, baths and sanitary conveniences, but not other fixtures, fittings and appliances for making use of the supply of water, gas or electricity); and

(c)   to keep in repair and proper working order the installations for space heating and heating water.

The covenants implied by s.11 do not require the tenant:

(a)   to do repairs for which the tenant is liable either under duty to use the premises in a tenant-like manner (see page 332) or under an express covenant to the same effect; or

(b)   to rebuild or reinstate the premises in the case of destruction or damage by fire, or by tempest, flood or other inevitable accident; or

(c)   to repair tenant's fixtures.

The standard of repair required by the covenant implied by s.11 is to be assessed having regard to the age, character and prospective life of the dwelling-house and the locality in which it is situated.

Section 11 entitles the landlord to enter the premises at reasonable times of the day to view their state and condition on giving 24 hours' written notice to the occupier. The landlord is not liable under s.11 until he has notice of the disrepair and is not in breach of covenant unless he fails to carry out the required repairs within a reasonable time of receiving such notice.

As a general rule, s.11 applies to leases of dwelling-houses granted on or after 24 October 1961 for a term of less than seven years.

Section 14 provides that s.11 does not apply to:

(a)   leases granted to existing tenants, or to former tenants remaining in possession, where s.11 did not apply to the previous lease;

(b)   tenancies of agricultural holdings;

(c)   leases granted on or after 3 October 1980 to various specified bodies, such as local authorities;

(d)   leases granted to a government department or to the Crown (unless the lease is managed by the Crown Estate Commissioners).

The landlord is not merely obliged to carry out repairs on structure, installations, etc. which form part of the property let to the tenant. Rather, he is required to repair the structure of all parts of the building in which he has an estate or interest. Equally, he is required to maintain installations which serve the dwelling-house, provided that they form part of the building in which he has an estate or interest or are owned by him or under his control. For example, where a tenant's flat is heated by radiators, but the boiler is in the basement, which does not form part of the property let to the tenant, the landlord is obliged to maintain both the radiators and the boiler.

Section 12 provides that an agreement or term of a lease is void to the extent to which its effect is to modify or exclude the provisions of s.11 unless it was authorized by the county court. The county court may authorize such an agreement or term with the consent of the parties if it is reasonable to do so. Unless such an agreement or term is authorized, any covenant made by the tenant to carry out repairs is of no effect to the extent that it requires the tenant to carry out repairs which the landlord is obliged to carry out under s.11.

4    *Repair and maintenance of common parts of blocks of flats.* Subject to express provision to the contrary in the lease, there is an implied condition at common law that landlords will take reasonable steps to repair and maintain common parts of the premises which remain in their control, such as lifts and stairs. This implied condition basically appears only to apply in relation to residential blocks of high-rise flats, though it may apply in other special situations in which maintenance of the common parts by the landlord is essential in order to make the tenancy agreement effective.

## Relating to the tenant

1    *Obligation not to commit waste.* The law of tort imposes upon tenants an obligation not to commit waste.

Voluntary waste comprises damage to or destruction of premises resulting from deliberate or negligent acts (such as demolition or the making of alterations), but does not encompass damage or destruction which results from their reasonable use.

Ameliorating waste comprises acts which improve the value of land or premises (such as constructing an extension of existing premises). The courts will not normally be prepared to award any remedy in respect of ameliorating waste.

Permissive waste comprises damage to or destruction of premises resulting from omission to undertake necessary repairs (such as failure to replace slates resulting in the loss of a roof). Periodic tenants (except, perhaps, yearly tenants) are not liable for permissive waste.

Where a tenant commits waste, the landlord may bring an action for damages and (except in the case of permissive waste) may claim an injunction to prevent continuing or repeated acts of waste.

2    *The tenant's duty to use the premises in a tenant-like manner.* Where the lease does not expressly impose repairing obligations, it is implied that the tenant will use the premises in a tenant-like manner.

A fixed-term tenant, being liable for permissive waste, should hand the premises over in repair at the end of the tenancy, though the tenant is not required to carry out repairs in respect of 'fair wear and tear'. A yearly tenant should carry out minor repairs necessary to keep the premises wind and water tight, but is not required to carry out repairs in respect of 'fair wear and tear'. A periodic tenant for a period of less than a year should perform small tasks which any reasonable tenant would perform (such as cleaning the windows).

Whatever the length of the tenancy, the tenant should carry out repairs required in consequence of voluntary waste.

# Security of tenure: termination of a tenancy at common law

At common law, a tenancy may terminate in a variety of ways. Until it terminates, both parties are bound by its terms.

The purpose of the present section is to examine briefly these common law methods of termination. It should be noted at this point, however, that the termination of business tenancies is regulated by Part II of the Landlord and Tenant Act 1954. Similarly, the ability of a landlord to obtain possession of a dwelling-house let under a residential tenancy is restricted by the Rent Act 1977 and Part I of the Housing Act 1988. These statutory provisions are examined later, after a discussion of the common law principles.

## *Methods of termination at common law*
### *Expiration of a fixed term*

A fixed-term tenancy terminates when the term expires. Thus, for example, a seven-year lease expires after seven years have passed.

### *Service of notice under a break clause*

A break clause is a proviso which empowers one or both parties to a lease to terminate the lease at certain specified dates by serving notice on the other. For example, a 14-year lease might contain a proviso empowering the tenant to terminate the lease at the end of the seventh year by prior service of notice.

### *Forfeiture*

Where a tenant is in breach of a covenant of a lease, the landlord can only forfeit the lease if it contains a proviso which gives him a right of re-entry on breach of that covenant. The option of forfeiture might, typically, be available to a landlord where a tenant has been in arrears of rent for 21 days.

If a landlord is aware of a breach which entitles him to forfeit the lease, and then expressly or by implication affirms the continued existence of the lease, he waives his right to forfeit it, unless it is a continuing breach (such as a state of disrepair). Acceptance of, or demand for, rent due in respect of a period following the breach will, for example, give rise to such waiver.

Other than in the case of forfeiture for non-payment of rent, s.146 of the Law of Property Act 1925 requires the landlord, before forfeiting the lease, to serve on the tenant a notice which:

(a)  specifies the breach;
(b)  if it can be remedied, requires the tenant to remedy it; and
(c)  requires the tenant to make monetary compensation for it.

Following service of the s.146 notice, the landlord can proceed to enforce the forfeiture unless the tenant remedies the breach and provides reasonable compensation within a reasonable time. Other than where the tenant peaceably vacates the premises, the landlord will normally be required to enforce his right of re-entry by bringing proceedings in court.

Where s.146 applies to a forfeiture, it entitles the tenant to apply to the court for relief against forfeiture. Taking into account both the conduct of the parties and all other relevant circumstances, including the fact that the tenant has remedied the breach or has promised to do so, the court can grant or refuse relief as it thinks fit. Special rules apply under s.146 to subtenants, including mortgagees.

In the case of forfeiture in respect of non-payment of rent, whilst s.146 does not apply, a right of re-entry will not arise until the landlord has formally demanded the rent unless, as is normally the case, this requirement is excluded by a proviso of the lease or by operation of s.210 of the Common Law Procedure Act 1852. The tenant may apply for relief from forfeiture. The extent of the tenant's entitlement to relief, and the powers of the court in respect of an application for relief, vary depending upon whether the action is brought in the High Court or the county court. Special rules also exist in relation to forfeiture where the tenant disputes the amount of a service charge (Housing Act 1996).

## Merger

Merger occurs when a tenant acquires a landlord's reversion and intends the two estates to merge into one, terminating the tenancy. Such intention will normally be expressed in the assignment of the landlord's reversion to the tenant.

## Surrender

Surrender may occur in either of two situations:

1   *Surrender by express agreement of the parties.* Express surrender occurs when landlord and tenant agree to end a lease before it is due to expire. Express surrender will normally be by deed, though an agreement in writing is sufficient in the case of terms of three years or less.

2   *Surrender by operation of law.* Surrender by operation of law occurs when both parties to a lease act in such a way that the surrender of the lease before it is due to expire is implied. This is the case where the conduct of the parties is such as permits of no explanation other than that the lease has been surrendered.

   For example, surrender may be implied where a tenant abandons possession and his landlord re-lets the premises to a new tenant. Again, it may be implied when a landlord grants his tenant a new lease before the old one has expired. It will probably not be implied where a tenant abandons possession and the landlord leaves the premises vacant in anticipation of the tenant's return.

Surrender, whether express or by operation of law, does not invalidate a lawful sublease.

## Service of notice to quit

A periodic tenancy may be terminated by service of notice to quit. The terms of a tenancy may expressly indicate the length of notice required in order to terminate it. Where this is not the case, the basic rule is that a tenancy may be terminated by service of one period's notice and, consequently:

- to terminate a weekly tenancy, one week's notice is required;
- to terminate a monthly tenancy, one month's notice is required;
- to terminate a quarterly tenancy, one quarter's notice is required.

   In the case of a yearly tenancy, or a periodic tenancy the period of which exceeds one year, six months' notice is required.

   In general, notice must be served so as to terminate a tenancy on an anniversary of its creation. In the case of a yearly tenancy, this would be one or more years from the date of grant. In the case of a weekly or monthly tenancy, this would be, respectively, one or more weeks or months from the date of grant. A quarterly tenancy must normally be terminated on a quarter day (25 March, 24 June, 29 September or 25 December).

   Notice to quit a dwelling-house must be in writing and must be served at least four weeks before the date of termination (Protection from Eviction Act 1977, s.5 – though s.5 does not apply to certain tenancies which are excluded by s.3A).

## Frustration

The doctrine of frustration is applicable, though in limited circumstances, to leases.

## Offences related to the recovery of possession of premises

Section 6 of the Criminal Law Act 1977 provides that it is an offence, without lawful authority, to use or threaten violence for the purpose of securing entry into premises, knowing that someone opposed to the entry is present on the premises.

Section 1(2) of the Protection from Eviction Act 1977 provides that it is an offence to unlawfully deprive or attempt to deprive the residential occupier of premises (namely, the person who occupies those premises as a residence under contract, statute or rule of law) of his or her occupation of all or part of the premises in the absence of belief, with reasonable cause, that the residential occupier has ceased to reside there.

The only lawful method of obtaining possession of premises occupied by a residential tenant is by obtaining a court order, unless either the tenant peaceably vacates the premises or the tenancy is a Housing Act 1988 excluded tenancy.

Section 1(3) provides that it is an offence for any person to commit acts likely to interfere with the peace or comfort of a residential occupier of premises or members of a household or to persistently withdraw or withhold services reasonably required for the occupation of the premises as a residence, intending in either case to cause the residential occupier to give up possession of the premises or to cause him or her to refrain from exercising rights or pursuing remedies in respect of the premises.

Section 1(3)(A) provides that it is an offence for the landlord of a residential occupier or his agent to commit such acts or withdraw or withhold such services, knowing or having reasonable cause to believe that the conduct is likely to result in one of those consequences, unless the person who did the acts or withdrew the services had reasonable grounds for so doing.

A tenant who gives up possession of premises in consequence of conduct falling within s.1(2) or (3) may be entitled to recover damages from the landlord.

Section 2 provides that where a person resides in premises which are let as a dwelling, a right of re-entry or forfeiture can only be lawfully enforced by proceedings in court. Section 3 provides that where a lease of a dwelling-house comes to an end and the occupier continues to reside in it, the landlord can only lawfully regain possession by court proceedings unless the tenancy was a protected tenancy or an excluded tenancy.

# Business tenancies (Landlord and Tenant Act 1954, Part II)

Where the provisions of Part II of the Landlord and Tenant Act 1954 apply to a tenancy, the Act may entitle the tenant to a new lease or to the continuation of an existing tenancy when the contractual term expires or the landlord serves notice to quit.

The Act only applies to leases, not to licences (s.23). It should be noted, however, that the principles stated by the House of Lords in *Street v. Mountford* (see page 326) are applicable to business leases. Thus, an agreement which is described as a licence but gives exclusive possession of premises for a fixed or periodic term will amount to a lease, not a licence, unless the agreement is one of the exceptional types recognized by the House of Lords in *Street v. Mountford*.

Under s.23, the tenant must occupy at least part of the premises. This does not require the tenant to occupy the premises personally. Rather, it may be sufficient if the premises are occupied by the tenant's agents or employees. The tenant need not be in sole occupation of the premises. The tenant must, however, retain a sufficient degree of control over the premises. For example, where a business tenant sublets the premises and acts as manager, providing services to the subtenants, this may be sufficient to amount to occupation of the premises by the tenant. The premises may include land without a building, such as an undeveloped lot used as a car park.

Occupation of the premises must, at least in part, be for the purposes of a business carried on by the tenant (s.23). Business includes a trade, profession or employment. The expression thus covers a wide variety of enterprises, from shops and offices to hospitals and clubs.

The Act does not apply to a tenancy if the lease prohibits business use by the tenant unless the landlord has agreed to the business use. The Act also does not apply where premises are mainly used for residential purposes and any business use is subsidiary to this. This may be so, for example, where a residential tenant takes in a lodger.

Where the purpose of a letting is a mixed business and residential purpose, the business purpose not being subsidiary to the residential purpose, then the Act may apply to the tenancy.

## Types of tenancy to which the Act does not apply

- Agricultural holdings (s.43).
- Mining leases (s.43).
- Service tenancies (s.43) – tenancies granted to the tenant for the purpose of the tenant's employment or an office or appointment which the tenant holds.
- Residential tenancies (s.23).
- Leases of premises required for public purposes (s.57) or regional development (s.60).
- Leases of premises required for national security purposes (s.58).
- Fixed-term tenancies not exceeding six months (s.43). If the tenant has been in occupation for more than 12 months or the lease contains a term permitting extension or renewal of the tenancy beyond six months, then the Act does apply.

- Extended tenancies granted under the Leasehold Reform Act 1967 (s.16 Leasehold Reform Act 1967).
- Tenancies of premises where the parties have agreed to a new tenancy of the premises (s.28).

## Contracting out of the Act

Section 38 of the Act allows both the landlord and tenant to agree that the provisions of the Act which relate to termination of the tenancy and entitle the tenant to a new tenancy will not apply to the tenancy. The landlord must serve notice on the tenant before the tenancy is completed, informing the tenant of the rights he or she is giving up. The tenant must either make a declaration or a statutory declaration, depending on the circumstances, that he or she has received the notice and understands it.

## Protection under the Act for the business tenant

Section 24 provides that a business tenancy to which the Act applies can only be terminated by forfeiture, by surrender, by merger, by service of notice to quit by the tenant, or by satisfying special requirements laid down in the Act.

Thus, where a fixed-term tenancy expires or the landlord serves notice to quit (in relation to a periodic tenancy or under a break clause), then, unless the special requirements laid down by the Act have been satisfied, the tenancy does not terminate. Rather, the tenancy continues until terminated in one of the five ways referred to above.

A tenancy which continues in this way is described as a continuation tenancy. Basically, it amounts to a continuation of the contractual tenancy and, consequently, it is possible for both landlord and tenant to assign their interests. The terms of the continuation tenancy are the same as the terms of the contractual tenancy, with the exception of any terms permitting the landlord to terminate the tenancy or creating personal obligations which do not run with the land. Whilst the tenancy continues in this way, the tenant continues to have the right to remove tenant's fixtures and to use easements given to him or her by the lease.

Under a continuation tenancy, the landlord cannot increase the rent except by applying to the court (s.24A). He can only do this if either he has commenced the special procedure which the Act lays down for determining the tenancy, or the tenant has commenced the special procedure which the Act lays down for applying for a new tenancy. The court may be prepared to grant an interim rent, which will remain in force until the continuation tenancy ends. A tenant can also apply to the court for an interim rent.

The court is not bound to grant such a rent but if it chooses to do so the rent must be an open market rent fixed in accordance with principles laid down by the Act. The principles are basically those which the Act lays down in relation to determining the rent payable under a new tenancy (s.34). Additionally, the court is required to take into account the rent payable under the terms of the tenancy and to determine the rent upon the basis of the assumption that the tenancy is a yearly tenancy (s.24A).

If a continuation tenancy ceases to be a business tenancy, the landlord may terminate it by giving the tenant between three and six months' notice or in accordance with the terms of the tenancy (s.24).

## Termination by the tenant

The Act permits termination by surrender and merger. The Act also allows the tenant to terminate the tenancy by serving notice to quit (in relation to a periodic tenancy or under a break clause) (s.24).

In relation to the expiration of a fixed-term tenancy, the tenant must give the landlord written notice that he or she does not wish the tenancy to continue either at least three months before the date when the tenancy would expire if the Act did not apply or at least three months before any day following that date (s.27).

## Termination by the landlord

Subject to the discretion of the court to grant relief, the Act allows the landlord to forfeit the tenancy where breach of a condition or covenant gives rise to a right of re-entry (s.24). Other than this, the landlord can only terminate the tenancy if he satisfies special requirements which the Act lays down (see below). In order to rely upon expiration of a fixed term or service of notice to quit (in relation to a periodic tenancy or under a break clause), the landlord must satisfy these special requirements. If he does not do so, then a continuation tenancy will result (s.24). Similarly, where a landlord wishes to terminate a continuation tenancy, his options are either to forfeit, if appropriate, or to satisfy the special requirements laid down by the Act.

The *special requirements* which the Act lays down for termination of a tenancy by the landlord are as follows:

1    The landlord must serve written notice (a 's.25 notice') on the tenant in the form laid down by regulations made under the Act. The notice must specify the date of termination and must be served on the tenant between six and 12 months before the date which it specifies (s.25). If it has not already terminated, the contractual tenancy will terminate on this date (though the current tenancy may continue as a continuation tenancy beyond this date).

   In relation to the expiration of a fixed-term tenancy, the date of termination cannot be earlier than the date when the tenancy would have expired had the Act not applied to it.

   With regard to termination of a tenancy by notice to quit (in relation to a periodic tenancy or under a break clause) the date of termination cannot be earlier than the earliest date upon which the tenancy could have been brought to an end, had the Act not applied to it, by notice to quit served by the landlord. Provided that the s.25 notice conforms with the requirements of the tenancy agreement, the landlord need not serve a separate notice to quit on the tenant.

In relation to the termination of a continuation tenancy, the date of termination can be any date chosen by the landlord.

If the landlord does not wish the tenant to be granted a new tenancy, the notice must state this, and must also indicate the grounds upon which the landlord objects to the grant of a new tenancy. The landlord may only oppose the grant of a new tenancy on grounds laid down by s.30. If the landlord is not opposing renewal, he must set out in his s.25 notice his proposals for the new tenancy.

2  The landlord who serves the notice will be the tenant's immediate landlord if he either owns the freehold or has a tenancy which will not expire within 14 months (s.44). If this is not the case, the landlord competent to serve the notice will be the first landlord in the chain above the tenant's immediate landlord who either owns the freehold or has a tenancy which will not expire within 14 months.

3  If the tenant does not want a new tenancy, he or she need do nothing and the tenancy will terminate on the date specified by the notice.

4  If the tenant does want a new tenancy, he or she must apply to the court for a new tenancy if the parties cannot agree upon one themselves (s.29). If the tenant wishes to apply to the court he or she must do so by the date specified in the s.25 notice as the termination date, otherwise the right to apply is lost. The parties may agree to extend this deadline. The relevant court will be the county court for the district in which the property is situated.

Once the tenant applies to the court for a new tenancy, s.64 provides that the current tenancy cannot terminate within three months of the disposal of the tenant's application by the court. This is so whether or not the court grants a new tenancy. Thus, if the termination date specified by the landlord's s.25 notice falls within this three-month period, the current tenancy will not terminate until the three months have passed.

The court will order the grant of a new tenancy unless the landlord establishes one or more of the grounds upon proof of which the court may or must refuse the tenant's application (s.29). If the tenant has a change of mind he or she may apply to the court for revocation of the order within 14 days of its making (s.36). If the tenant does this the current tenancy will continue for a period agreed by the parties or determined by the court, thus giving the landlord a reasonable opportunity to re-let or dispose of the premises (s.36).

If the court grants a new tenancy it will determine the terms of the new lease if the parties cannot reach agreement upon them. In the absence of agreement, the Act specifies in ss.32–35 rules with regard to: the length of the tenancy, up to a maximum of 15 years; the property to be included in the tenancy; the rent payable, which will be an open market rent fixed in accordance with statutory principles; and other terms of the new tenancy.

A landlord also has the right to apply to the court for either the renewal of the tenancy or for an order terminating the tenancy.

5   Under s.30 there are statutory grounds on which landlords can object to the grant of a new tenancy. Proof of certainty of these grounds means that a court *must* refuse the tenant's application for a new tenancy. In other cases, the court *may* refuse to grant a new tenancy.

## Mandatory grounds for refusing the tenant's application

There are three grounds upon proof of which by the landlord the court must refuse the tenant's application:

- **Ground d** – the landlord has offered and is willing to provide the tenant with alternative accommodation which is suitable for the tenant's requirements on terms which are reasonable.

  In assessing whether the alternative accommodation is suitable for the tenant's requirements (including the preservation of goodwill), the court should take into account the time when the new accommodation will be available for the tenant, the nature and class of the tenant's business and the situation, extent and facilities of the accommodation provided by the current tenancy.
- **Ground f** – the landlord intends to demolish or reconstruct the premises or to undertake substantial construction work, and could not reasonably do so without obtaining possession of the premises. The landlord must prove at the time of the court hearing that he has the necessary intention. He will do so if, on the evidence, a reasonable person would think that the landlord had a reasonable prospect of carrying out his intention. (Note that the tenant may be entitled to compensation under s.37 where the landlord relies on this ground – see page 349.)
- **Ground g** – the landlord intends to occupy the premises either as his residence or for the purposes of a business to be carried on by him. This ground is not normally available to a landlord who acquired the freehold or a tenancy of the premises by purchase within five years of the date of termination specified by his s.25 notice. Again, the landlord must prove at the time of the court hearing that he has the necessary intention. (Note that the tenant may be entitled to compensation under s.37 where the landlord relies on this ground – see page 349.)

## Discretionary grounds for refusing the tenant's application

There are four grounds upon proof of which by the landlord the court may, but is not obliged to, refuse the tenant's application:

- **Ground a** – the tenant ought not to be granted a new tenancy in consequence of the state of repair of the premises resulting from the tenant's failure to comply with his or her obligations under the current tenancy to repair and maintain them. The breach must be serious enough to persuade the court not to grant a new tenancy, an undertaking made by the tenant to remedy the breach being a relevant factor.
- **Ground b** – the tenant ought not to be granted a new tenancy in consequence of his or her persistent delay in paying rent due under the current tenancy. This requires either a failure to pay rent due on a number of occasions or a very long delay in paying rent due. The court will take into account factors such as the reason for the delay and how, if a new tenancy is granted, the landlord may be protected against breaches of the tenant's covenant to pay rent.
- **Ground c** – the tenant ought not to be granted a new tenancy in consequence of other substantial breaches of his or her obligations under the current tenancy, or for any other reason connected to use or management of the premises. The court will take into account factors such as the seriousness of the breach and whether the tenant has remedied or proposes to remedy it.
- **Ground e** – the tenant ought not to be granted a new tenancy where: the tenant is a subtenant of part of the property let by the landlord under a superior tenancy; and the landlord might reasonably expect to achieve a substantially better rent by letting the property as a whole; the landlord requires possession of the premises in order to let or dispose of the property let under the superior tenancy as a whole. The court will take into account factors such as the fact that the landlord consented to the subletting. The ground is only of value to the landlord where the superior tenancy is due to terminate shortly after the subtenancy. (Note that the tenant may be entitled to compensation under s.37 where the landlord relies on this ground – see page 349.)

## Tenant's request for a new tenancy

A tenant who has not been served a s.25 notice by the landlord may wish to request a new tenancy. The requirements which must be satisfied are as follows:

1   The current tenancy must have been granted for a term exceeding one year. This includes a tenancy granted for a fixed term and thereafter from year to year.
2   The tenant must serve an s.26 notice, requesting a new tenancy, on the 'competent landlord'. A tenant cannot serve an s.26 notice if the landlord has served an s.25 notice on him or her, or he or she has given the landlord notice to quit, or notified the landlord (under s.27)

that he or she does not wish the fixed-term tenancy to continue when its expiration date arrives.

The notice must be in writing and in the form laid down by regulations made under the Act. It must state the date upon which the new tenancy is to commence. The current tenancy will automatically terminate on this date. The date must be between six and 12 months after the making of the s.26 request to the landlord, and cannot be earlier than the date on which, had the Act not applied to it, the tenancy would have expired or could have been terminated by notice to quit served by the tenant. The notice must contain the tenant's proposed terms of the new tenancy, including the property to be comprised in the tenancy and the rent to be paid.

3    If the landlord does not wish to oppose the tenant's application for a new tenancy, he need not respond to the s.26 notice served on him by the tenant. He thus loses his right to oppose the tenant's application.

If the landlord does wish to oppose the tenant's application, he must serve a written counter-notice on the tenant indicating that he opposes the application. He must serve this counter-notice on the tenant within two months of the serving of the s.26 notice on him by the tenant, otherwise he loses his right to oppose the tenant's application.

The notice must indicate the grounds upon which the landlord objects to the grant of a new tenancy. The grounds upon which the landlord may so object, laid down by s.30, were outlined on page 346.

4    Once the tenant serves the s.26 notice he or she becomes entitled to apply to the court for a new tenancy if the parties cannot agree upon one themselves (s.29). If the tenant wishes to apply to the court, he or she must do so before the date specified in the s.26 notice. The parties can agree to extend this deadline. The landlord also has the right to apply to the court for renewal or an order terminating the tenancy.

Once the tenant applies to the court for a new tenancy, s.64 provides that the current tenancy cannot terminate within three months of the disposal of the tenant's application by the court. This is so whether or not the court grants a new tenancy. Thus, if the termination date specified by the tenant's s.26 notice falls within this three-month period, the current tenancy will not terminate until the three months have passed.

The court will order the grant of a new tenancy, unless the landlord establishes one or more of the grounds upon proof of which the court may or must refuse the tenant's application (s.29). If the tenant has a change of mind he or she may apply to the court for revocation of the order within 14 days of its making (s.36). If the tenant does this, the current tenancy will continue for a period agreed by the parties or determined by the court, thus giving the landlord a reasonable opportunity to re-let or dispose of the premises (s.36).

If the court grants a new tenancy it will determine the terms of the new lease if the parties cannot reach agreement upon them. In the absence of agreement: the length of the tenancy will be that which the court considers to be reasonable in the circumstances, to a maximum of 15 years (s.33); the property included in the new tenancy will be that part of the property included in the current tenancy which the court designates in the circumstances (s.32); the rent payable will be an open market rent, fixed in accordance with principles laid down by the Act (s.34); the other terms will be determined by the court having regard to the terms of the current tenancy and all other relevant circumstances (s.35).

5    If the landlord establishes one of the statutory grounds upon which he is entitled to object to the grant of a new tenancy (see page 346), it may be that the tenant will not be granted a new tenancy. Upon proof of certain grounds by the landlord, the court must refuse the tenant's application for a new tenancy. Upon proof of other grounds, the court has discretion to grant or refuse to grant a new tenancy.

## Tenant's compensation
### Under s.37 Landlord and Tenant Act 1954, Part II

Compensation under this section is available in two cases:

1    Where the tenant's application to the court for a new tenancy is refused because the landlord intends to demolish the premises, or because the landlord intends to occupy the premises himself, or because the landlord wants to amalgamate a subtenant's property with other property in order to get a better rent by letting as a whole (s.30).

2    Where either the landlord's s.25 notice or his counter-notice served in response to the tenant's s.26 notice states any or all of these three grounds of opposition (but no others), and the tenant either fails to apply for a new tenancy or makes and then withdraws an application. (Note: the leave of the court is required to withdraw an application.)

The amount of the compensation is determined by reference to a formula laid down from time to time by statutory instrument.

It may be, however, that landlord and tenant have agreed in writing to exclude the tenant's right to compensation under s.37. Such an agreement may be valid if the tenant and his or her predecessors in the business occupied the premises for business purposes for less than five years by the time he or she leaves. Similarly, such an agreement may be valid if made after the right to compensation has accrued. Otherwise, however, s.37 provides that such an agreement will be void.

### Compensation for improvements under s.1 Landlord and Tenant Act 1927, Part I

*Entitlement to compensation.* Part I of the 1927 Act applies to premises held under a lease which are used for trade or business purposes, but does not apply to mining leases, leases of agricultural holdings or leases which express in writing that the premises are let to the tenant as holder of an office, appointment or employment which he or she continues to hold (s.17). An agreement depriving a tenant of the right to compensation under Part I is invalid unless made for valuable consideration before 10 December 1953 (s.9 as amended by s.49 Landlord and Tenant Act 1954).

Where Part I of the 1927 Act applies, it entitles the tenant, on vacating a holding at the end of the tenancy, to compensation from the landlord in respect of improvements made by the tenant or his/her predecessors in title (s.1). Where premises are used only partly for trade or business purposes, compensation is only payable in relation to improvements if and to the extent to which they relate to the trade or business (s.17). It is the tenant's immediate landlord who is required to pay compensation. If the tenant is not the freeholder, then when the tenancy comes to an end and the premises are vacated, he or she is entitled to claim compensation from the landlord (s.8).

In order to give rise to this right to compensation, an improvement must not be a fixture which the tenant is entitled to remove, and it must add to the letting value of the holding at the end of the tenancy (s.1). Moreover, the improvement must not have been made for valuable consideration (s.2).

Further, a tenant is not entitled to compensation under Part I unless the requirements laid down by s.3 have been satisfied. Prior to making the improvement, the tenant is required to notify the landlord of his or her intention, providing the landlord with a specification and plan of the improvement. Unless the making of the improvement is required by statute, the landlord is entitled to serve notice of objection on the tenant within three months of receiving the tenant's notice.

If the landlord does not serve a notice of objection within three months, he loses his right to do so. The tenant can lawfully go ahead with the improvement even if it is prohibited by the terms of the lease.

If the landlord does serve a notice of objection within three months, the tenant can apply to the court for a certificate that the improvement is proper. When such an application is made, all superior landlords must be notified of the tenant's intention and have a right to be heard before the court. The court must give the tenant a certificate if it is satisfied that the improvement will add to the letting value of the holding at the end of the tenancy, is reasonable and suitable to the character of the property, and will not diminish the value of any other property belonging to the landlord.

The landlord is entitled to offer to make the proposed improvement himself in return for a reasonable increase in rent or an increase determined by the court. In this case the court cannot grant a certificate unless it is subsequently shown to the court's satisfaction that the landlord has failed to do so.

Before giving the tenant a certificate, the court is entitled to make such modifications to the specification or plan as it thinks fit and to impose such conditions as it thinks reasonable. Once the certificate has been issued, the tenant can lawfully carry out the improvement, even if it is prohibited by the terms of the lease. He or she must comply with conditions imposed by the court and must complete the work within time limits agreed with the landlord or fixed with the court.

### Claiming compensation

The tenant's claim for compensation, made in the prescribed form, must be served on the landlord within time limits laid down by s.47 of the Landlord and Tenant Act 1954. Where the claim relates to the giving of notice to quit, whether by the landlord or the tenant, the claim must be made within three months of the giving of the notice.

If the claim for compensation is in relation to termination following the service of an s.26 notice by the tenant, it must be made within three months of the date upon which the landlord served the counter-notice. If the claim is in relation to termination by expiration of time, it must be made between three and six months before the end of the tenancy. Where the claim is in relation to forfeiture, it must be made within three months of the effective date of a court order for possession or, in the case of re-entry without a court order, within three months of the date of re-entry.

### Amount of compensation

In the absence of agreement between the parties to the lease, the amount of compensation is determined by the court (s.1). The amount of compensation cannot exceed either the net addition to the value of the holding which directly results from the improvement, or the reasonable cost of carrying out the improvement at the end of the tenancy less the cost of putting the improvement into a reasonable state of repair (unless or to the extent to which the lease obliges the tenant to carry out such repairs).

In determining the net addition to the holding's value, the intended use of the premises after the end of the lease, including an intention to demolish or reconstruct, and the likely time lapse between the date of termination and any change of use, demolition or reconstruction must be taken into account. If the landlord does not carry out his intentions within the time specified, the tenant is entitled to make a further application for compensation. The amount of compensation will be reduced in accordance with the value of any benefits received by the tenant or predecessors from the landlord or predecessors in consideration of the making of the improvement (s.2(3)).

## Residential tenancies

In relation to a residential tenancy, where the provisions of the Housing Act 1988 or the Rent Act 1977 apply, the tenant may be provided with security of tenure.

## Housing Act 1988

A tenancy to which the Act applies is known as an *assured tenancy*. The Act only applies to leases and not to licences. However, as was noted on page 326, an agreement which is described as a licence, but which gives exclusive possession of premises for a fixed or periodic term may in fact create a tenancy (*Street v. Mountford*).

The tenancy must be of a dwelling-house, the purpose of the letting being residential. A dwelling-house may be a house or part of a house (s.45) and thus can include a flat. It must be let as a separate dwelling, which must provide the tenant or joint tenants with exclusive possession of rooms in which the tenant can conduct the main activities of living, especially cooking, eating and sleeping. Other rooms, such as bathrooms, may be shared (s.3).

In order to be within the protection of the Act, the tenant or one of the joint tenants must be an individual (i.e. not a company) and must occupy the dwelling-house as their only or principal home. Furthermore, the tenancy *must not* be one of the following, which are examples of tenancies incapable of being an assured tenancy:

- A tenancy entered into before 15 January 1989.
- A tenancy of a dwelling-house the rent for which (or rateable value if the tenancy was entered into prior to 1 April 1990) exceeds specified limits.
- A tenancy under which either no rent is payable or the rent payable falls below specified limits.
- A business tenancy within Part II of the Landlord and Tenant Act 1954.
- A tenancy under which agricultural land exceeding two acres is let with the dwelling-house.
- A tenancy granted to a student at a specified educational institution by a specified body.
- A tenancy the purpose of which is to let the dwelling-house to the tenant for a holiday.
- A tenancy of a dwelling-house which forms part of a building, where the landlord is an individual who occupies another dwelling-house in the same building as his only or principal home. The tenancy is not necessarily rendered incapable of being an assured tenancy by the presence of a resident landlord if the dwelling-house forms a flat in a purpose-built block of flats. (Note that the complex provisions of Sched. I, para. 10, relating to resident landlords, fall outside the scope of this book.)
- A tenancy under which the interest of the landlord belongs to a local authority or other specified body.
- A protected tenancy (see discussion of the Rent Act 1977 below); a housing association tenancy (see Rent Act 1977); a secure tenancy (see Housing Act 1985).

## Security of tenure under the 1988 Act

1  *Periodic tenancies.* A landlord who wishes to terminate a periodic tenancy which falls within the protection of the Act (a periodic assured tenancy) may only do so by obtaining a court order (s.5).

Before taking proceedings in court, the landlord must normally serve on the tenant a notice which satisfies the requirements of s.8. The notice must state the landlord's statutory grounds for possession (see page 354), indicate the earliest date on which the proceedings can begin (which must be at least two weeks from the date of service of the notice) and indicate that the proceedings will not begin later than 12 months from the date of service of the notice.

Where the notice specifies, with or without other grounds, any of statutory grounds 1, 2, 5, 6, 7, 9 or 16 (see page 354–357), the proceedings cannot begin earlier than the date at which the landlord could have terminated the tenancy at common law by serving notice to quit and, in any event, not earlier than two months from the date of service of the notice.

The court can dispense with the s.8 notice requirement where it is just and equitable to do so, except where the landlord seeks to recover possession under statutory ground 8.

2  *Fixed-term tenancies.* Unless the tenant surrenders a tenancy, a landlord who wishes to terminate a fixed-term tenancy which falls within the protection of the Act (a fixed-term assured tenancy) may only do so either by obtaining a court order or by exercising the power to terminate the tenancy under a break clause (s.5). In order to obtain a court order, the landlord must comply with the s.8 notice procedure (outlined above).

If a fixed-term assured tenancy comes to an end other than by surrender or by court order, the tenant is entitled to remain in possession under a periodic assured tenancy which arises under s.5 (a statutory periodic tenancy), the period of which is the same as that for which rent was payable under the fixed-term tenancy. Basically, the terms of the statutory periodic tenancy are the same as those of the fixed-term tenancy except that terms permitting termination of the tenancy cease to have effect so long as the tenancy is assured.

At the end of a fixed-term tenancy, if the landlord grants the tenant a new fixed-term or periodic tenancy of substantially the same dwelling-house, a statutory periodic tenancy will not arise.

3  *Possession by court order.* The court may only grant the landlord an order for possession of a dwelling-house let on an assured tenancy if the landlord establishes one or more of the statutory grounds for possession (s.7). In relation to a fixed-term tenancy, the court cannot make an order if the tenancy has not expired or otherwise been terminated unless at least one of grounds 2, 8, 10, 11, 12, 13, 14 or

15 is established and the lease permits termination on the ground in question.

Where an order is made in relation to a fixed-term tenancy which has come to an end, this is sufficient to terminate a statutory periodic tenancy which has arisen under s.5 without further notice being required (ss.7 and 8).

There are two types of grounds for possession: mandatory grounds and discretionary grounds. Upon proof of a mandatory ground, the court is required to make an order for possession (subject both to what was said immediately above with regard to the making of orders in relation to fixed-term tenancies and to satisfaction of the s.8 notice requirements examined above).

Upon proof of a discretionary ground, the court may make an order for possession if it considers it reasonable to do so. The court possesses discretion to stay or suspend the execution of the order or postpone the date of possession for such period as it thinks fit.

## Mandatory grounds (Part I of Schedule 2, Housing Act 1988)

In order to obtain possession on grounds 1–5, the landlord must have notified the tenant not later than the beginning of the tenancy that possession might be recovered on the ground in question. In relation to grounds 1 and 2, the court can dispense with the notice requirement where it is just and equitable to do so.

**Ground 1** applies either where the landlord at some time prior to the beginning of the tenancy occupied the dwelling-house as his only or principal home or where the landlord requires the dwelling-house as his or his spouse's only or principal home.

**Ground 2** applies where the dwelling-house is subject to a mortgage granted before the tenancy began and the mortgagee requires vacant possession in order to exercise a power of sale.

**Ground 3** applies to fixed-term tenancies for a term not exceeding eight months if at some time within the 12 months prior to commencement of the tenancy the dwelling-house was let as a holiday home.

**Ground 4** applies to fixed-term tenancies for a term not exceeding 12 months if at some time within the 12 months prior to commencement of the tenancy the dwelling-house was let to a person pursuing or intending to pursue a course of study provided by an educational institution specified by the Secretary of State and the tenancy was granted by that institution or by another specified institution or body of persons.

**Ground 5** applies where a dwelling-house is held for the purpose of being available for occupation by a minister of religion as a residence from which to perform his duties and the court is satisfied that the dwelling-house is required for this purpose.

**Ground 6** applies where the landlord intends to demolish or reconstruct the whole or a substantial part of the dwelling-house or to carry out substantial works. In order to obtain possession on this ground, it must be established that:

(a) the work cannot reasonably be carried out without the tenant giving up possession because: the tenant will not agree to such a variation of the terms as would permit the work to be carried out; or the nature of the work is such that no such variation is practicable; or the tenant is not willing to accept an assured tenancy of such part of the dwelling-house as would leave in the landlord's possession so much of the dwelling-house as would reasonably enable the work to be carried out; or the nature of the work is such that such a tenancy is not practicable; and

(b) the landlord seeking possession either acquired his interest in the dwelling-house before the grant of the tenancy or the interest was in existence at the time of the grant and neither the landlord nor any other person who has acquired the interest since that time acquired it for money or money's worth; and

(c) the assured tenancy on which the dwelling-house is let did not come into being in succession to a Rent Act 1977 or Rent (Agriculture) Act 1976 statutory tenancy.

Section 11 provides that where the court makes an order for possession on ground 6 or ground 9 (but not any other ground) the landlord must pay to the tenant the reasonable expenses likely to be incurred by the tenant in moving from the dwelling-house. The amount will either be determined by agreement between landlord and tenant or by the court.

**Ground 7** applies to a periodic tenancy which has devolved under the will or intestacy of the former tenant (but not by statutory succession) if proceedings for recovery of possession are commenced either not later than 12 months after the death of the former tenant or, if the court so directs, after the date on which the landlord became aware of the former tenant's death. For the purpose of this ground, acceptance by the landlord of rent from a new tenant after the former tenant's death does not create a new periodic tenancy unless the landlord agrees in writing to change a term or terms of the tenancy.

**Ground 8** applies when both at the date of service of the s.8 notice and at the date of the court hearing:

- At least eight weeks' rent is unpaid in cases where the rent is paid weekly or fortnightly.
- At least two months' rent is unpaid in cases where the rent is paid monthly.
- At least one quarter's rent is more than three months in arrears in cases where the rent is paid quarterly.
- At least three months' rent is more than three months in arrears in cases where the rent is paid yearly.

## *Discretionary grounds (Part II of Schedule 2, Housing Act 1988)*

**Ground 9** applies where suitable alternative accommodation is available for the tenant or will be available when the order for possession takes effect.

Part III of Schedule 2 provides that a certificate of the local housing authority for the district in which the dwelling-house is located, certifying that the authority will provide suitable alternative accommodation for the tenant by a specified date, is conclusive evidence that suitable alternative accommodation will be available by that date.

Otherwise, Part III of Schedule 2 provides that alternative accommodation is suitable if:

(a)   it will either be let on an assured tenancy (other than an assured shorthold tenancy or an assured tenancy in relation to which notice is given that possession may be required on any of grounds 1–5), or the terms of the letting will give the tenant security of tenure reasonably equivalent to that provided by such an assured tenancy; and

(b)   the accommodation is reasonably suited to the needs of the tenant and family in relation to proximity to place of work; and

(c)   the accommodation is similar in relation to rent and extent to that provided in the neighbourhood by any local housing authority for persons whose needs in relation to extent are similar to those of the tenant and family or is reasonably suited to the means of the tenant and to the needs of the tenant and family in relation to extent and character; and

(d)   if any furniture was provided under the assured tenancy, furniture is provided which is either similar to that formerly provided or is reasonably suited to the needs of the tenant and family.

Accommodation is not suited to the needs of the tenant and family if their occupation of it would render it an overcrowded dwelling within the meaning of Part X of the Housing Act 1985.

Section 11 provides that where the court makes an order for possession on ground 6 or ground 9 (but not any other ground) the landlord must pay to the tenant the reasonable expenses likely to be incurred by the tenant in moving from the dwelling-house. The amount will either be determined by agreement between landlord and tenant or by the court.

**Ground 10** applies where rent is unpaid on the date on which the proceedings for possession are begun and (except where the court considers it to be just and equitable to dispense with the requirement that notice be served under s.8) was in arrears at the date of the service of the s.8 notice.

**Ground 11** applies where (whether or not rent is in arrears on the date on which proceedings for possession are begun) the tenant has persistently delayed in paying rent.

**Ground 12** applies where any obligation of the tenancy, other than one related to the payment of rent, has been broken or has not been performed.

**Ground 13** applies where the condition of the dwelling-house or common parts has deteriorated due to acts of waste by, or the neglect or default of, the tenant or any other person residing in the dwelling-house. Where the waste, neglect or default is that of a lodger or subtenant, it must be shown that the tenant has not taken such steps as ought reasonably to have been taken in order to remove the lodger or subtenant.

**Ground 14** applies where the tenant or any other person residing in the dwelling-house has been guilty of conduct which is or is likely to cause a nuisance or annoyance to persons residing in, visiting or engaging in a lawful activity in the locality, or has been convicted of using the dwelling-house or allowing its use for immoral or illegal purposes or an indictable offence has been committed in it or in its locality.

**Ground 14A** applies where the landlord is a registered social landlord or a charitable housing trust and the tenant or spouse or partner has left the dwelling-house in consequence of violence or threats of violence by his/her spouse or partner towards him/her or a member of his/her family and is unlikely to return.

**Ground 15** applies where the condition of any furniture provided under the tenancy has deteriorated owing to ill-treatment by the tenant or by any other person residing in the dwelling-house. In the case of ill-treatment by a lodger or subtenant, it must be shown that the tenant has not taken such steps as ought reasonably to have been taken in order to remove the lodger or subtenant.

**Ground 16** applies where the dwelling-house was let to the tenant in consequence of his or her employment by the landlord (or a previous landlord under the tenancy) and the tenant is no longer in that employment.

**Ground 17** applies where the tenant or a person acting at the tenant's instigation induced the landlord to grant the tenancy by a false statement made knowingly or recklessly.

4    *Possession under an assured shorthold tenancy granted before 28 February 1997.* To be an assured shorthold tenancy, a tenancy must be a fixed-term assured tenancy, granted for a term of not less than six months, which the landlord has no power to determine within six months of its commencement (s.20). Further, before the assured tenancy was entered into, the landlord must have served on the tenant a notice which stated that the tenancy was to be a shorthold tenancy.

Where an assured shorthold tenancy comes to an end, a statutory periodic tenancy arises and any new assured tenancy of the same premises with the same landlord and tenant will be a shorthold tenancy, even though it does not satisfy these requirements, unless the landlord serves notice on the tenant before the tenancy is entered into that it is not to be a shorthold tenancy. Thus, a periodic tenancy may be an assured shorthold tenancy if it follows a fixed-term assured shorthold tenancy.

A tenancy cannot be an assured shorthold tenancy if, immediately before the tenancy was granted, one of the tenants was tenant under a tenancy granted by the landlord which was an assured tenancy but was not a shorthold tenancy.

Where a tenancy is an assured shorthold tenancy, the Act gives the landlord a specific right to recover possession in addition to the grounds for possession which have already been examined. In relation to a fixed-term assured shorthold tenancy, s.21 provides that the court must order possession when the tenancy comes to an end, even if a statutory periodic tenancy has come into existence, provided that the landlord has given the tenant at least two months' notice in writing stating that he requires possession and no further assured tenancy (other than a statutory periodic tenancy) is in existence. Notice must be given to the tenant before or on the day when the tenancy comes to an end.

In relation to a periodic assured tenancy, s.21 provides that the court must order possession if the landlord has notified the tenant in writing that after a specified date, possession of the dwelling-house is required under s.21. The date cannot be earlier than the earliest day on which the tenancy could have been terminated at common law by notice to quit given on the same day as the s.21 notice.

5    *Possession under an assured shorthold tenancy granted on or after 28 February 1997 (Housing Act 1996).* All assured tenancies granted after 28 February 1997, whether fixed term or periodic, are presumed to be assured shorthold tenancies other than in certain specified exceptional circumstances. It is no longer necessary for the landlord to notify the tenant that the tenancy will be a shorthold tenancy, though the tenant is, upon request, entitled to a written statement of certain terms. Moreover, it is no longer necessary that an assured shorthold tenancy be granted for a term of at least six months.

Examples of circumstances in which an assured tenancy granted after 28 February 1997 will not be an assured shorthold tenancy are provided by those in which:

- a periodic tenancy arises under s.5 of the Housing Act 1988 when a fixed-term assured tenancy comes to an end; or
- the landlord has notified the tenant that the tenancy will not be an assured shorthold tenancy; or
- the lease contains a provision stating that it is not an assured shorthold tenancy; or
- the landlord offers an assured tenant a new tenancy (unless the tenant notifies the landlord that the tenancy will be an assured shorthold tenancy).

A fixed-term assured shorthold tenancy may be terminated during the fixed term by obtaining an order for possession under grounds 2 or 8 (examined above). When the fixed term expires, the court may grant an order for possession if the requirements of s.21 of the 1988 Act (examined above) have been satisfied. It should be noted, however, that an order for possession made under s.21 may not take effect until at least six months after the commencement of the tenancy if the tenancy is an assured shorthold tenancy granted on or after 28 February 1997.

## Rent control under the Housing Act 1988

*Fixed-term assured tenancies.* The Act does not regulate the amount of rent which may be charged or the extent of rent increases under a fixed-term tenancy. When the fixed term expires, then, if a statutory periodic tenancy commences, the provisions of s.13 (considered immediately below) will apply.

*Periodic assured tenancies.* Section 13 provides for rent increases under assured periodic tenancies (including statutory periodic tenancies). The minimum period between such increases is one year. The s.13 procedure is not applicable where the lease contains a rent review clause which is binding upon the tenant, and the section does not prevent the parties from reaching an agreement of their own. Section 13 also applies to periodic assured shorthold tenancies granted on or after 28 February 1997.

*Assured shorthold tenancies.* Section 22 provides that a tenant under an assured shorthold tenancy may apply to a rent assessment committee for determination of the rent which the landlord might reasonably be expected to obtain under the tenancy, provided that the requirements of the section are satisfied. Where the tenancy was granted on or after 28 February 1997, the tenant cannot make an application if the tenancy commenced more than six months prior to the making of the application.

## Succession by tenant's spouse to an assured periodic tenancy

Where the tenant under an assured periodic tenancy dies, the spouse or partner of the tenant can take over the tenancy under s.17, provided that the spouse or partner occupied the dwelling-house as his or her only or principal home immediately before the death of the tenant. This rule can only operate where the deceased tenant was not already a successor, e.g. by virtue of being the survivor of joint tenants, or by inheriting under the will or intestacy of a previous tenant.

## Rent Act 1977

Basically, the Rent Act 1977 applies to private residential tenancies entered into before 15 January 1989, though certain tenancies created after that date do still fall within the provisions of the 1977 Act. In relation to tenancies created after that date, the appropriate legislation is normally the Housing Act 1988.

A tenancy to which the Rent Act applies is known as a *protected tenancy*. Like the Housing Act 1988, the Rent Act applies to tenancies but not to licences. The tenancy must be of a dwelling-house, which may be a house or part of a house, and thus can include a flat. The dwelling-house must be let as a separate dwelling, which must provide the tenant or joint tenants with exclusive possession of rooms in which the tenant can conduct the main activities of living, especially cooking, eating and sleeping. Other rooms, such as bathrooms, may be shared (s.22).

The tenancy *must not* be one of the types which are incapable of being protected tenancies. The following are examples of types of tenancy which are incapable of being protected tenancies:

- Certain tenancies where the rent/rateable value exceeds specified limits (ss.4 and 25).
- A tenancy under which either no rent is payable or the rent payable falls below specified limits.
- A shared ownership lease granted under Part V of the Housing Act 1985 (s.5A), or a lease granted by a housing association which satisfies the requirements of s.5A(2).
- A dwelling-house let with land other than the site of the dwelling-house (s.6) – though where land is let with a dwelling-house and the primary purpose of the agreement is the letting of the dwelling-house, the Act may apply.
- A dwelling-house let at a rent which includes payments for genuine board and attendance (s.7).
- In relation to a student studying a course provided by a specified educational institution, a tenancy granted to the student by a specified body (s.8).
- A tenancy the purpose of which is to let the dwelling-house to the tenant for a holiday (s.9).
- A tenancy under which the dwelling-house is comprised in an agricultural holding (see Agricultural Holdings Act 1986) and is occupied by the person who controls the farming of the holding (s.10).
- A tenancy of premises with an 'on-licence' for the sale of intoxicating liquors (s.11).
- A tenancy of a dwelling-house which forms part of a building, where the landlord from the time the tenancy was granted has occupied another dwelling-house in the same building as his residence (s.12). The tenancy is not rendered incapable of being a protected tenancy by the presence of a resident landlord if the dwelling-house forms a flat in a purpose-built block of flats. (Note that the complex provisions of s.12, relating to resident landlords fall outside the scope of this book.)
- A tenancy under which the interest of the landlord belongs to certain types of public body such as a county council, development corporation or housing action trust (s.14).
- Business tenancies which fall within Part II of the Landlord and Tenant Act 1954 (including premises with mixed business/residential use) (s.24(3)).

## Security of tenure under the Rent Act

A protected tenancy comes to an end when the term expires (if the lease is for a fixed term) or when the tenancy is otherwise terminated at common law. Following the termination of the protected tenancy, s.2 provides that the former protected tenant becomes a statutory tenant of the dwelling-house provided that and for as long as the dwelling-house remains his or her residence.

A statutory tenancy is not an interest in land; rather, it is a statutory right to remain in possession of a dwelling-house. The terms of the statutory tenancy are those of the protected tenancy except to the extent to which they are inconsistent with provisions of the Act (s.3). Thus, terms of the protected tenancy entitling the landlord to recover possession of the dwelling-house will not become terms of the statutory tenancy.

1   *Termination of a statutory tenancy by act of the tenant.* A statutory tenancy will not arise, or will come to an end, if the dwelling-house ceases to be the tenant's residence (s.2). The fact that the tenant is frequently absent from the dwelling-house for lengthy periods does not necessarily mean that the dwelling-house has ceased to be his or her residence.

Similarly, where a tenant has two houses, the fact that he or she occupies one house more regularly than the other will not necessarily make the former house a residence for this purpose. In each case, the primary consideration in determining whether a dwelling-house is the tenant's residence is the intent.

A statutory tenant may terminate his or her statutory tenancy by serving notice on the landlord (s.3). If the protected tenancy was a fixed-term tenancy, the tenant must give the landlord at least three months' notice for this purpose. If the protected tenancy was a periodic tenancy, the length of notice required is either that which was required under the protected tenancy or four weeks, whichever is the greater.

A statutory tenant may surrender statutory tenancy expressly or impliedly (see page 340).

2   *Termination of a statutory tenancy by the landlord.* A landlord can only obtain possession of a dwelling-house which is subject to a statutory tenancy by obtaining a court order (s.98). In order to obtain a court order, the landlord must prove one or more of the discretionary or mandatory grounds for possession which are laid down by the Act. If the tenant can prove a discretionary ground, the court may then order possession if it considers it reasonable to do so. The court possesses discretion to stay or suspend the execution of the order or postpone the date of possession for such period as it thinks fit. On proof of one or more of the mandatory grounds, the court must order possession.

The landlord also needs a court order to obtain possession of a dwelling-house which is subject to a protected tenancy. Even though the landlord can establish grounds for possession under the Act, the court will not grant such an order in respect of a protected tenancy unless the landlord is entitled to possession of the dwelling-house at common law.

## Discretionary grounds for possession (Rent Act 1977)

Section 98(1)(a) applies where suitable alternative accommodation is available for the tenant. The phrase 'suitable alternative accommodation' is extensively defined in Part IV of Schedule 15 to the Act in terms similar to those already examined in the context of the Housing Act 1988 on page 356.

Schedule 15, Part I of the Rent Act 1977 sets out the following discretionary grounds for a landlord obtaining possession.

**Case 1** applies where the tenant has failed to pay rent due or has broken or failed to perform an obligation of the tenancy.

**Case 2** applies where the tenant has been guilty of conduct which is a nuisance or annoyance to adjoining occupiers, or has been convicted of using the dwelling-house or allowing its use for immoral or illegal purposes.

**Case 3** applies where the condition of the dwelling-house or common parts has deteriorated due to acts of waste by, or the neglect or default of, the tenant, a lodger or a subtenant.

**Case 4** applies where the condition of any furniture provided under the tenancy has deteriorated owing to ill-treatment by the tenant, a lodger or a subtenant.

**Case 5** applies where the tenant has given his landlord notice to quit and, consequently, the landlord has contracted to sell or let the dwelling-house.

**Case 6** applies where the tenant has, without the landlord's consent, either assigned or sublet the whole of the dwelling-house.

**Case 7** was repealed by the Housing Act 1980.

**Case 8** applies where the dwelling-house was let to the tenant in consequence of his or her employment by the landlord or a former landlord and the tenant is no longer in that employment, provided that the dwelling-house is reasonably required by the landlord for a person engaged in full-time employment by him or by one of his tenants.

**Case 9** applies where the dwelling-house is reasonably required by the landlord as a residence for himself, for a child of his over 18 years old, for his father or mother or for his father-in-law or mother-in-law. This case does not apply if the landlord became landlord by purchasing the dwelling-house or an interest in it after 23 March 1965.

Part III of Schedule 15 provides that the court shall not make an order under case 9 if greater hardship would be caused by granting the order than by refusing to grant it.

**Case 10** applies where the tenant has sublet part of the dwelling-house under a protected or statutory tenancy at a rent in excess of the maximum rent recoverable under Part III of the Rent Act 1977.

## Mandatory grounds for possession under Part II of Schedule 15, Rent Act 1977

In order to obtain possession under cases 11–15 and 20, the landlord must have notified the tenant not later than the beginning of the tenancy that possession might need to be recovered under the appropriate case (though, in relation to cases 11, 12 and 20, the court may dispense with these requirements where it is just and equitable to do so).

**Case 11** applies where the owner, prior to letting the dwelling-house, occupied it as his residence and:

(a) the house is required as a residence by him or by a member of his family who lived with him when he last occupied the dwelling-house; or

(b) he has died and the house is required as a residence for a member of his family who was living with him at the time; or

(c) he has died and the house is required by a successor in title (but not a purchaser) either as a residence or in order to dispose of it with vacant possession; or

(d) the house is subject to a mortgage granted before the tenancy and the mortgagee requires vacant possession in order to exercise a power of sale; or

(e) the house is not reasonably suited to his needs, having regard to his place of work, and he requires it in order to dispose of it with vacant possession, the proceeds to be used to acquire a house as his residence which is better suited to his needs.

**Case 12** applies where the owner intends to occupy the dwelling-house as his residence when he retires from regular employment and:

(a) he has retired and requires the house as his residence; or

(b) he has died and the house is required as a residence for a member of his family who was living with him at the time; or

(c) he has died and the house is required by a successor in title (but not a purchaser) either as a residence or in order to dispose of it with vacant possession; or

(d) the house is subject to a mortgage granted before the tenancy and the mortgagee requires vacant possession in order to exercise a power of sale.

**Case 13** applies to fixed-term tenancies for a term not exceeding eight months if at some time within the 12 months prior to commencement of the tenancy the dwelling-house was let as a holiday home.

**Case 14** applies to fixed-term tenancies for a term not exceeding 12 months if at some time within the 12 months prior to commencement of the tenancy the dwelling-house was let to a person pursuing or intending to pursue a course of study provided by a specified educational institution and the tenancy was granted by a specified body.

**Case 15** applies where a dwelling-house is held for the purpose of being available for occupation by a minister of religion as a residence from which to perform his duties and the court is satisfied that the dwelling-house is required for this purpose.

**Cases 16, 17 and 18** relate to premises once occupied by agricultural workers which are required for agricultural employees, provided that notice was served prior to commencement of the tenancy that possession might be required under the relevant case, and that the other requirements of the relevant case are satisfied.

**Case 19** may be applicable where a dwelling-house was let under a protected shorthold tenancy (see consideration of the similar provisions of the Housing Act 1988).

**Case 20** applies where a dwelling-house was let after 28 November 1980 by a person who was a member of the regular armed forces when he acquired it, provided that he can satisfy requirements similar to those in case 12.

### Succession to a statutory tenancy (s.2 and Part I of Schedule 1, Rent Act 1977 as amended by s.39 and Schedule 4, Housing Act 1988)

Where a statutory tenant who was the original tenant under the protected tenancy dies, the tenant's spouse (including a common law spouse) succeeds to the statutory tenancy provided that the spouse resided in the dwelling-house immediately prior to the tenant's death. If there is no resident spouse, then a family member who lived with the tenant in the dwelling-house immediately before the tenant's death and for the two years prior to it succeeds to an assured periodic tenancy under the Housing Act 1988.

If a surviving spouse has succeeded to a statutory tenancy, then on the death of that person, any member of his or her family who was also a member of the original tenant's family may be able to succeed to an assured periodic tenancy.

In *Mendoza v. Ghaidan* (2002) EWCA Civ 1533; [2002] 4 All E.R. 1162, the Court of Appeal decided that a person who was in a same-sex relationship with a deceased tenant who held a Rent Act tenancy was entitled to succeed to a statutory tenancy as the 'surviving spouse' of the tenant. In reaching this decision, the Court departed from a unanimous ruling of the House of Lords in *Fitzpatrick v. Sterling Housing Association* (2001) 1 A.C. 27, where it had been decided that a surviving partner of the same sex could not be treated as a surviving spouse (though the Court also decided, by a majority, that the surviving partner was entitled to be treated as a member of the deceased tenant's family).

### Premiums (Rent Act 1977, Part IX)

It is an offence to require a premium in addition to rent in relation to the grant of a protected tenancy. Any part of a contract requiring the payment of such a premium is void.

Where an excessive price is charged for the purchase of furniture, and the grant of the tenancy is conditional upon such payment, the excess part of the payment above that which is reasonable is treated as a premium.

A requirement that rent be paid prior to commencement of the period to which it relates is void, and it is an offence to require or receive such payments.

Taking a deposit is allowed provided that it does not exceed one-sixth of the annual rent and is reasonable.

## Protected shorthold tenancies (Housing Act 1980 ss.51–55 and Housing Act 1988 s.34 and Schedule 18)

A tenancy may only be a protected shorthold tenancy if it is a fixed-term tenancy for a term of between one and five years which cannot be brought to an end by the landlord before expiration of the term other than in consequence of breach by the tenant of his or her obligations under the lease. Further, prior to commencement of the tenancy, the landlord must have notified the tenant that the tenancy was to be a protected shorthold tenancy. If a tenant was a protected or statutory tenant of a dwelling-house immediately prior to the grant of a new protected tenancy, the new tenancy cannot be a protected shorthold tenancy.

Where a shorthold tenancy is created after 15 January 1989 it will be an assured shorthold tenancy under the Housing Act 1988 unless granted under a contract made before that date.

## Rent control (Rent Act 1977, Part IV)

Either or both parties to a protected tenancy may apply to the rent officer to have a fair rent registered for two years. Usually, no further application can be made within the two-year period. This registered rent is the maximum which the landlord can recover, though if it exceeds the current rent no increase in rent is authorized unless permitted by the terms of the tenancy. If no such application is made, the rent payable is that which is authorized by the lease or agreed by the parties.

Similarly, either or both parties to a statutory tenancy may apply to the rent officer to have a fair rent registered for two years. This is the maximum which the landlord can recover, though if it exceeds the current rent no increase in rent is authorized unless permitted by the terms of the tenancy. If no such application is made, the maximum rent payable is that which was payable under the last rental period of the protected tenancy (or, if a fair rent was registered under the protected tenancy, that which was so registered).

In determining a fair rent, regard is to be had to all relevant non-personal circumstances, such as: the age, character, locality and state of repair of the dwelling-house; the quantity, quality and condition of furniture; and any premium which has lawfully been required or received on the grant or renewal of the tenancy.

In determining a fair rent, certain matters must be disregarded, namely: scarcity in the neighbourhood; disrepair or defects attributable to the failure of the tenant or a predecessor in title to comply with the obligations under the tenancy; improvements carried out by the tenant or a predecessor in title other than under the terms of the tenancy; and improvement of or deterioration of furniture due to the conduct of the tenant, a lodger or a subtenant.

In practice, the best method of assessment is normally comparison with registered fair rents of comparable dwelling-houses.

## Long leases of residential premises: the Leasehold Reform Act 1967

The Act applies to tenancies which satisfy the following conditions, laid down by s.1:

1   The tenancy must be a long tenancy. This basically means that the tenancy must be for a fixed term exceeding 21 years (s.3).
2   The tenancy must be a tenancy of a house (but does not include a flat created by horizontal division of a building).
3   The rateable value of the house must fall within specified limits (if the tenancy was granted after 1 April 1990 a statutory formula is applicable). (Note that this requirement only applies where the tenant wants an extended lease, not where he or she wishes to purchase the freehold: Leasehold Reform, Housing and Urban Development Act 1993.)

Where the requirements of the Act are satisfied, the tenant may claim the freehold of the house (s.8) or a 50-year extension to the term of the lease (s.14) by serving notice on the landlord. A tenant may, however, lose the right to make a claim in a number of ways, for example by forfeiture, or by failing to make a claim within two months of receiving notice to terminate the tenancy from the landlord, or by giving the landlord notice to terminate the tenancy or being granted a new tenancy under Part II of the Landlord and Tenant Act 1954 (business tenancies).

The landlord may apply to the court for possession. The court may grant possession if the landlord can prove statutory grounds, for example that he or one of his family wishes to occupy the house.

Where the freehold is claimed, the purchase price will be calculated upon the basis of factors laid down by the Act (s.9). If the tenant decides not to go ahead with the purchase when the price is calculated, he or she loses his right to make another claim for three years.

Where an extension to the term is claimed, the rent will be calculated upon the basis of factors laid down by the Act (s.14). The other terms of the extended lease will basically be those of the original lease, subject to modifications relating to options and rights of termination (s.15).

A tenant who claims an extended tenancy under the Act is not subsequently entitled to claim a further extension. He or she may purchase the freehold before the date when the original tenancy would have ended. After that date, there are no rights to enfranchisement under the Act.

Where the Leasehold Reform Act 1967 is not applicable to a long tenancy at a low rent or the tenant does not claim the freehold or an extended tenancy under it, he or she may be entitled to security of tenure under Part I of the Landlord and Tenant Act 1954 or Schedule 10 of the Local Government and Housing Act 1989. The 1954 Act essentially applies to tenancies granted prior to 1 March 1990 provided that the tenancy is a tenancy of a separate dwelling in which the tenant resides. When the term of the long lease expires, the tenancy continues until:

- the tenant terminates it; or
- the landlord proposes a Rent Act 1977 statutory tenancy on terms which the tenant accepts or the court determines; or
- the court grants an order for possession, the landlord having established statutory grounds for possession.

The 1989 Act essentially applies to tenancies granted on or after 1 March 1990, provided that the tenancy is a tenancy of a separate dwelling in which the tenant, who is an individual, resides as his or her only or principal home. When the term of the long lease expires, the tenancy continues until:

- the tenant terminates it; or
- the landlord proposes a Housing Act 1988 assured tenancy; or
- the court grants an order for possession, the landlord having established statutory grounds for possession.

Where a long lease at a low rent expires on or after 15 January 1999, the provisions of the 1989 Act and not those of the 1954 Act will apply to it, even if the lease was granted prior to 1 March 1990.

# Residential tenancies of flats: Landlord and Tenant Act 1987

## *Tenants' right to purchase interest of landlord where landlord is in breach of repairing covenants*

Part III of the Act applies to qualifying residential tenants of flats under leases for terms exceeding 21 years. Its effect is that a majority of the qualifying tenants are entitled to compulsorily purchase the landlord's interest in premises to which Part III applies, if he fails to remedy the breach of his repairing or management covenants within a reasonable period after receiving notice from the tenants and is unlikely to do so, or a manager has been in place for three years (s.29). The landlord's interest is transferred to a nominated person on terms either agreed by the parties or determined by a rent assessment committee.

## Tenants' right to purchase interest of which landlord intends to dispose

Part I of the Act applies to qualifying residential tenants of flats. Its effect is that if the landlord decides to dispose of an interest in premises to which Part I applies, qualifying tenants are given the right to purchase the interest. Part I applies to disposals such as the sale or lease of the landlord's freehold reversion or, if he is a tenant, the assignment of his lease. It does not apply to disposals such as disposals by mortgage, or compulsory purchase, or family or charitable gifts, or where required by an option or under a will (s.4).

Where the landlord proposes to dispose of an interest in the premises, he must notify the qualifying tenants of his intention, offering the interest to them (s.5). Failure to serve notice without reasonable excuse is now a criminal offence (Housing Act 1996). Moreover, the purchasing landlord may be required to comply with a similar notice requirement (Landlord and Tenant Act 1985). The tenants must be given time to accept the offer (s.6) which requires a majority of the qualifying tenants. Moreover, the purchasing landlord may be required to comply with a similar notice requirement (Landlord and Tenant Act 1985).

If the tenants accept offer, or the landlord accepts their counter-offer, they must be given another two months to nominate a person to take the landlord's interest. If they fail to nominate such a person within the two-month period, the landlord has 12 months to dispose of the premises for consideration not less than that specified in the notice.

If they do not accept the offer within the specified period, the landlord has 12 months to dispose of the premises, though the consideration must not be less than that specified in the notice.

## Tenants right to apply to the court for variation of the terms of the lease

Part IV of the Act permits a residential tenant (or residential tenants) of a flat (or flats) under a lease (or leases) for a term exceeding 21 years to apply to the court for variation of the terms of the lease or leases. The court may order variation if grounds are established (s.35). Basically, variation may be ordered where the terms of the lease are inadequate with regard to matters such as repair and maintenance of the building, of the flat or of installations, provision of insurance or computation of service charges.

# Rights of tenants of flats with long leases to purchase the freehold or an extended lease (Leasehold Reform, Housing and Urban Development Act 1993)

Qualifying tenants of flats under leases the terms of which exceed 21 years may be entitled to purchase the freehold of the premises in which the flats are contained if a number of conditions are satisfied and the tenants follow the detailed procedures laid down by the 1993 Act as amended by the Commonhold and Leasehold Reform Act 2002. For example: the premises must be structurally detached; they must contain at least two flats occupied by qualifying tenants; and the qualifying tenants amount to at least one-half of the total number of flats. Moreover, the right to purchase does not apply if more than 25% of the internal floor area is occupied for non-residential purposes.

The procedural steps which the tenants must follow include the creation of an 'RTE company', which is a private company limited by guarantee, and its memorandum of association must state that its object is to exercise the right of collective enfranchisement and the service of notice by the RTE company upon the landlord. The landlord may, by serving a counter-notice, dispute the tenants' right to purchase. In particular, in certain circumstances he may be entitled to defeat the tenants' attempt to purchase by asserting his intention to redevelop. The end result of the process, if successful, is that the freehold is transferred to the RTE company, the landlord receiving an open market price (determined by reference to statutory provisions).

Alternatively, a qualifying tenant of a flat who has a lease the term of which exceeds 21 years may be entitled to an extended lease. Again, certain conditions must be satisfied and procedures followed (e.g. notice must be served on the landlord). Again, in certain circumstances, the landlord may be entitled to defeat the tenant's attempt to purchase by asserting his intention to redevelop. The end result of the process, if successful, is that the tenant obtains a fixed-term tenancy for a term of 90 years, the new lease commencing when the existing lease expires. The tenant pays a premium calculated in accordance with statutory provisions.

## The right to manage: Commonhold and Leasehold Reform Act 2002

The Commonhold and Leasehold Reform Act 2002 gives tenants of flats the right to take over the management of the building in which their flat is situated without having to prove that there is any fault on the part of the landlord. No compensation is payable to the landlord on exercise of this right.

The right may be exercised if the premises consist of a self-contained building which contains two or more flats held by qualifying tenants and the total number of flats held by qualifying tenants is not less than two-thirds of the total number of flats (s.72). A qualifying tenant is a tenant under a lease the term of which exceeds 21 years.

The procedural steps which the tenants must follow include the creation of an 'RTM company', which is a private company limited by guarantee, and its memorandum of association must state that its object is to exercise the right to manage the premises. There are also a number of requirements regarding notice (ss.78 and 79).

# Council and other public sector tenants: Housing Act 1985

The Housing Act 1985 applies to a tenancy of a house let as a separate dwelling (s.79), provided that the tenant occupies the dwelling-house as his or her only or principal home (s.81) and the landlord is a local authority or other specified public body (s.80). It also applies to a licence to occupy a dwelling-house (s.79), other than a temporary licence given to an occupier who entered the premises as a trespasser.

Tenancies to which the Act applies are known as *secure tenancies.* The Act does not apply to tenancies for terms exceeding 21 years, nor where the tenant occupies the premises for the purposes of employment where the local authority (or another specified public body) is the employer, nor where the house is used for temporary accommodation, because it is on land acquired for the purposes of development. The Act is also inapplicable where the house is let to the tenant as temporary accommodation whilst work is carried out on a home if the tenant is not a secure tenant of this home. There are also exclusions from the protection in respect of tenants of agricultural holdings and on-licensed premises, and in the case of certain student lettings, and in relation to the provision of housing for homeless persons and persons moving into the district in order to take up employment. The Act is also inapplicable where Part II of the Landlord and Tenant Act 1954 applies to the tenancy.

A tenancy ceases to be secure if the tenant gives up possession of the house or sublets the entire house (s.93) or (in the case of a fixed-term tenant) if he or she dies (s.90). Secure tenancies cannot normally be assigned (s.91).

## Succession

When a secure periodic tenant dies, a spouse or a member of the tenant's family who had resided with the tenant for the 12 months before death and had occupied the house as a principal home may succeed to the tenancy (s.89). Only one succession is allowed.

## Possession

A secure fixed-term tenancy may end by expiration of the term or by court order in certain circumstances (s.86). Once the fixed term ends (even if by forfeiture), a periodic tenancy will automatically arise (s.86) unless the tenant has been granted another tenancy of the dwelling-house which commences as soon as the original tenancy ends. The period of the periodic tenancy will be determined by reference to how the rent was paid under the fixed-term tenancy. The terms will be a modified form of those of the original tenancy.

A secure periodic tenancy may only be terminated by court order. The court can only make an order if grounds for possession are established by the landlord (s.84). The grounds relied upon must be contained in a notice which the landlord serves on the tenant prior to bringing proceedings. The date specified by the notice for commencement of possession proceedings cannot be earlier than the first possible date of common law termination. The grounds are contained in Schedule 2.

On **grounds 1–8**, the court will only make an order for possession if it is reasonable to do so, the court possessing the power to adjourn proceedings, postpone or suspend an order, or impose conditions in relation to rent. The grounds are as follows:

1   The tenant has failed to pay rent or is in breach of covenant.
2   The tenant or persons residing with or visiting him or her have caused or been guilty of conduct likely to cause a nuisance or annoyance to persons residing, visiting or engaged in a lawful activity in the locality or have been convicted of using or allowing the use of the house for illegal or immoral purposes or of an indictable offence committed in its locality.
2A  The tenant or a spouse or partner has left because of the violence of or threats of violence by his/her spouse or partner towards him/her or a member of his/her family and is unlikely to return.
3   The house has deteriorated in consequence of acts of waste by the tenant or persons residing with him or her. If these are the acts of a lodger or subtenant, the tenant must have failed to take reasonable steps to remove this person.
4   Furniture provided by the landlord has deteriorated in consequence of ill-treatment by the tenant or persons residing with him or her. If these are the acts of a lodger or subtenant, the tenant must have failed to take reasonable steps to remove this person.
5   The tenant was granted the tenancy in consequence of a false statement made by the tenant intentionally or recklessly.
6   The tenant acquired the tenancy by exchange and either paid or received a premium.
7   The building which contains the house is mainly used for non-residential purposes, the tenancy was granted because the tenant is employed by the landlord and the nature of the tenant's behaviour or that of a person who resides with him or her has been such that, taking into consideration the use of the rest of the building, it is not proper for the tenant to remain in occupation.

8   The tenant was a secure tenant of a different house and accepted the present house on the basis that he or she would leave when work on the original house was complete, that work now being complete.

On **grounds 9–11**, suitable alternative accommodation must be available at the time when the order is to take effect. The grounds are as follows:

9   The occupier is liable for an offence in respect of the overcrowding of the house.
10   The landlord intends to demolish, reconstruct or carry out work on the building and cannot reasonably do so without possession.
11   The landlord is a charity and the tenant's continued occupation of the house would conflict with its objectives.

On **grounds 12–16**, the court will only make an order for possession if it is reasonable to do so. The court has the power to adjourn proceedings, postpone or suspend the order, or impose conditions in relation to rent. Further, suitable alternative accommodation must be available at the time when the order is to take effect. The grounds are as follows:

12   The building which contains the house is mainly used for non-residential purposes, the tenancy was granted because the tenant is employed by the landlord, the employment has ended and the landlord needs the house for a new employee.
13   The house is adapted for the disabled, no disabled person is living there and the landlord needs it for a disabled person.
14   The landlord is a housing association or trust which only lets houses to people who are difficult to house, no such person is living in the house and the landlord needs it for such a person.
15   The house is part of a group for people with special needs, there being special facilities in the area for such persons, no such person is living in the house and the landlord needs it for such a person.
16   The house is more extensive than the tenant needs, the tenant succeeded to the tenancy, not being the deceased tenant's spouse, and notice was served on the tenant between six and 12 months after the former tenant's death.

## Right to buy

A person who has been a secure tenant (not necessarily of the same house or under the same landlord) for two years or more at any time may be entitled to purchase the house or take a long lease (125 years) of a flat (s.119). A tenant cannot claim this right if an order for possession of the house or flat has been made (s.121). Certain dwellings (such as dwellings particularly suitable for the elderly or disabled) are excluded (Schedule 5). The price is calculated taking account of factors specified by the Act (s.127), subject to a discount based upon the length of the period during which the tenant has been a secure tenant. The tenant also has the right to a mortgage from the landlord (s.132).

## The introductory tenancy (Housing Act 1996)

Instead of granting secure tenancies to new tenants, local authorities may now elect to grant introductory tenancies. Essentially, an introductory tenancy is a periodic tenancy which will become a secure tenancy after a trial period of one year unless one of a number of specified events occurs within the year (e.g. if the local authority commences possession proceedings within the year and obtains a court order for possession, whether or not possession is obtained within the year). In order to obtain possession, it is not necessary to prove statutory grounds for possession but the landlord must notify the tenant of the landlord's reasons for seeking possession. The tenant may require the landlord to review the reasons for its decision.

# 14
## Planning law

## Central and local planning responsibilities

If people are to enjoy and maintain a reasonable quality of life in the place where they live, then issues relating to land use planning should obviously concern them. The very word 'planning' suggests change, and the policies developed should aim to maintain and improve the environment as an attractive place in which to live and work.

Any coherent system of planning control is necessarily dependent on a proper administrative structure. The local government structure was established by the 1890s and the first Planning Act followed in 1909. The emphasis in that early Act was directed towards planning being used as a tool to secure better sanitary conditions but it introduced for the first time the important issues of 'amenity and convenience'. Early planning Acts were based on the idea of local councils preparing planning schemes to control land use in suburban areas. General control of land use was introduced in 1947, creating the modern system of planning law. The main principles are now contained in the Town and Country Planning Act 1990 (the 1990 Act), the Planning and Compulsory Purchase Act 2004 (the 2004 Act) and the Planning Act 2008 (the 2008 Act).

The overall structure for administering the rules on planning is headed by the Secretary of State for Communities and Local Government, who has power over local planning authorities – county councils, district councils, metropolitan authorities and other bodies vested with planning responsibilities, such as National Park Authorities.

## Development plans

The system of planning in the UK is often referred to as a 'plan-led' system. This is because the role of the development plan is central to the determination of planning applications. Under s.70 of the 1990 Act, local planning authorities must have regard to the provisions of the development plan when determining planning applications. Under s.38(6) of the 2004 Act, their determination must be made in accordance with the plan unless there are other material considerations which indicate otherwise.

Galbraith's Building and Land Management Law for Students. DOI: 10.1016/B978-0-08-096692-2.00014-2
Copyright © 2010 by Elsevier Ltd

For each region of the country (excluding Scotland and Wales) there is a regional spatial strategy (RSS), setting out the Secretary of State's policies in relation to the development and use of land within the region. The purpose of the RSS is to provide a broad development strategy for the region over a 15- to 20-year period. The RSS will identify the regional and subregional priorities for issues such as housing, the environment, transport, infrastructure, economic development, agriculture, minerals extraction, and waste treatment and disposal. Since April 2009 the RSS must also include policies designed to secure that the development and use of land in the region contributes to the mitigation of and adaption to climate change.

Local planning authorities produce a local development framework (LDF), which must be in general conformity with the RSS for the region within which the local planning authority is situated. The LDF consists of a portfolio of local development documents, which are made up of development plan documents, supplementary planning documents and a statement of community involvement which, when taken together, set out the local planning authority's policies for meeting the local community's economic, environmental and social aims in relation to the development and use of land. As such, the framework will not only be restricted to matters that can be implemented through the planning system, but will include policies on issues that relate to the use and development of land such as regeneration, economic development, education, housing, health, waste, energy, recycling, protection of the environment, transport, and cultural and social issues.

The position is slightly different in Wales and London. In Wales, instead of RSSs there is a single Wales Spatial Plan, adopted in 2004, setting the context for local and community planning over a 20-year period. Similar in scope to an RSS, it sets out national priorities, providing the context for the application of national and regional policies for specific sectors, such as health, education, housing and the economy. Instead of LDFs, local planning authorities in Wales produce Local Development Plans. Unlike the 'portfolio' approach of LDFs, a Local Development Plan is a single document setting out area-wide polices for development, such as the allocation of land for various types of development (i.e. housing, employment and business), as well as specific policies for key areas.

In London, the Mayor has significant planning powers, and is responsible for strategic planning through the mechansim of a Spatial Development Strategy (known as the London Plan). Similar to an RSS, the Spatial Development Strategy addresses issues of strategic importance such as transport, economic development, housing, the built environment, major cultural and commmunity facilities, and the River Thames. The London borough councils must ensure that their LDFs are in compliance with the Spatial Development Strategy.

## Development

The key concept in planning control is development. It is defined by the 1990 Act, s.55 as: 'The carrying out of building, engineering, mining or other operations in, on, over or under the land, or the making of any material change in the use of any building or other land.' If an activity comes within this definition, then as a broad general rule, planning permission is required. The usual way to seek planning permission is to apply to the local planning authority. However, there are instances where no individual planning permission is necessary. This may be because:

1    The operation planned is not within the meaning of the word 'development'.
2    The development is permitted within the terms of the Town and Country Planning (General Permitted Development) Order, so that no specific application need be made.

The definition of the term 'development' has two aspects – 'operations' and 'material changes of use'. An operation results in changes to the physical form of the land which will have some degree of permanence. Use consists of the activities done in, or on, the land which do not change its physical form.

## Operations

'Building' is defined to include structure or erection, but does not include the plant and machinery within the building. 'Building operations' are defined to include 'demolition of buildings, rebuilding, structural alterations of, or additions to, buildings, and other operations normally undertaken by a person carrying on business as a builder'. This at first sight seems to be virtually all-embracing, but s.55 of the Act does say that development does *not* include: 'The carrying out of work for the maintenance, improvement or other alteration of any building, being works which affect only the interior of the building or which do not materially affect the external appearance of the building.'

For many years, it was not altogether clear where demolition work stood within the s.55 definition of development. The position was resolved in 1991 by the insertion of 'demolition of buildings' to the statutory definition of development. In practice, however, following the publication of a direction by the Secretary of State in 1995, the demolition of certain types of building (e.g. a building of less than 50 cubic metres or a building which is not a dwelling-house and does not adjoin a dwelling-house) is not to be taken to involve development. Moreover, the General Permitted Development Order grants permission for the demolition of most dwelling-houses. It should be noted, however, that planning permission is required for partial demolition of a building and may be required in circumstances in which demolition amounts to an engineering operation.

Engineering operations are not specifically defined in the 1990 Act except to the extent that they include the formation or laying out of means of access to a highway, whether for vehicles or pedestrians. The definition, however, specifically excludes works by a highway authority on land within the boundaries of a road for the maintenance or improvement of the road. Similarly, works by local authorities or statutory undertakers for the purpose of inspecting, repairing or renewing sewers, pipes, cables and other apparatus are excluded. Mining operations include matters such as the removal of materials from mineral-working deposits and the extraction of minerals from disused railway embankments. Other operations include matters such as installing metal grilles over the windows and doors of a shop.

## Change of use

Only those changes of use which are 'material' require planning permission. There is no definition of 'material change of use' in the 1990 Act and inevitably what is material is a question of fact and degree. The 1990 Act does aim to give some assistance by laying down certain specific instances of material change of use. These are:

1    Where a single dwelling-house is converted into two or more separate dwellings.
2    Where there is to be a display of advertisements on any external part of a building not previously used for that purpose.
3    Where refuse or waste materials are to be deposited, including extending an existing tip or raising its height above the height of the adjoining land.

Conversely, certain changes of use are specified not to involve development. These are:

1    Using buildings or land within the curtilage of a dwelling-house for a purpose incidental to the enjoyment of the house.
2    Using land for purposes of agriculture or forestry.
3    Changing between uses within the same 'use class'.

The Town and Country Planning (Use Classes) Order 1987 specifies 14 different use classes, divided into four parts, dealing with:

● shops, financial services, restaurants, cafes, drinking establishments and hot food takeaways;
● business, industrial and storage, and distribution;
● hotels and residential uses;
● non-residential and leisure uses.

For example, use as a travel agency or for hairdressing both fall within the same class and, consequently, planning permission would not be required to change the use of premises from a travel agency to a barber's shop.

Whilst changing between uses in the same use class does not amount to development, changing between uses within different use classes (i.e. from use as a travel agency to use as a bank) will constitute a change of use that requires an application for planning permission. Further, it should be noted that some uses are specified not to fall within the Use Classes Order (e.g. use as a theatre, taxi business, scrapyard or for the sale of fuel for motor vehicles).

If the use for which permission was originally granted is not listed in the Use Classes Order, then its rules cannot assist and a normal planning application will be necessary. Again, planning permission for the original use may have been subject to conditions which override the provisions of the Use Classes Order. It should also be noted that if the change of use requires new buildings, or alterations or extension to existing buildings, an individual planning permission is still likely to be required because the building work will presumably amount to a 'building operation'.

# Permission under the General Permitted Development Order 1995

Instead of seeking permission for development by applying to the local planning authority, there are cases, covered by the General Permitted Development Order (GPDO) made by the Secretary of State, where permission is automatically given for all classes of development set out in the GPDO. In addition, since 2006 local planning authorities can promote 'local development orders' for their area, giving additional permitted development rights over and above those set out in the GPDO for certain types of development in certain specified areas. Some commonly encountered examples of development which are permitted by the GPDO include:

1    Development within the curtilage of a dwelling-house. Where it is proposed to extend a dwelling-house, then provided that the extension (and any previous extensions, sheds or outbuildings) does not exceed 50% of the curtilage, excluding the footprint of the original house, such development is permitted. There are other conditions to be satisfied, e.g. that the height of the building when extended must not exceed the height of the original dwelling-house, the extension must not front on to a highway and must not extend more than a certain distance beyond the rear of the original house.

2    Construction of porches, garages and the installation of satellite antennae, subject to size, location and height limitations. It should be noted that the paving over of a front garden to provide off-street parking is now not covered by the GPDO unless the surface is made of porous materials or provision is made to direct run-off water from the hard surface to a permeable or porous area.

3   The installation of certain types of domestic microgeneration equipment, such as solar panels and ground source heat pumps. At the moment this does not include wind turbines, due to issues of visual impact, noise and possible electrical interference.

4   Minor operations, e.g. erecting, constructing, maintaining, improving or altering gates, fences or walls.

5   Temporary buildings. Where building operations for which planning permission has been granted are in progress, this class allows temporary buildings needed in connection with the building work, such as site huts or workmen's lavatories, to be erected. But they must be removed at the end of the building operations.

6   Temporary uses. Land can be used for most temporary purposes for up to 28 days each year. This rule allows land to be used for an occasional agricultural show (or boot sale), or point-to-point, or as a site for a summer fête. The permission also covers erection of movable structures (e.g. marquees) for these purposes. If the purpose is to hold a market, or motor sports, temporary use is limited to 14 days each year.

7   Building or engineering operations required for agricultural purposes may be carried out on agricultural land, e.g. erection of barns and steadings. This permission is restricted in terms of size and height limitations, and distances from highways.

8   Additions to industrial buildings and warehouses, limited in size and height.

It should be noted that permitted development under certain classes of the GPDO can only operate once. For example, whilst extending a house by up to 50% of the curtlilge may be authorized by the GPDO, this does not authorize a further extension of the same house by another 50% of the remaining curtilage at a later date.

Even if work looks as though it is covered by the GPDO, the local planning authority may have made an 'Article 4 Direction'. Such directions are made in the interests of good planning in a particular area. The local planning authority will impose restrictions on development in that area, so that an application for permission must be made, even though it appears that permission already exists under the GPDO. There is likely to be an 'Article 4 Direction' if the place is a conservation area, or if a particular street has a special character or appearance. Landowners in such areas should have been served with a copy of the direction. A condition imposed upon the grant of planning permission may also restrict development which would otherwise be permitted under the GPDO.

## Simplified planning zones

A further situation in which planning consent may be automatically granted is where a 'simplified planning zone' is established by a local planning authority. Within such zones, designated by a local planning authority only after extensive consultative processes, certain developments or classes of development are permitted without the need for formal application. Examples of appropriate schemes are to grant permission for the development of large-scale residential estates or the redevelopment of large derelict industrial sites.

# Restrictions upon development imposed by covenants relating to freehold land or in leases

It should be noted that planning permission can do nothing to help in those cases where the type of development is prevented by a restrictive covenant, or where a tenant is precluded from carrying out the development in question by the express terms of the lease.

# Making a planning application

One immediate advantage of development which is permitted under the GPDO is that it costs the developer nothing to get permission. Since 1980, fees have been payable in respect of planning applications, and the level of these can be varied from time to time. The application will not be considered until the fee is paid. The government has increased the fees so that they now cover all of a local authority's development control costs. If there is any doubt as to whether planning permission is required for a development an application can be made to the local planning authority for a certificate of lawfulness of proposed use or development.

Applications for planning permission are made to the local planning authority, and may be made by anyone, whether or not he or she owns the land or any interest in it. The actual owner does not need to give consent, but the applicant has to give notice of the application to owners or tenants of the relevant land (but not to tenants whose leases have less than seven years to run). The necessary forms (which are standard across the country) are supplied by the local planning authority, and applications can now be made online. It may be sensible to appoint a surveyor, architect, planning consultant or solicitor to make the application. A full application will require plans so that the site can be adequately identified and the development proposals fully understood. Engaging professional assistance may be costly, and at an early stage a developer may simply want to discover the attitude of the planning authority in principle to what is proposed. In such a case, an outline planning application is appropriate. A developer may still have to acquire the land and commission detailed plans, but can then go ahead with more confidence in the business venture, armed with the outline permission. This is not usually the best way for private individuals to proceed, as the outline permission, if granted, must be followed by a second application for 'approval of reserved matters'. Once an application has been submitted, a copy will be placed in the planning register, which is a public document available for inspection by anyone.

Since April 2010 applications for what are known as 'nationally significant infrastructure projects' must be made to the Infrastructure Planning Commission, rather than the local planning authority. As the name implies, these applications are for large-scale infrastructure projects such as railways, wind farms, power stations, airports and sewage treatment works.

The General Development Procedure Order requires that all planning applications must be publicized by the local planning authority, usually by a site notice or by serving notice on the owners or occupiers of adjoining land. In some instances, a newspaper advertisement and a site notice may be required, for example where an 'environmental statement' accompanies an application, or where development involves a departure from the development plans or affects a public right of way. Where there is a major development, such as mineral extraction, a newspaper advertisement and either a site notice or the serving of written notice upon all adjoining owners and occupiers is required. The planning authority may also be required to notify specific bodies of specific types of development (e.g. the Secretary of State or a parish council). There are also specific provisions relating to the publication of certain types of development (e.g. development affecting the character or appearance of a conservation area or affecting the setting of a listed building).

The planning authority may also be required to consult certain bodies in certain circumstances (e.g. the local highway authority, where development involves the creation or alteration of access to the highway). Where a development would involve a radical departure from the local plan, the planning authority must send a copy of the application and various other documentation to the Secretary of State. The planning authority must then wait for 21 days before granting planning permission, though the planning authority will have to refer the application direct to the Secretary of State if he or she uses the power to 'call in' the application. This is likely to happen only in the case of very large-scale controversial development proposals.

Regulations in 1988 introduced the concept of environmental impact assessment into the planning process, to implement a European Community Directive. Those regulations were replaced by new regulations in 1999, which were further amended in 2008. Circular 02/99 states that: 'Environmental Impact Assessment is a means of drawing together, in a systematic way, an assessment of a project's likely significant environmental effects. This helps to ensure that the importance of the predicted effects, and the scope for reducing them, are properly understood by the public and the relevant competent authority before it makes its decision.'

An environmental impact assessment is required in the case of developments such as oil refineries, chemical installations, nuclear power stations, railway lines for long-distance traffic and aerodromes with runways exceeding a specified length. An assessment may be required in relation to developments for industries such as agriculture, mining, civil engineering, rubber, paper and textiles, if the development is likely to have significant environmental effects because of its nature, size or location. The local planning authority can be asked to provide a 'screening opinion' if there is any doubt as to the need for an environmental impact assessment.

The planning authority must now make a decision on the application before it. It will have regard to the development plan and any other material considerations. The decision must be made in accordance with the development plan unless material considerations indicate otherwise. Examples of material considerations are provided by planning circulars, decisions from previous planning appeals, the demands of the proposed development upon roads and other services, and its impact upon adjacent land. The planning authority will also take account of the views of organizations it has consulted and objections it has received.

A decision by the planning authority should be taken within eight weeks for a minor development, 13 weeks for a major development (more than 10 houses or 1000 square metres of commercial floor space) or within 16 weeks if an environmental impact assessment is required. That decision may be:

- Unconditional permission or approval.
- Conditional permission.
- Refusal of permission.
- Refusal to take a decision.

In the vast majority of cases permission is conditional. The conditions imposed can be whatever the planning authority 'thinks fit', so long as they fairly and reasonably relate to the permitted development, are imposed for a planning purpose and are not so unreasonable that no reasonable planning authority would have imposed them.

For instance, permission may be given for building operations to take place only during a limited period, or subject to work being commenced within a particular period. Equally, conditions may relate to the materials to be used, or the location of a building on the site.

A planning authority may refuse to take a decision where an application is substantially the same as one which the Secretary of State has refused within the last two years or as one in relation to which the Secretary of State has dismissed an appeal within the last two years. The planning authority may only do so if, in the opinion of the authority, there has been no significant change in the development plan (in a sense material to the application) or in any other material consideration. Once granted a planning permission usually has a lifespan of three years within which time the development authorized by the permission must be commenced. A grant of outline planning permisison requires a further application for the approval of 'reserved matters' to be made within three years of the grant of the outline permission and work to commence within two years of the approval of the reserved matters.

# Planning obligations and the Community Infrastructure Levy

Sections 106, 106A and 106B of the Town and Country Planning Act 1990 enable developers to enter into planning obligations, to apply to the planning authority to have planning obligations modified or discharged, and to appeal to the Secretary of State if the planning authority refuses to modify or discharge a planning obligation. Essentially, a planning obligation arises where a person interested in land enters into an obligation which:

- restricts the use of the land; or
- requires specified operations or activities to be carried out in, on, under or over it; or
- requires it to be used in a specific way; or
- requires specified sums of money to be paid to the planning authority.

A planning obligation may be enforced by the planning authority whether the planning obligation takes the form of an agreement between the developer and the local authority or whether it amounts to an undertaking by the developer which the Secretary of State has approved upon appeal following refusal of planning permission.

It may be that entering into a planning obligation will, by removing objections to a proposed development, persuade the planning authority, or the Secretary of State on appeal, to grant planning permission. Essentially, a planning authority should only seek planning obligations which are both necessary to the granting of permission and relevant to the specific development. For example, it might be appropriate for the developer of a new football stadium to enter into an obligation to provide a specific number of car parking places on land close to the prospective stadium or to pay specified periodic sums of money to the planning authority in respect of the provision by them of an adequate car park.

In addition to planning obligations the Planning Act 2008 introduced the Community Infrastructure Levy (CIL). A local planning authority can set a rate of CIL through the development plan process which will have to be paid on all developments that are granted planning permission, save for small-scale 'householder' development. The aim of the CIL is to cover the cost of new infrastructure required as a result of the development. This includes roads and other transport facilities, flood defences, schools and medical facilities, sporting and recreational facilities, and open spaces.

# Appeals

If planning permission is refused or the applicant is aggrieved by any conditions imposed or the planning authority has failed to make a decision within, as appropriate, eight, 13 or 16 weeks, the applicant may appeal to the Secretary of State. In most cases, the decision is actually made on behalf of the Secretary of State by an inspector appointed by the Secretary of State. An appeal must be lodged within six months of the date of the planning authority's decision. There are three methods for dealing with appeals – written representations, hearings or public inquiries. The Secretary of State will determine which method will be used, although he or she will take account of the parties' preferences.

1   *Written representations.* This procedure avoids the expense of an inquiry. An inspector considers written representations made by both parties and then makes a site visit. This written representations procedure is used in the majority of cases. There is an expedited version of the written representations procedure for householder appeals through the Householder Appeals Service. The deadline for lodging an appeal under this process is 12 weeks from the date of the local planning authority's decision. No further representations are allowed beyond the documentation lodged with the initial application and such appeals should be determined within eight weeks.
2   *Hearings.* These take the form of informal discussions between the inspector and the parties.
3   *Public inquiry.* Inquiries tend to be formal and legalistic. The procedure is governed by detailed procedural rules. As with all planning appeals, the rules of natural justice must be observed so that both parties are given a fair hearing. The proceedings will be held in public, and they give an opportunity to persons closely affected by the development to put their objections and cross-examine the other party's witnesses.

Following an appeal to the Secretary of State, it is possible for a person aggrieved with the decision to apply to the High Court for review of the decision within six weeks of its making. Such an application is made under s.288 of the 1990 Act. The statutory grounds of review are that an order or action was not empowered by the Act or that requirements of the Act were not complied with.

If an applicant for planning permission does not appeal to the Secretary of State (for example, because he or she has been granted unconditional permission), a person aggrieved may be able to challenge the planning authority's decision by applying to the High Court for judicial review. Such applications must be made promptly (which essentially means within three months or preferably sooner).

## Planning enforcement

The planning rules need to provide for those cases where development is carried out without valid permission, or where a developer breaches the conditions imposed on the permission. Breaches of the planning rules can come to light in a number of ways. Some are reported, often anonymously, by neighbours or persons affected by the development. Others become obvious when a subsequent permission is applied for. The planning officers are regularly out and about making site inspections and they tend to know what is going on and what has been authorized within their 'patch'. They liaise with the building inspectors, who have the opportunity to observe 'unauthorized' development in the course of their work. Breach of the rules is not a criminal offence as such. The planning authority uses an enforcement notice where some action is necessary. Before serving an enforcement notice, the planning authority may serve a planning contravention notice, requiring the developer to provide information about operations on the land, the use of the land, etc. It is also possible for a landowner to establish whether any existing use of a building or land or whether any operation carried out on the land is lawful by applying to the local planning authority for a certificate of lawfulness of existing use or development.

It must be emphasized that, where possible, the planning authority will seek to resolve matters without the need to resort to enforcement procedures. For example, the development may be of a type which would clearly have been permitted, if permission had been sought at the appropriate time. In such a case, the developer is encouraged to make a late application, which will be heard in the ordinary way by the planning authority. Only as a last resort, when persuasion has failed, will an enforcement notice be issued. The time limits for issuing an enforcement notice are essentially within four years of the completion of unauthorized operations or of unauthorized change of use to a single dwelling-house or within 10 years of any other change of use or breach of condition. The notice must specify the alleged breach of the planning laws, and the steps which must now be taken to remedy the breach, together with the time limit for taking the necessary steps. The measures to be taken can vary considerably. For example, where a building has been erected without permission, at one extreme the step may consist of ordering it to be demolished, or at the other, may simply require some tree planting or landscaping to disguise the building.

A developer may appeal to the Secretary of State against an enforcement notice on a number of grounds, e.g. the matters specified in the notice do not constitute a breach of planning laws, that planning permission ought to be granted, or the time specified to take the necessary steps is unreasonably short. The appeal has the effect of suspending the enforcement notice until the appeal has been resolved. This means that the operation or change of use can go on as before until the appeal is resolved. That could be disastrous where, by allowing the operation to continue, it would be impossible to subsequently correct matters. In such circumstances it may be appropriate to issue a stop notice. Failure to comply with a stop notice is a criminal offence. If the enforcement notice is quashed on appeal on certain grounds, the local planning authority must compensate an owner or occupier of land for losses occasioned by the stop notice. This can include a contractor who has been held up on a site. In addition to the stop notice, the Planning and Compensation Act 1991 introduced the right for a local planning authority to apply for an injunction to restrain any actual or apprehended breach of planning control.

If an enforcement notice is upheld on appeal, or if no appeal is made against it, then it must be complied with within the prescribed time limits. Non-compliance with an enforcement notice is a criminal offence, punishable by a fine. Moreover, in cases of non-compliance, the local planning authority may enter the land and carry out the necessary steps at the expense of the owner.

Finally, it should be noted that, in respect of the breach of a planning condition, instead of serving an enforcement notice the planning authority may serve a breach of condition notice. If the developer fails to comply with the condition within the specified time he is guilty of a criminal offence. There is no appeal to the Secretary of State.

# Planning and the preservation of amenity

A vital part of the planning process is the preservation and enhancement of amenity. That term is not defined by the planning legislation, but the dictionary defines it as 'pleasantness, as in situation and characteristics'. The ordinary planning rules can do much to preserve amenity, by refusing permission in appropriate cases, or attaching conditions to preserve or improve amenity in others. Additionally, certain specific forms of control are particularly geared to protect amenity.

### Trees

Local planning authorities may make tree preservation orders in the interests of amenity. These may be designed to protect individual trees or areas of woodland. In such cases, once an order is made, permission must be sought to fell or lop a tree. If such permission is granted, the local planning authority will often require replacement planting. Ordinary planning permission may be granted subject to the preservation of trees on the site, or subject to the planting of additional trees.

Interference with protected trees is a criminal offence, punishable by a fine. Before felling or lopping a tree in a conservation area (see below for details on conservation areas), the local planning authority must be notified, even if the tree is not protected by a tree preservation order. The local planning authority can consent to the works, make a tree preservation order to protect the tree or do nothing. Under the lost option the works to the tree can be carried out six weeks after the date of the notice to the local planning authority.

## Hedgerows

Subject to exceptions, it is necessary to serve a hedgerow removal notice on the planning authority before removing a hedgerow to which the Hedgerows Regulations 1997 apply (e.g. a hedgerow in or adjacent to agricultural land with a continuous length of 20 metres or more unless within the curtilage, or forming part of the curtilage of a dwelling-house). Within 42 days the planning authority may serve a hedgerow retention notice if the hedgerow is an important hedgerow (as defined by the regulations), thus preventing its removal. An appeal against such a notice may be made to the Secretary of State. Removal of a hedgerow in breach of such a notice is a criminal offence.

## Buildings of special interest and conservation areas (Planning (Listed Buildings and Conservation Areas) Act 1990)

Buildings of special interest are usually referred to as 'listed buildings'. They become listed by the Secretary of State for Culture, Media and Sport after consultation with experts such as English Heritage if they are of special architectural or historic interest. There are three grades of listing to indicate a building's relative importance: Grade I, Grade II* and Grade II. Buildings that are listed are often proposed to the Secretary of State by the local planning authority. The owner of the building has no formal right to object to its listing. He or she may know nothing of a decision to list the building until notified by the local planning authority. The listing will also be registered in the local land charges register. The local planning authority may protect a building which they consider is of special architectural or historic interest and is in danger of being demolished or substantially altered by the use of a 'building preservation notice'. This gives the authority six months to initiate the procedures for listing with the Secretary of State.

If a building is listed, any work to alter or demolish it must have 'listed building consent'. Demolishing or altering a listed building without consent may amount to a criminal offence. Further, the planning authority may serve a listed building enforcement notice, requiring the restoration of the building or the alleviation of the effects of alteration. If such consent is refused or is conditional, the owner can then appeal to the Secretary of State and argue *inter alia* that the building should never have been listed. This argument is significant because the owner had no right initially to object to the listing. Many owners do not welcome listing of their property. Not only is the planning control much more rigorous, but it may also cause considerable loss if it was hoped to sell the building to realize its development value. This was the problem facing the parties in the case of *Amalgamated Investment Co. Ltd v. John Walker and Sons Ltd*, where a property with development potential was agreed to be sold for £1.7 million. The building on the land was listed before the sale went through, and the value of the land then slumped to £250,000!

Where listed building consent is refused or is conditional, an owner may serve a 'listed building purchase notice' on the planning authority, requiring them to purchase the listed building if he or she can show that it has become incapable of reasonably beneficial use. Planning authorities also possess limited powers to repair listed buildings at the expense of the owner and, if authorized by the Secretary of State, to acquire such buildings by compulsory purchase.

The concept of conservation areas stems from the Civic Amenities Act 1967. Prior to 1967 the emphasis was on the preservation of individual buildings, not areas. Local planning authorities are now required to determine those areas which are of special historical or architectural interest, where it is desirable to preserve or enhance their appearance. Effectively these are areas which are rich, or potentially rich, in listed buildings, or of scenic beauty. Where an area is to be designated as a conservation area, the local planning authority must give notice in the local press, and enter appropriate notices in the land charges register.

When the local planning authority is considering designating a conservation area it will take account not only of individual buildings but groupings of buildings, their relationship to each other, and the quality and character of the space between buildings. The group of houses and shops in a marketplace of a small town may not be particularly worthy of note individually, but the overall impact of the area may be sufficiently pleasing to warrant creating a conservation area. New development within a conservation area is not forbidden but, in determining whether to grant planning permission, the planning authority must pay special attention to the desirability of preserving or enhancing the character or appearance of the area. Further, demolition in a conservation area may require conservation area consent.

## Advertisements

The Secretary of State can, by regulations, control the display of advertisements so far as it appears to be expedient in the interests of amenity or public safety. This is generally designed to control the display of outdoor advertising, which includes any model, sign, placard, board, notice, device or representation, whether illuminated or not. Various classes of advertisements are excluded from the regulations, however (e.g. traffic signs). The power of the planning authority relates to issues of amenity (e.g. historic and architectural considerations) and public safety (e.g. road safety); it could not therefore refuse permission on grounds of social desirability.

Under Regulation 1(3) of the Control of Advertisements Regulations 2007, certain advertisements are exempted from control. These include advertisements on vehicles and advertisements incorporated into the fabric of a building or displayed inside a building.

Under Regulation 6, deemed consent is granted for such advertisements as the functional advertisements of local authorities and statutory undertakers, nameplates for businesses or companies, and temporary advertisements, including signs relating to the sale or letting of premises. If an advertisement is not exempt from control under Regulation 3 or there is no deemed consent under Regulation 6, planning permission is required from the planning authority, though there is a right of appeal to the Secretary of State. Displaying an advertisement in breach of the Regulations is a criminal offence. Moreover, the local planning authority may require the removal of such an advertisement or may remove such an advertisement.

## Financial problems – compensation and betterment

When a planning decision has been made (e.g. to create a new motorway, or build a new airport runway) there will be owners of property adjacent to the development whose property needs to be acquired by compulsory purchase for the purposes of the development. For some owners the value of their property will be significantly reduced by the development, while for others the value of the property will be increased by the planning decision.

Local authorities and development corporations have power to acquire land compulsorily for planning purposes. In such a case the appropriate authority or corporation will have initiated the necessary legal steps to acquire the land. There are two situations, however, where a landowner can initiate proceedings to force the local authority to purchase the land:

1 *Purchase notices* – where planning permission has been refused or made subject to onerous conditions, an owner may argue that the land is incapable of reasonably beneficial use. He or she may then serve a purchase notice on the local authority, requiring it to buy the property.

2 *Blight notices* – where an owner is unable to sell the land at a fair market price because the prospect that it may be compulsorily purchased (as revealed by a development plan) has rendered it virtually unsaleable, he or she may serve a blight notice on the appropriate public authority requiring it to buy the land.

Further, where a revocation or modification of planning permission results in wasted expenditure, loss or damage, compensation may be payable.

# 15
# Highways

## Introduction

Even a short journey by road will serve to convince the traveller that the law relating to highways must be complex. Workmen are to be seen digging up portions of the road; part of the road is blocked off for repair; notices indicate the building of a new road; trees adjacent to the road have become a hazard with their overhanging branches; the road has a bad surface which needs repair; builders have deposited a skip and piles of sand and stones on the road by a building site; demonstrators have blocked the road with a march; animals have strayed on to the road; snow is making the road dangerous. Many of these incidents can cause annoyance and, more seriously, delay, injury or loss. Who will be responsible? Against whom can a complaint be made? When can damages be claimed? The answer to these questions can be learned from an analysis of highway law.

The first question to be posed is what is a highway? Simply, the answer is any portion of land over which the public has the right to pass. When it is remembered that many highways have existed for over 1000 years, and that the law on this subject has been developing for almost as long, this may explain some of the complexity. Most of the relevant statutory law on the point is now contained in the Highways Act 1980 (the 1980 Act). The main aim of the Act was to bring together many rules previously contained in Highways Acts passed between 1959 and 1971, and in the Private Street Works Act 1961, but account was also taken of recommendations made by the Law Commission. The enormity of the subject can be grasped when it is seen that the 1980 Act runs to 432 pages.

Although the mental picture which most people have of a highway is a busy road full of motor traffic, the term is far wider than that. Highways can be subdivided, and further definitions of some other common expressions will be useful:

1   *Footpath* – a highway over which the public have a right of way on foot only.
2   *Footway* – a part of a highway which is also a carriageway (see below), where the public have a right of way on foot only. A footway is what is commonly called the pavement at the side of a road.

Galbraith's Building and Land Management Law for Students. DOI: 10.1016/B978-0-08-096692-2.00015-4
Copyright © 2010 by Elsevier Ltd

3   *Carriageway* – a highway over which the public have a right of way for the passage of vehicles, i.e. what is commonly called a road.

4   *Public bridleway* – a highway over which the public have a right of way on foot or on horseback.

5   *Restricted byway* – a highway over which the public have a right of way on foot, horseback or in non-mechanically propelled vehicles (i.e. horse-drawn carriages and cycles).

6   *Byways open to all traffic (BOAT)* – a highway over which the public have a right of way for vehicular and all other types of traffic, but which is used mainly for the purpose for which footpaths and bridleways are used. BOATS are usually used by 4 × 4 vehicles and off-road motorbikes, and are often referred to as 'green lanes'.

7   *Cycle track* – a way constituting or comprised in a highway over which the public have a right of way on pedal cycles (not being motor vehicles) with or without a right of way on foot.

8   *Walkways* – a footpath created in pursuance of an agreement between a local highway authority or district council and a building owner to provide ways over, through or under the building 'for the dedication … of those ways as footpaths subject to such limitations and conditions, if any … as may be specified in the agreement'. (This right of way may apply to shopping centres. The reference to 'limitations and conditions' allows for the centres to be closed at night without infringing any right of way.)

9   *Toll roads* – toll roads have not existed in this country since the abolition of turnpikes in the late nineteenth century. The first modern toll road, the Birmingham Northern Relief Road (M6 Toll Road), opened in December 2003.

Certain obvious questions arise in connection with highways:

- How does a highway come into existence?
- Who owns the highway?
- Who is responsible for the upkeep of the highway?

# Acquisition and ownership of highways

There are two ways in which a highway can be created: by dedication or by statute.

## Dedication

A highway can come into existence if the owner of the land concerned dedicates the right to cross it to the general public, which right is then accepted by the public actually using it. Dedication sounds like a very formal act, but in fact it can occur impliedly. Several points about dedication should be noted:

1   The owner of the land must intend to dedicate it as a highway. This intention is shown in the case where a formal dedication is made, e.g. if a formal document is used. Often, the intention will be implied from the circumstances, e.g. people are passing and repassing over the land, and the owner takes none of the steps available to stop the creation of a highway (see below).

2   The highway must be dedicated to the public generally.

3   The person dedicating the land as a highway must be granting a right for all time. Effectively, therefore, only an outright owner of land can dedicate.

4   The right to pass and repass must be accepted by the public actually using the highway.

Some assistance in proving the existence of a highway created by dedication is given by statute. Section 31 of the 1980 Act provides that where a way has been enjoyed by the public as a right, without interruption for 20 years (this does not apply to use by mechanically propelled vehicles), then it is deemed to be dedicated as a highway unless the owner can prove that the intention was not to dedicate the way. The owner could best prove lack of intention to dedicate by showing that notices had been displayed to that effect, e.g. 'No public right of way' or 'This land is not dedicated as a public highway'. Alternatively the owner can lodge a map of the land with the local authority showing the ways (if any) which he or she admits as being dedicated as public highways and can thereafter make a statutory declaration at any time within 10 years of lodging the map (or a previous declaration) that no other ways over the land have been dedicated.

The rule created by s.31 is important because it places responsibility on the owner to show that there is no highway, rather than on the public to show that there is. Even if a 20-year period has not elapsed, it would still be open to a member of the public to try to prove the existence of a highway. It would then be necessary to prove that the owner intended to dedicate it. This might be possible, for example, by showing that the owner had allowed repairs to be undertaken by the highway authority. (Merely because a highway is acknowledged to exist by the landowner it does not necessarily follow that it is maintainable at public expense – see below.)

The most common method of dedicating a new highway is under s.38 of the 1980 Act. A developer, usually of a new housing estate or industrial/retail park, will enter into an 's.38 Agreement' with the local highway authority to build the roads in the development to the standard specified by the authority. After a short period, usually one year, the highway authority agrees to adopt the roads as highway maintainable at the public expense (see below for details).

## Creation of highways by Act of Parliament

The main rules are now consolidated in the 1980 Act, s.24. Under its provisions highway authorities can build highways including particular types of road (e.g. motorways where access and passage are restricted to certain classes of vehicle).

## Who owns the highway?

Although the highway authority, when building new roads, may have acquired the land on which the highway is built, it is quite usual for the land under a highway to remain in the ownership of a private person. Often the landowners on either side of a highway own the land to the centre line of the road. Naturally, the use to which they can put such land is very limited. At common law, owners did have rights, e.g. to tunnel under the land, but many of the common law rights have been eroded by controls in Acts of Parliament. Ownership of a highway may still be important, however, if it is one which is not maintainable at public expense (see below).

# Upkeep of the highway

The position with regard to repair and maintenance of highways is more easily understood if a brief mention is made of the historical development of the rules. At common law (before the introduction of any statutory changes), the duty and expense of repair rested with parishes, of which there were originally more than 10,000. The drawbacks of such a fragmented system were obvious, especially once methods of communication and transport were quicker. Changes made by statutes gradually produced the result that 'highway authorities' were created with responsibility for repair and maintenance. Today the position is that highway authorities are either:

1    The Secretary of State for Transport or
2    County councils, London borough councils and unitary authorities.

The responsibility of the Secretary of State is for trunk roads, i.e. the principal roads constituting the national system of routes for through traffic in Great Britain. However, the work of repair and maintenance is commonly delegated to the councils as agents. The councils are the highway authorities responsible for all other roads maintainable at public expense. In turn, repair work is commonly delegated by a county council to district councils.

Section 36 of the 1980 Act deals with the question of which highways are maintainable at public expense. Effectively, these are:

1    Those highways in existence before 31 August 1835.
2    Highways created since then which have been adopted by the highway authority. Adoption can occur in a number of ways, e.g. by the dedicator serving notice on the highway authority which is accepted within the terms of s.37, or by agreement (s.38). Most important, however, are the adoption procedures with regard to private streets provided for by Part XI of the 1980 Act.

By s.41 the local highway authority is under a duty to maintain the highway. This means, according to *Rider v. Rider* (1973), that an authority must 'reasonably maintain and repair the highway so that it is free of dangers to all users who use that highway in the way normally to be expected of them – taking account of the traffic reasonably to be expected on the particular highway'. The duty is not one to ensure that a highway has to be free from surface water, snow and ice at all times or take preventative steps to stop ice forming or snow settling. Rather it is limited to the taking of reasonable steps such as preventive or clearance measures which are sufficient to keep the surface reasonably safe (*Goodes v. East Sussex CC* (2000)). However, it is to be noted that if the surface water or consequent ice in cold weather is due to defective design of the highway or drainage then there may be breach of duty on the part of the local highway authority. A highway includes any verge, and a difference in height between the carriageway and verge could, in the event of an accident, lead to a finding that the highway authority had breached their duty of care (*West Sussex CC v Russell* (2010)).

## Standard of the highway

When considering the state of repair of a highway, all highway authorities should bear in mind the words of a judge in the House of Lords:

> It is the duty of the road authorities to keep their public highways in a state fit to accommodate the ordinary traffic which passes or may be expected to pass along them. As the ordinary traffic expands or changes in character, so must the nature of the maintenance and repair of the highway alter to suit the change.

In fact, most highway authorities will wish to improve highways under their care in so far as their budget allows them to do so. The 1980 Act makes extensive provision for such improvements. Part V of the Act contains rules, e.g. with regard to dual carriageways and roundabouts, cycle tracks, footways, guardrails, refuges, subways and footbridges, levelling of highways, improvement of corners, fencing and lighting of the highway, and roadside planting.

Where a person is affected by a failure to repair the highway, he or she may wish to exercise the enforcement procedure provided by s.56 of the 1980 Act. To use this, a notice is served on the highway authority (or other party alleged to be responsible for maintenance) requiring the repair to be done. If the highway authority disputes liability, the case is referred to the Crown Court, which may order the highway to be put in proper repair within a specified period. In assessing whether a highway meets the requisite standard of repair, the court will consider the character and ordinary use of the highway (*Kind v. Newcastle upon Tyne City Council* (2001)). If the authority admit liability but dispute the extent of the repairs necessary the matter is settled by the magistrates' court.

The failure to repair the highway may have led to injury or damage being caused. Until 1961 it was very difficult to bring a civil action for damages against a highway authority, as the rule was that the authority owed no liability for injury caused by failure to repair, only for injury caused by negligent carrying out of the repairs. That rule is now changed; the present law is that the highway authority can be fully liable for injury caused by its failure to maintain or repair the highway. However, by way of defence, the authority may plead that it took such care as was reasonably required to secure that the highway was not dangerous to traffic (s.58). Naturally, this will involve factors such as the type of highway involved, the usual nature of the traffic using it, and the state of knowledge of the highway authority. People frequently seek to bring actions under these rules when they have been injured as pedestrians by a fall on an uneven or broken pavement. As the Court of Appeal held in the leading 'tripping' case of *Winterhalder v. Leeds CC* (2000):

> It is quite impossible and very misleading to state that a depression or gap or a dip of any particular dimension forms a danger to pedestrians. Any case such as this is fact sensitive and will be resolved on the basis of its own facts.

An interesting example of the application of s.56 can be seen in *Wentworth v. Wiltshire County Council* (1993). A farmer had a long-running dispute with the council about the status of the road to his farm, which he argued was a highway maintainable at public expense. The council did not agree. After the dispute had raged for a considerable number of years, matters came to a head when the Milk Marketing Board withdrew the tanker to his farm because of the state of the road. On application to the Crown Court, the farmer was successful in obtaining an s.56 order, under which the council were given nine months to effect the repairs. Unfortunately, as a result of the withdrawal of the milk tanker, the farmer had had to abandon his dairy business. He now wanted to seek financial compensation. The High Court refused to make such an order, as he had a remedy under s.56, which he ought to have pursued earlier, and his losses in relation to his dairy business were economic loss.

## Heavy traffic

The damage caused to highways is often due to excessively heavy traffic. It seems very unjust that the public should bear the cost of repair and maintenance when the damage is perhaps caused by the vehicles of one person, company or organization. Rules to redress this injustice are contained in s.59 of the 1980 Act. They provide that a highway authority may recover 'excess expenses' for maintaining the highway from any person causing excessive weight or other extraordinary traffic to pass along the highway. The excess expenses are those incurred over and above the average expenses for maintaining that or a similar highway which have resulted from damage arising from the extraordinary traffic.

Where a person knows that he will be operating traffic likely to cause such damage, he can agree in advance a payment to the highway authority. The advantage of this is that the payment will be a fixed amount, whereas if the operator waits for recovery procedures (proceedings may be taken in a county court) to be used by the highway authority, the 'excess expenses' are an unknown quantity. There is a significant amount of case law, particularly on the meaning of the phrases 'excessive weight' and 'extraordinary traffic', which is not defined in the Act. Given the loads carried by builders and the type of vehicles being brought to sites, it will be appreciated that the section may be relevant to them.

# Streets and street works

Although the word 'street' may conjure up a mental picture of a road with buildings on either side of it, the definition provided by the 1980 Act is much more vague. 'Street' is given the same meaning for both the 1980 Act and the New Roads and Street Works Act 1991 (the 1991 Act) and means, irrespective of whether it is a thoroughfare, any highway, road, lane, footway, alley or passage, square or court and any land laid out as a way whether it is for the time being formed as a way or not. The 1991 Act goes on to define a street as including one which is not a highway maintainable at public expense. In order for a private street to be 'adopted', i.e. to become maintainable at public expense, street works as specified by the 1980 Act must be executed, and then the adoption rules applied (s.228). These rules are important to landowners who have done the work specified and now want to rid themselves of responsibility for repair and maintenance. This will be particularly significant for an estate developer.

## The Private Street Works Code

This applies to a private street, meaning a street that is not a highway maintainable at the public expense. It allows Street Works Authorities to carry out work on such streets and recover the expense for so doing. Street Works Authorities (generally, outside London, these are the county councils or unitary authorities) and inside London the relevant London borough council have power to execute street works in private streets (s.205). Where this power is used, it is generally referred to as the Private Street Works Code. This involves the council preparing:

- A specification of the street works.
- An estimate of the probable expense.
- A provisional apportionment of those probable expenses between the premises liable to be charged – this means premises fronting the street.

The definition of 'fronting' in the 1980 Act includes premises that adjoin the street and therefore premises with side or rear frontages that have no access to the street are liable to be charged.

Once these details are approved by the Street Works Authority, it must:

1  Publish notices containing relevant details in local newspapers.
2  Post notices in the street affected.
3  Serve individual notices on the owners of the premises liable to be charged for the work, stating the sum provisionally apportioned to their premises. Generally, this will be decided according to the length of frontage of the premises to the street, but s.207 provides for the provisional sum calculated to be increased or decreased, depending on factors such as the degree of benefit to be derived by the owners of the premises, and any work already carried out by the owners.

Once all the required notices have been given, the owners of premises affected may wish to object. Section 208 sets out possible grounds of objection. These include arguments that:

- The street is not a private street.
- The estimated expenses are excessive.
- The proposed works are unreasonable or insufficient.
- The provisional apportionment is unacceptable.

Hearings of these objections take place before the magistrates' court, which has authority to quash the proposals in whole or in part, or amend them.

Once any objections are resolved and the work is completed, the Street Works Authority can then recover the costs in the agreed proportions from the owners, who may be given up to 30 years to pay. The actual amount due becomes a charge on the premises. This means that the agreed cost passes as a liability to any new owner of the premises.

Once the street works have been executed, it is then possible to 'adopt' the street and make it a highway maintainable at public expense under s.228. This is done by the Street Works Authority displaying a notice in the street declaring it to be a highway maintainable at public expense. If no objection is made within a month, the street becomes such a highway. But a majority of owners in the street may object and in that case the Street Works Authority must apply to the magistrates' court for an order overruling their objections, before the street can be turned into a highway maintainable at public expense.

## Advance Payments Code

The standards imposed by Street Works Authorities in the making up of streets mean that this work can prove very costly. The authority operating under the Private Street Works Code is responsible for recovering the sums due after completion of the work and money may be outstanding for long periods. The drawback of this system can be avoided in cases where the Advance Payments Code applies. The purpose of this code is to secure payment of the expenses of executing street works in private streets adjacent to new buildings. Section 219 provides that where (a) it is proposed to erect a building for which plans must be deposited with the local authority and (b) the building fronts on to a private street where the Street Works Authority could execute street works under the Private Street Works Code, then no building can take place until the owner of the land has paid or secured the payment of the sums required to the Street Works Authority under the Advance Payments Code. The sanction for disobeying this rule is a fine. However, no offence will be committed if the Street Works Authority does not notify the developer of the sum they are required to pay within six weeks of the plans being deposited. Section 219, however, is subject to a number of important exceptions, e.g. if the building to be erected is in the grounds of an existing building or where an agreement under s.38 has been made with the Street Works Authority, whereby the person undertaking street works at his own expense will, on completion of the works, dedicate the street as a highway.

If the Advance Payments Code does apply, then the Act provides the machinery for establishing the amount to be deposited – this is 'such sum as, in the opinion of the Street Works Authority, would be recoverable under the Private Street Works Code if the Street Works Authority were to carry out the work'. Provision is made for appeal to the Secretary of State for Transport against the amount proposed. Once an advance payment has been made, it opens the way for the 'adoption' procedure under s.229. As long as one frontager to the street has paid under the Code then a majority of the frontagers (either in terms of numbers of owners, or length of frontage owned) may request the Street Works Authority to secure the carrying out of appropriate street works and then declare the street to be a highway maintainable at public expense.

Two final points with regard to street works should be noted:

1    Where a private street is in need of repair to obviate danger to traffic, the Street Works Authority can require owners fronting on to the street to undertake specified repairs within a given time. If the owners fail to carry out the repairs, the Street Works Authority may proceed with the repairs and recover the cost proportionately from the owners. Where a notice is served requiring urgent repairs, it may be in the interests of the owners to ask the authority to undertake the work of properly making up the street under the Private Street Works Code, as it will, thereafter, be a highway maintainable at public expense (s.230).

2    Section 236 gives power to a Street Works Authority to resolve to bear all or any part of the costs of works under the Private Street Works Code. This then discharges or reduces the liability of the owner.

## New Roads and Street Works Act 1991

Various utilities may need to dig up a street to install or maintain apparatus beneath the surface, e.g. gas, electricity, water and telecommunications. Authority to carry out work of this type is given by Parliament to utility companies who are known as 'undertakers'. The process of doing such work is regulated by the New Roads and Street Works Act 1991. The Act applies, subject to certain restrictions, to both highways maintainable at the public expense and private roads.

By s.51 it is a criminal offence, punishable by a fine, for a person, e.g. a company, other than the street authority (usually the highway authority), amongst other things, to place apparatus in or break up or open up a street otherwise than in pursuance of a statutory right or a street works licence. An undertaker who proposes to carry out street works must give not less than seven days' notice (three months in the case of major works) to the street authority and to any other company having apparatus likely to be affected by the works. The requirement of notice does not apply to emergency works, which are defined in the Act as works whose execution is required to put an end to circumstances that are likely to cause danger to persons or property.

Various obligations are placed upon an undertaker: s.66 provides that street works shall be carried on and completed 'with all such dispatch as is reasonably practicable'; Regulations made under s.74A provide for an undertaker executing street works in a highway maintainable at the public expense to pay a charge, calculated by reference to the duration of the works, to the local highway authority; s.70 provides that it is the duty of the undertaker to reinstate the street; s.59 places a duty on the street authority to coordinate street works and a duty on undertakers to cooperate. To lessen the disruption caused by street works, s.58 provides that where 'it is proposed to carry out substantial road works in a highway, the street authority may by notice … restrict the execution of street works during the 12 months following the completion of those works'. This notice allows for coordination of street works, e.g. if a highway authority digs up a street other undertakers may take advantage of the street being opened up to effect any necessary repairs to their apparatus.

# Closing and obstruction of highways

It may be necessary, from time to time, to close or divert a highway, e.g. while work is in progress on it. Alternatively, the need for the highway may have disappeared and the authority may wish to close it permanently. The procedure for 'stopping up' is contained in the 1980 Act, s.116. Generally, this involves an application to the magistrates' court by the highway authority. (Where the closure is sought by a person other than the highway authority, he or she must ask the highway authority to take the necessary steps. There is no appeal if the highway authority refuses to activate the procedure but it must not unreasonably withhold consent.) The magistrates can make an appropriate order once the highway authority has followed through the necessary steps. These include notifying adjoining owners and occupiers, advertising the proposed closure or diversion in the press, and fixing notices at each end of the highway.

It is also possible for a council to stop up or divert a footpath, bridleway or restricted bridleway if it is considered that it is expedient to do so. This can be either on the grounds that it is not needed for public use (stopping up) or that the path should be diverted in the interests of either the owner of the land or the public (ss.118 and 119). Both possibilities are, however, subject to strict controls that make the relevant stopping up or diversion orders difficult to obtain.

Under s.247 of the Town and Country Planning Act 1990, the Secretary of State has the power to stop up or divert a highway if he or she is satisfied that it is necessary to do so in order that a development for which planning permission has been granted can be carried out. A local planning authority has similar powers in respect of footpaths, bridleways and restricted bridleways (ss.257 and 258).

Any obstruction of, or interference with, the highway, unless authorized in some way, may constitute a civil and/or criminal offence. The Highways Act 1980, Part IX, contains numerous offences (e.g. in relation to straying animals and unlawful deposits) and also creates various duties and powers.

The highway authority is under a general duty to assert and protect the rights of the public to the use and enjoyment of any highway, and to prevent as far as possible the obstruction of the highway. It may enforce these rules in legal proceedings. Some specific examples of interference with the highway which are relevant to construction work should be noted:

1    Section 133 – if excavation or other work on land adjoining a street causes damage to a footway (pavement), the highway authority can recover the cost of repair from the landowner, or from the person causing the damage.
2    Section 131 – it is an offence, without lawful authority, to make an excavation on the highway, or to deposit anything on a highway that will cause it damage. The penalty imposed in respect of such offences is a fine.
3    Section 137 – it is an offence to wilfully obstruct free passage along the highway.
4    Section 139 – builders' skips may not be deposited on a highway without permission of the highway authority. When granting such a permission, the authority may specify the size of the skip, conditions as to its siting, lighting and guarding, means for making it visible to oncoming traffic, disposal of its contents and its ultimate removal. The owner must make sure that the skip is:
     (a)   properly lit;
     (b)   clearly marked with his name and address;
     (c)   removed as soon as practicable when full;
     (d)   complying in all respects with any conditions attached to the permission.

Breaches of these rules are punishable by fines. The section specifically states that 'nothing in this section is to be taken as authorizing the creation of a nuisance'. Where the skip is deposited on a highway, even when permission has been granted, s.140 provides that the highway authority or a police constable can require it to be removed or repositioned, or may themselves remove or reposition it and recover from the owner the expense of doing so.

5   Section 168 deals with building operations affecting public safety. If building operations in, or near, a street cause an accident which gives rise to a risk of injury to persons in the street then the owner of the land on which the work is taking place or any other person liable for causing such a risk (e.g. the building contractors) is liable to a fine.

6   Section 169 provides that where scaffolding will obstruct the highway, it can only be erected if a licence is obtained from the highway authority. There is a right of appeal against refusal of a licence, or against any conditions attached to it. If a licence is granted, then the scaffolding must be adequately lit during darkness and any necessary traffic signs must be erected and maintained.

7   Section 170 controls the mixing of mortar and cement on the highway in such a way that it may stick to the highway or enter drains and sewers and solidify. The section permits mixing in a receptacle or on a plate, but any breach of the rules is an offence carrying a fine.

8   Section 171 controls deposits of building materials and the making of excavations in streets. Such activities can be carried on with the consent of the highway authority, which may impose conditions, particularly if there is a danger of damage or obstruction to the apparatus of statutory undertakers (e.g. the gas, electricity and water boards). Appeal against refusal of permission, or the conditions imposed, is to the magistrates' court. The highway authority may levy a charge in the case of depositing building materials, erecting scaffolding under section 169 or mixing mortar and cement on the highway under section 170 if the works are not completed in a reasonable period.

9   Section 172 regulates when building works must be close boarded by a hoarding or fence. This section operates whenever a building is to be erected or taken down in a street, unless the local authority dispenses with the requirement. Where a hoarding is required, conditions are likely to be imposed regarding the creation and lighting of a footway and the maintenance of the hoarding. Under s.173, it is an offence to use such a hoarding which is not securely fixed to the satisfaction of the local authority.

It should be noted that some of the foregoing sections apply to highways whereas others apply to streets. Many create statutory offences punishable by a fine. But the behaviour constituting the statutory offence may also cause damage or injury to an individual, who may be able to bring a civil claim for damages in the law of tort. A number of torts may be relevant, e.g. a person falling into an unfenced and unlit excavation on the highway may succeed in negligence, whereas a person injured by the collapse of an unsafe hoarding around building work in a street may succeed in negligence or public nuisance. An accumulation of rubbish in a builder's skip may constitute a nuisance. For the relevant rules in tort, see Chapter 10.

The tort of greatest significance in relation to activities on or near the highway is public nuisance. Claimants have succeeded in cases where crowds blocked the highway, where smoke obscured visibility in the highway, where a dangerous lamp overhanging the highway fell on to it, where there was unlit scaffolding, where the danger consisted of broken cellar flaps and where the danger arose from a heap of slates piled at the side of the road.

## Building and improvement lines

Part V of the 1980 Act contains a number of rules, powers and duties with regard to the improvement of highways. It may be obvious to the highway authority that, because of changing or increasing traffic volume, it will sooner or later become necessary to widen a particular highway. The need may not yet be urgent or funding may not be available, but it would be absurd if in the meantime the local authority continued to allow building work to take place, thereby making ultimate widening of the highway more difficult and costly. To prevent this situation occurring, s.73 of the Act provides that the highway authority may prescribe an improvement line for the street in question. This is the line to which it is ultimately intended to widen the street. Naturally, such a decision can affect the value of adjacent properties and the development which can be undertaken in relation to them. It would not be surprising, therefore, if the owners wished to object to the prescription of the improvement line. This can be done by any aggrieved person appealing to the Crown Court. If the appeal fails, it should be noted that s.73(9) provides for compensation to be paid to persons whose property is thus injuriously affected.

Section 74 of the Act provides for the highway authority to prescribe a building line. Once such a building line is operative, no new building shall be erected beyond the building line without the consent of the highway authority. Rather like the rules in s.73, provision is made to compensate owners whose property is injuriously affected, but unlike s.73 there is no provision in s.74 for appealing against the imposition of the building line. Schedule 9 merely provides that the highway authority must consider objections.

Once either type of line is properly prescribed, it must be shown on a duly authenticated plan which is available for inspection by interested parties. Both ss.73 and 74 provide means whereby these lines can be removed if they are subsequently seen to be unnecessary.

# 16
# Building Regulations

## Purpose and scope of the Building Regulations

Since 1875, the law has imposed controls to ensure that buildings are constructed in a way which is conducive to good health. Originally, local authorities made bye-laws on matters such as the materials to be used, ventilation and sanitation. Relevant legislation is now consolidated in the Building Act 1984 which empowers the Secretary of State for the Environment to make Regulations. The current regulations are the Building Regulations 2000. Many specific points previously contained in the Regulations are now covered by guidance documents. Section 7 of the 1984 Act provides that a failure to comply with guidance documents does not in itself render a person liable to civil or criminal proceedings, but in any such proceedings a failure to comply may be relied on as tending to establish liability.

Under s.1 of the Building Act 1984, the Secretary of State may make regulations:

1   To secure the health, safety, welfare and convenience of persons in or about buildings and of others who may be affected by buildings.
2   To further the conservation of fuel and power.
3   To prevent waste, undue consumption, misuse or contamination of water.
4   To further the protection or enhancement of the environment.
5   To facilitate sustainable development.
6   To further the prevention or detection of crime.

The regulations may relate to the design and construction of buildings, and the provision of services, fittings and equipment in, or in connection with, buildings. Such regulations are to be known as Building Regulations. The Secretary of State has power to dispense with or relax a Regulation in particular cases where it is considered, after consultation with the local authority, that its operation would be unreasonable (s.8). When making regulations, the Secretary of State is advised by the Building Regulations Advisory Committee, appointed under s.14. The 2000 Regulations lay down requirements in relation to matters such as structure, fire safety, site preparation, toxic substances, soundproofing, ventilation, drainage, stairways, energy conservation and access for the disabled. In 2007 the Regulations were

Galbraith's Building and Land Management Law for Students. DOI: 10.1016/B978-0-08-096692-2.00016-6
Copyright © 2010 by Elsevier Ltd

amended to include the energy performance of buildings and continuing requirements relating to fuel, power and emissions from buildings.

## Enforcement procedures

The Regulations apply to any 'building work' or 'material change of use' of a building. The work done must be supervised, either by the local authority or an approved inspector, i.e. a person approved by the Secretary of State.

When work is to be supervised by the local authority, the builder will provide the authority with a building notice, or will deposit full plans. Building notices are only appropriate for small and minor works. The local authority can ask the builder for such extra detail as it needs, and will make the same site inspections as in other cases. On a deposit of full plans, these must be passed subject to conditions or rejected by the authority within five weeks. Where the plans show that work will be carried out in accordance with the Building Regulations, the authority must approve them. Local authorities are authorized to charge fees in relation to matters such as the passing or rejection of plans and the carrying out of inspections.

Where the local authority is supervising, there are prescribed time limits for notifying the authority about aspects of the work:

1    At least two days' notice of commencement.
2    At least one day's notice before covering up foundations, or a dampcourse, or any concrete material.
3    At least one day's notice before covering up any drain.
4    At least five days' notice if the drain is a foul water drain.
5    At least five days' notice of completion of the work.

Many local authorities issue pro-forma notices for a builder to post off to them at relevant times.

Under s.35 of the 1984 Act, contraventions of the Building Regulations render a person liable to a fine on summary conviction in the magistrates' court. A local authority may also serve notice on an owner, requiring him to pull down or remove the work, or make the alterations necessary so that it does comply with the Regulations. If necessary, the local authority may pull down or remove the work themselves, and charge the owner (s.36). No s.36 notice can be served after the expiration of 12 months from the date of completion of the work in question.

## Defective and dangerous buildings

Sections 76–83 of the Building Act 1984 created several important controls regarding defective and dangerous premises and demolition. These are in addition to rules which exist in other legislation (e.g. the Town and Country Planning Act 1990). Where a building is in such a defective state that it is prejudicial to health or a nuisance, the local authority may serve a notice on the owner indicating that it intends to remedy the defective state of affairs. Nine days after serving the notice, the authority may execute the work and recover its expenses from the owner. An owner can serve a counter-notice within seven days, saying that he intends to remedy the situation himself. The local authority can then only intervene if the owner does not make a start within a reasonable time, or does not make reasonable progress. The advantage of using this s.76 procedure is speed. It is only available in cases where it appears to the authority that the procedures under the Environmental Protection Act 1990 would cause unreasonable delay. The drawback of the procedure is that an authority may need to bring proceedings to recover its expenses and, at that point, the court can enquire whether the authority was justified in the assumptions it made and the steps it took. If not justified, the authority cannot recover the expenses or any part of them. Similar rules exist with regard to dangerous buildings (ss.77 and 78), and dilapidated buildings and neglected sites (s.79).

Demolition work is now subject to a measure of control by local authorities by virtue of ss.80–83. Where the whole or part of a building is to be demolished, then no work should commence until a notice of intention to demolish has been given to the local authority and either the local authority has served an s.81 notice, or six weeks have elapsed since it received the notice of intention. If the authority decides to serve an s.81 notice, it may require the demolition contractor to take any, or all, of the following steps:

- Shore up adjacent buildings.
- Weatherproof surfaces of adjacent buildings exposed by demolition.
- Repair and make good any damage to adjacent buildings caused by the demolition.
- Remove rubbish or material resulting from the demolition.
- Disconnect and seal sewers or drains.
- Make good the surface of the ground.
- Make arrangements for disconnection of services.
- Make such arrangements with regard to the burning of structures or materials on the site as may be reasonably required by the fire and rescue authority.
- Take such steps as the authority considers necessary for the protection of the public and preservation of public amenity.

A contractor can appeal to the magistrates' court against an s.81 notice, particularly against the shoring-up and weatherproofing requirements, where he may argue that the adjacent owner ought to pay or contribute to the costs.

For the effective operation of many of these rules, it is essential that authorized officers of local authorities have power to enter premises. Powers are granted by s.95, and must sometimes be exercised with the authority of a justice's warrant. Officers must ensure that they have appropriate authority to enter, otherwise their actions constitute the tort of trespass.

# Index